Multiple myeloma

Multiple myeloma

Edited by

GÖSTA GAHRTON MD

Professor of Medicine, Karolinska Institutet, Huddinge
University Hospital, Huddinge, Stockholm, Sweden

BRIAN G M DURIE MD

Head of the Research and Treatment Center, Division of Hematology and Oncology,
Sinai-Cedars Medical Center, Los Angeles, USA

A member of the Hodder Headline Group
LONDON • SYDNEY • AUCKLAND
Co-published in the USA by Oxford University Press, Inc., New York

First published in Great Britain 1996 by
Arnold, a member of the Hodder Headline group
338 Euston Road, London NW1 3BH

Co-published in the United States of America by
Oxford University Press, Inc.,
198 Madison Avenue, New York, NY 100016
Oxford is a registered trademark of Oxford University Press

Whilst the advice and information in this book is believed to be true and
accurate at the date of going to press, neither the authors nor the publisher
can accept any legal responsibility or liability for any errors or omissions
that may be made. In particular (but without limiting the generality of the
preceding disclaimer) every effort has been made to check drug dosages;
however it is still possible that errors have been missed. Furthermore,
dosage schedules are constantly being revised and new side-effects
recognized. For these reasons the reader is strongly urged to consult the
drug companies' printed instructions before administering any of the drugs
recommended in this book.

British Library Cataloguing in Publication Data
A catalogue record for this book is available from the British Library

Library of Congress Cataloging-in-Publication Data
A catalog record for this book is available from the Library of Congress

ISBN 0 340 576030 (hb)

Typeset in Times and Optima by
J&L Composition Ltd, Filey, North Yorkshire

Printed and bound in Great Britain by
The Bath Press, Somerset

Contents

Contributors   vii
Preface   ix

1. **History of multiple myeloma**   *Robert A Kyle*   1

2. **Epidemiology and incidence of plasma cell neoplasms**   *Daniel E Bergsagel*   12

3. **Basic concepts – the plasma cell in multiple myeloma**   *Federico Caligaris-Cappio,*
Maria Grazia Gregoretti   22

4. **Immunoglobulins in multiple myeloma**   *Anders Österborg, Lars-Olof Hansson, Håkan*
Mellstedt   36

5. **Human myeloma-induced bone changes**   *Régis Bataille*   51

6. **Chromosome abnormalities in multiple myeloma**   *Gunnar Juliusson, Gösta Gahrton*   55

7. **Oncogenes related to multiple myeloma**   *Paolo Corradini, Mario Boccadoro, Alessandro Pileri*   64

8. **Growth factors in the pathogenesis of multiple myeloma**   *Bernard Klein*   73

9. **Monoclonal gammopathies of undetermined significance**   *Philip R Greipp, John A Lust*   83

10. **Clinical features and staging**   *Anders Österborg, Håkan Mellstedt*   98

11. **Principles of chemotherapy and radiotherapy**   *Diana Samson*   108

12. **Treatment of refractory multiple myeloma with cytotoxic therapy and radiation therapy**
Francis J Giles   130

13. **Use of interferon in the treatment of multiple myeloma**   *Giuseppe Avvisati*   148

14. **Cytokines in the treatment of multiple myeloma**   *Heinz Ludwig, Elke Fritz*   159

15. **Autologous transplantation in multiple myeloma**   *Michel Attal, Jean Luc Harousseau*   182

16. **Bone marrow transplantation with syngeneic or allogeneic marrow in multiple myeloma**
Gösta Gahrton, Per Ljungman   194

17. **Complications and supportive treatment**   *Dietrich Peest, Helmuth Deicher*   205

Index   215

Contributors

Michel Attal MD
Professor of Hematology, University of Toulouse, Toulouse, France

Giuseppe Avvisati MD, PhD
Consultant Hematologist, Department of Human Biopathology, University of La Sapienza, Rome, Italy

Régis Bataille MD, PhD
Professor of Hematology and Head, Hematology Laboratory, Institute of Biology, Nantes, France

Daniel E Bergsagel CM, MD, DPhil
Professor of Medicine, University of Toronto and Consultant Physician, Ontario Cancer Institute and Princess Margaret Hospital, Toronto, Ontario, Canada

Mario Boccadoro MD
Professor, Division of Hematology, University of Torino, Torino, Italy

Federico Caligaris-Cappio MD, PhD
Professor of Clinical Immunology, University of Turin, Turin, Italy

Paolo Corradini MD
Assistant Professor, Division of Hematology, University of Turin, Turin, Italy

Helmuth Deicher MD
Emeritus Professor of Medicine and Head, Division of Immunology and Blood Transfusion Medicine, Hannover Medical School, Hannover, Germany

Elke Fritz
First Department of Medicine and Medical Oncology, Wilhelminenspital, Vienna, Austria

Gösta Gahrton MD
Professor of Medicine and Head of the Department of Medicine, Karolinska Institutet, Huddinge University Hospital, Huddinge, Sweden

Francis J Giles MD
Director of the Myeloma Research and Treatment Center; Director of Stem Cell/Bone Marrow Transplantation Unit; UCLA Assistant Professor of Medicine, Division of Hematology and Oncology, Cedars-Sinai Medical Center, Los Angeles, California, USA

Maria Grazia Gregoretti MD
FIRC Fellow, University of Turin, Turin, Italy

Philip R Greipp MD
Consultant Physician, Division of Hematology and Internal Medicine and Hematopathology, Mayo Clinic and Mayo Foundation; Professor of Medicine and Laboratory Medicine, Mayo Medical School, Rochester, Minnesota, USA

Lars-Olof Hansson MD, PhD
Protein Laboratory, Department of Clinical Chemistry, Karolinska Hospital, Stockholm, Sweden

Jean Luc Harousseau MD, PhD
Professor of Hematology, Hotel Dieu Hospital, Nantes, France

Gunnar Juliusson MD
Head, Department of Hematology, University Hospital, Linköping, Sweden

Bernard Klein PhD
Director of Research, Institute of Molecular Genetics, Montpellier, France

Robert A Kyle MD
Consultant, Devision of Hematology and Internal Medicine, Mayo Clinic and Mayo Foundation; Professor of Medicine and of Laboratory Medicine, Mayo Medical School, Rochester, Minnesota, USA

Per Ljungman MD, PhD
Department of Medicine, Karolinska Institutet at Huddinge University Hospital, Huddinge, Sweden

Heinz Ludwig MD
Professor, First Department of Medicine and Medical Oncology, Wilhelminenspital, Vienna, Austria

John A Lust MD, PhD
Consultant Physician, Division of Hematology and Internal Medicine and Laboratory Genetics, Mayo Clinic and Mayo Foundation; Assistant Professor of Medicine, Mayo Medical School, Rochester, Minnesota, USA

Håkan Mellstedt MD, PhD
Associate Professor, Department of Oncology, Karolinska Hospital and Karolinska Institutet, Stockholm, Sweden

Anders Österborg MD, PhD
Assistant Professor, Department of Oncology, Karolinska Hospital and Karolinska Institutet, Stockholm, Sweden

Dietrich Peest MD
Professor of Medicine and Clinical Immunology, Hannover Medical School, Hannover, Germany

Alessandro Pileri MD
Professor and Chairman, Division of Hematology, University of Turin, Turin, Italy

Diana Samson MD, FRCP, FRCPath
Senior Lecturer and Consultant Haematologist, Royal Postgraduate Medical School, Hammersmith Hospital, London, UK

Preface

Multiple myeloma is one of the most studied malignancies since its discovery in the mid-nineteenth century. One reason for the interest in multiple myeloma is probably the protein abnormalities, later recognized as immunoglobulins, that made it possible very early on to have a direct marker of the disease in the peripheral blood and urine. In the late 1940s it was seen that these marker proteins could be affected by treatment with urethane, but it was not until the early 1960s, when melphalan became available, that multiple myeloma was more regularly treated with the aim of affecting the tumor cells. Melphalan was the backbone of myeloma treatment for two decades. Combination with prednisolone was a step forward, but it was not until the early 1980s that high-dose treatment combined with autologous or allogeneic bone marrow transplantation was used in an attempt at finding a cure. The reason for this delay in comparison to similar earlier approaches in leukemia was, of course, that multiple myeloma patients are usually of old age. However, an increasing fraction of patients appear to benefit from these new treatment modalities.

The new methodologies, such as cytogenetics, molecular genetics and PCR, oncogene and suppressor gene research, as well as increasing knowledge about regulation of cells by cytokines, have affected multiple myeloma research. Although it has turned out that multiple myeloma in this respect is one of the most complex malignancies, the information available about it has increased tremendously in recent years. Attempts have been made to translate some of the discoveries into treatment, such as the use of cytokines, interferons, and anti-IL-6 with or without combination with chemotherapy or transplantation.

The recent progress in stem cell research has made it possible to use peripheral blood stem cells instead of bone marrow for transplantation. The success in purification of the stem cell may rapidly improve results with transplantation. Recent successful attempts to introduce genes by retroviral vectors into myeloma cells may open a new field of treatment research in multiple myeloma.

This book has mainly focused on the new developments in multiple myeloma. However, it also covers the basic facts about the disease, the clinical features and the practical handling. Our hope is that it will be of value, not only for scientists in the field but also for the practitioner and the specialist who deals with the treatment of patients with multiple myeloma.

Gösta Gahrton
Brian Durie

CHAPTER 1

History of multiple myeloma

ROBERT A KYLE

The first recorded case	1	Multiple myeloma – some early cases	6
Henry Bence Jones	4	Multiple myeloma in the twentieth	
Other players in the history of Bence Jones		century	8
proteinuria	6	References	9

In 1850, Dr William Macintyre, a 53-year-old Harley Street consultant and physician to the Metropolitan Convalescent Institution and to the Western General Dispensary, St Marylebone,[1] described one of his patients:

> Mr. M_____, a highly respectable tradesman, aged 45, placed himself under my care on the 30th of October, 1845. He was then confined to the house by excruciating pains of the chest, back, and loins, from which he had been suffering, more or less, for upwards to twelve months.[2]

Because edema had been noted, Dr Macintyre examined the urine but found no evidence of sugar. The urine specimen was opaque, acid, and of high density, with a specific gravity of 1.035. When heated, it was found to 'abound in animal matter.' The precipitate dissolved on boiling but again consolidated on cooling.

The urine sample and following note were sent by Dr Macintyre and a leading physician of London, Dr Thomas Watson, to Dr Henry Bence Jones, a 31-year-old physician at St George's Hospital who had already established a reputation as a chemical pathologist.[3]

Saturday, Nov. 1st, 1845

Dear Doctor Jones

The tube contains urine of very high specific gravity. When boiled it becomes slightly opaque. On the addition of nitric acid, it effervesces, assumes a reddish hue, and becomes quite clear; but as it cools, assumes the consistence and appearance which you see. Heat reliquifies it. What is it?[4]

The first recorded case

Thomas Alexander McBean, the patient who contributed the urine discussed above, was a highly respectable grocer of 'temperate habits and exemplary conduct.' Having married early, he had numerous offspring, and, with the exception of two or three severe attacks of 'frontal neuralgia,' he had enjoyed good health. His family history was non-contributory: his father had died of the complications of gout, and his mother had died suddenly after operation for carcinoma of the breast. For slightly less than a year previous to the onset of symptoms, his family had noted that he fatigued easily and appeared to stoop while walking. He also had urinary frequency, and he was concerned that his 'body linen was stiffened by his urine' despite the absence of a urethral discharge.[2]

While vaulting out of an underground cavern on a country vacation in September 1844, McBean had 'instantly felt as if something had snapped or given way within the chest, and for some minutes he lay in intense agony, unable to stir.'[2] The pain abated, and he was able to walk to a neighboring inn. Soreness and stiffness of the chest persisted but were relieved by the application of a 'strengthening plaster to the chest.' Three or 4 weeks later, the pain recurred, and the patient was treated by removal of a pound of blood and the application of leeches. The pain resolved but, as might be expected, the bleeding was followed by considerable weakness for 2 or 3 months.

In the spring of 1845, the pain recurred with an episode of pleuritic pain in the right side, between the ribs and the hip, which was treated by cupping. Therapeutic bleeding produced much greater weakness than before. Wasting, pallor, and slight puffiness of his face and ankles led to consultation with Dr Thomas Watson. Steel and quinine therapy resulted in rapid improvement. By the middle of summer, he was able to travel to Scotland, where on the coast 'he was capable of taking active exercise on foot during the greater part of the day, bounding over the hills, to use his own expression, as nimbly as any of his companions.'[2] His appetite became ravenous – he expressed it as being so much that he dreamed of eating dogs and cats.[4] His recovery was interrupted by an episode of diarrhea, which proved to be obstinate, reducing his strength considerably. In September 1845, he returned to London in a very debilitated state but free of the excruciating pains that he had experienced during the spring and early part of the summer. In October, the lumbar and sciatic pains became severe. Warm baths, Dover's powder (ipecac and opium powder), acetate of ammonia, camphor julap, and compound tincture of camphor did not help.

The pain became 'fixed in the left lumbar and iliac regions, obliging the patient to observe a semi-bent posture, on account of the agony caused by every attempt at movement of the body upon the thighs.'[2] There was intermittent pain involving the chest and shoulders. Great care and cautious maneuvering enabled him to 'get in and out of bed on all-fours.' He became weaker and was confined to his bed. He had considerable flatulence and 'marked fulness and hardness in the region of the liver.'[2] Citrate of iron

and quinine produced no benefit. His urine became turbid and thick like pea soup. This change coincided with improvement, characterized by sleeping well for two nights and the ability to get up and walk around the room with little or no pain. The urine, however, contained the same amount of animal matter. Phlegm in the chest, a cough, and recurrence of the severe pain then developed. He also had an attack of diarrhea, which was precipitated by a dose of rhubarb and soda that had been given to correct the flatulence.

On 15 November, Dr Henry Bence Jones saw the patient in consultation and recommended alum as treatment 'with the view of checking the exhausting excretion of animal matter.' The specific gravity of the urine and animal matter in the urine decreased, and McBean was able to sit up daily for an hour or two and continued to enjoy his food. Unfortunately, on 7 December, he experienced a 'dreadful aggravation of lumbar pains,' which crude opium and morphine failed to relieve. He became weaker and died on 1 January 1846, exhausted, in full possession of his mental faculties (Fig. 1.1).

Post-mortem examination revealed emaciation; the ribs, which crumbled under the heel of the scalpel, were soft, brittle, readily broken, and easily cut by the knife. Their interior was filled with a soft 'gelatiniform substance of a blood-red colour and unctuous feel.' The sternum also was involved. The heart and lungs were not remarkable. 'The liver was voluminous, but of healthy structure.' The kidneys appeared to be normal on both gross and microscopic examination. They had 'proved equal to the novel office assigned to them' and had 'discharged the task without sustaining, on their part, the slightest danger.' The thoracic and lumbar vertebrae had the same changes as found in the ribs and sternum, but the humeri and

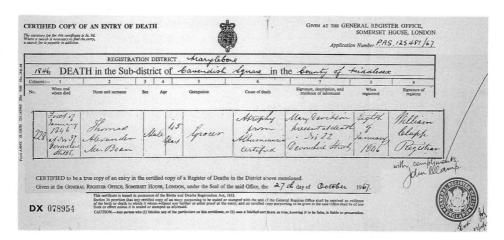

Fig. 1.1 Death certificate of Thomas Alexander McBean. (Courtesy of General Register Office, London, England.)

Fig. 1.2 Plasma cells (wood engravings made from drawings by Mr Dalrymple). (From ref. 5.)

femurs resisted 'all efforts to bend or break them by manual force.'[2]

John Dalrymple, surgeon to the Royal Ophthalmic Hospital, Moorfields, examined two lumbar vertebrae and a rib of Mr McBean. Dalrymple noted that the disease appeared to begin in the cancellous bone, then grew and produced irregularly sized round, dark-red projections that were visible through the periosteum. Nucleated cells formed the bulk of the gelatiniform mass that filled the cancellous cavities. Most of these cells were round or oval and about one-half to two times as large as an average blood cell. The cells contained one or two nuclei, each with a bright, distinct nucleolus. Wood engravings made from the accurate drawings of Mr Dalrymple were consistent with the appearance of myeloma cells (Fig. 1.2).[5]

Macintyre stated that his 'own share in this part of the inquiry, it must have been seen, was very humble.' He went on to say that the examination and course of the patient seemed to be 'deserving of a detailed account.' and that he 'shall be content if I have succeeded in pointing out to future observers, gifted with the requisite qualifications for conducting researches of a higher order, certain definite and distinctive characters by which a peculiar and hitherto unrecorded pathological condition of the urine may be recognised and identified.'[2]

The diarrhea, weakness, emaciation, hepatic enlargement, flatulence, dyspepsia, edema of the ankles, puffiness of the face, and large amounts of Bence Jones proteinuria all suggest the possibility of amyloidosis, in addition to multiple myeloma; but the autopsy findings of a normal heart and kidneys and 'voluminous liver of healthy structure' make the presence of amyloidosis unlikely. Because the waxy changes of amyloidosis in the liver were commonly recognized in this era, it is unlikely that amyloidosis would have been overlooked.

As with many so-called first cases, one can find an earlier example. It is almost certain that 39-year-old Sarah Newbury, the second patient described by Solly in 1844, had multiple myeloma.[6] She experienced fatigue and, 4 years before her death, was seized with a violent pain in her back when stooping. Rheumatic pains occurred a year later. Pain in her limbs increased after a fall in February 1842. She felt excruciating pain, 'just as if her thighs were being

Fig. 1.3 Sarah Newbury. Fractures of femurs and right humerus. (From ref. 6.)

broken into a thousand pieces,' while her husband was lifting her from the fireplace to carry her to bed. He felt her thighs give way, and she was unable to walk thereafter. Fractures of the clavicles, right humerus, and right radius and ulna occurred (Fig. 1.3). She was hospitalized at St Thomas's Hospital, London, and was treated with an infusion of orange peel and a rhubarb pill when necessary as well as an opiate at night. She died suddenly on 20 April 1844. Autopsy revealed that the cancellous portion of the sternum (Fig. 1.4) had been replaced by a red substance, which Macintyre reported was similar to the red substance seen in Mr McBean. The red matter had replaced much of both femurs (Fig. 1.5).

Examining the specimen of McBean's urine received from Watson and Macintyre on 1 November, Bence Jones corroborated Macintyre's finding that the addition of nitric acid produced a precipitate that was redissolved by heat and formed again on cooling. He calculated that the patient excreted 67 g/day and concluded that the protein was an oxide of albumin, specifically 'hydrated deutoxide of albumen.'[7]

There is some justification for changing the name 'multiple myeloma' to 'McBean's disease with Macintyre's proteinuria'. Although Macintyre described the heat properties of the urine, Bence Jones emphasized its place in the diagnosis of myeloma, for he said, 'I need hardly remark on the importance of seeking for this oxide of albumen in other cases of mollities ossium' (softening of the bone).[4]

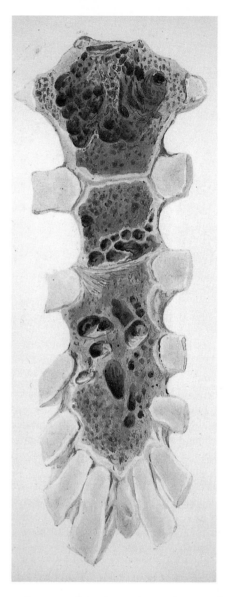

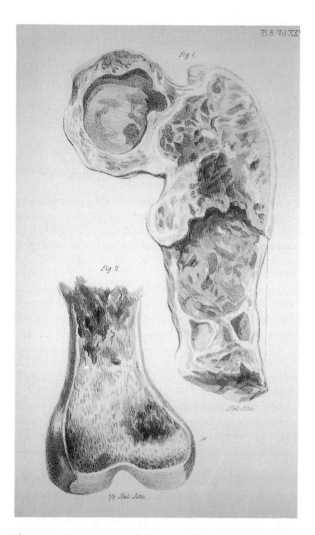

Fig. 1.5 Destruction of femurs of Sarah Newbury by myeloma tumor. (From ref. 6.)

Fig. 1.4 Sternum of Sarah Newbury, showing destruction of bone. (From ref. 6.)

Henry Bence Jones

Born on 31 December 1813, at Thorington Hall, Yoxford, Suffolk, England, at the home of his maternal grandfather, the Reverend Mr Bence Sparrow, Henry Bence Jones attended Harrow, where he excelled in sports and was on the cricket team. Proceeding from Harrow to Cambridge, he entered Trinity College, where he was a member of the boating crew. He attended the Divinity Lectures in preparation for ordination but decided against the career of a clergyman[8] after obtaining his arts degree in

January 1836. Instead, he became a pupil of Mr John Hammerton, apothecary at St George's Hospital. He enrolled as a full-time medical student after 18 months. During this time, he learned the use of the stethoscope, a relatively new instrument. To acquire a knowledge of chemistry, he became a private pupil of Professor Thomas Graham at University College, London. As a part of his studies he was required to examine a calculus from the University College Museum. This stone consisted of cystine and led to his first publication. After 6 months during 1841 spent studying chemistry with Justus von Liebig at Giessen, Germany, he returned to St George's Hospital, where he subsequently advanced rapidly to assistant physician and then to physician. In 1846, he was elected a Fellow of the Royal Society.

Dr Bence Jones was an accomplished physician

Fig. 1.6 Dr Henry Bence Jones. (From Snapper I, Kahn A. *Myelomatosis: fundamentals and clinical features.* Baltimore: University Park Press, 1971. By permission of S Karger.)

and soon acquired a large and remunerative practice (Fig. 1.6). His patients included the German chemist August Wilhelm Hofmann and the English biologist Thomas Huxley. Charles Darwin, the great naturalist, was another of his patients, whom he treated with a diet that 'half starved him to death.' His profits reached £7400 from 5 April 1864 to 5 April 1865.[9] About drugs, Bence Jones said that 'there is scarcely one which may not under different conditions produce opposite effects.'[10] Hermann von Helmholtz, the inventor of the ophthalmoscope, had great respect for Bence Jones, whom he described as a charming man, simple, harmless, cordial as a child, and extraordinarily kind.

Bence Jones was also well acquainted with Florence Nightingale and had a high opinion of her, seeking her advice about a project that he was considering for reform of nurses' training in the hospitals of London. She, in turn, regarded him as the best 'chemical doctor' in London. He served as one of the original members of the Council of the Nightingale Fund and was also influential in the

establishment of the Hospital for Sick Children on Great Ormond Street, on whose board he served.

As a student, Bence Jones attended lectures of the physicist Michael Faraday at the Royal Institution. He subsequently became a friend and physician to Faraday, and in 1870 he published a well-received two-volume biography of the prominent physicist.

Bence Jones was the first to describe xanthine crystals in the urine.[11] He emphasized the frequency of diabetes in the older population – 11 of his 29 patients were more than 60 years of age,[12] and noted that sugar was still found in the urine despite the withholding of sugar-containing foods.[13] He published a series of lectures on the 'applications of chemistry and medicine to pathology and therapeutics.' He became secretary of the Royal Institution in 1860 and subsequently wrote a history of it, including its first founders and first professors. He served as editor of the *Manual of Chemistry*. He believed that medical men would be better served if they spent time acquiring knowledge about chemistry and physics rather than learning Latin and Greek.

Bence Jones began his laboratory work each day at dawn and spent the afternoons and evenings doing hospital rounds. Medical students did not seek his clerkship because he was not adept at clinical teaching and had a well-known lack of punctuality. He made his diagnoses quickly and briefly. Irritable in manner and at times impetuous, he was sometimes too quick with criticism for those with opposing views. He was self-reliant and strong-willed and his chief characteristics were scientific truth, accuracy, and a dislike of empiricism. He always called for the 'medical facts.'

In 1861, Bence Jones experienced frequent heart palpitations and diagnosed rheumatic heart disease with his stethoscope on hearing a mitral systolic murmur. He had had an episode of rheumatic fever in 1839. A pleural effusion developed in 1866, and in early 1873 he gave up his practice because of hepatomegaly, ascites, and anasarca. He died at home at 84 Brook Street in London of congestive heart failure on 20 April 1873 at the age of 59 years and was buried at Kensall Green Cemetery.[9] Interestingly, although Bence Jones's obituary described his work on renal stones, diabetes mellitus, and malignant and tuberculous involvement of the kidney and his emphasis on the value of microscopic analysis of the urine, there was no mention of his articles on the unique urinary protein that bears his name.[13]

Incidentally, Henry Bence Jones did not use the hyphen in his name, and it does not appear in any of his more than forty papers and books; books published during his lifetime enter him under Jones. The hyphen was added by his descendants over half a century after his death.[14] The Bence Jones ward still exists at St George's Hospital in Tooting (Fig. 1.7)

Fig. 1.7 Bence Jones ward at St George's Hospital, Tooting. (From ref. 59.)

Other players in the history of Bence Jones proteinuria

Several other persons were involved in the story of Bence Jones proteinuria. J.F. Heller (Fig. 1.8) in 1846 described a protein in the urine that precipitated when warmed a little above 56°C and then disappeared on further heating. Although Heller did not recognize the precipitation of the protein when the urine was cooled, it is nearly certain that this was Bence Jones protein. He distinguished this new protein from albu-

Fig. 1.8 J.F. Heller. (From ref. 59.)

min and casein.[15] Bradshaw,[16] in 1898, found that meals had little or no influence on the amount of Bence Jones proteinuria. There was no nocturnal variation, and he believed that the rate of excretion was 'pretty constant throughout the 24 hours.' This finding was supported by Walters,[17] who reported that the amount of protein in a patient's diet had no effect on the amount of Bence Jones proteinuria.

Two distinct groups of Bence Jones proteins were recognized by Bayne-Jones and Wilson in 1922.[18] In 1956, Korngold and Lipari[19] demonstrated a relationship between Bence Jones protein and the serum proteins of multiple myeloma. It is in tribute to Korngold and Lipari that the two major classes of Bence Jones protein have been designated κ and λ.

One hundred and seventeen years after the description of the unique heat properties, Edelman and Galley[20] demonstrated that the light chains prepared from a serum immunoglobulin G (IgG) myeloma protein and the Bence Jones protein from the same patient's urine had the same amino acid sequence, similar spectrofluorometric behavior, the same molecular weight, identical appearance on chromatography with carboxymethylcellulose and on starch gel electrophoresis after reduction and alkylation, and the same ultracentrifugal pattern – as well as the same thermal solubility. The light chains precipitated when the protein was heated to between 40°C and 60°C, dissolved on boiling, and reprecipitated when cooled to between 40°C and 60°C.

Multiple myeloma – some early cases

In 1867, Hermann Weber[21] described a 40-year-old man with myeloma who suffered frequent colds in the spring and experienced sternal pain in May. The sternum was tender and then became deformed. The patient also had severe pain in the lumbar region. Movement of his head produced pain in his neck and arms. He died $3\frac{1}{2}$ months after the onset of pain. Post-mortem examination revealed that the sternum was almost entirely replaced by a grayish-red substance that had the microscopic appearance of a sarcoma. There were also two fractures of the sternum. Several round defects in the skull were replaced by the same morbid substance as found in the sternum. Many of the ribs, several vertebrae, and parts of the pelvis were involved. Amyloid was found in the kidneys and spleen. Five years later, William Adams[22] described a patient with 'acute rheumatism' characterized by bone pain, fractures, and fever. The left humerus and femur fractured while placing the body on the autopsy table. Lardaceous changes were found in the liver and kidneys. The cancellous portion of the bones was replaced by a homogeneous, soft, gelatinoid substance. Examined microscopically, the colloid substance filling the hollowed bone consisted of small spherical and oval cells that contained one oval nucleus (rarely, two) with a bright nucleolus.

The term 'multiple myeloma' was introduced by von Rustizky[23] (working in von Recklinghausen's laboratory) in 1873 when, during autopsy, he found eight separate tumors of bone marrow and designated them as multiple myeloma. The patient, a 47-year-old man, had presented with a gradually enlarging tumor in the right temple. Subsequently, thickening of the manubrium and seventh rib developed. This was followed by paraplegia. At autopsy, it was revealed that an apple-sized tumor in the right frontal region extended into the orbit and had produced ophthalmoplegia. Other findings included an apple-sized tumor in the right fifth rib, a tumor in the left seventh rib (producing a fracture), a tumor of the sternum, a tumor of the sixth to the eighth thoracic vertebrae (producing paraplegia), and three tumors of the right humerus. Although von Rustizky's description of the tumor cells is vague, he described round cells the size of white cells whose nucleus was located in the periphery near the cell membrane. He did not mention albumosurie (Bence Jones protein). In Russia, the term 'Rustizky's disease' is often used.

Willy Kühne, in 1883,[24] described a 40-year-old man from Amsterdam who he thought had acute osteomalacia. He had marked tenderness of the cervical and thoracic spine and had trouble sleeping at night because of pain and curvature of the spine. He was unable to lie on his back. His head was flexed forward. During the last weeks of his life, dysphasia and paralysis of the seventh cranial nerve developed, probably from an extradural plasmacytoma. He died on 27 August 1869, 9 months after the onset. There was no autopsy report. His brother was thought to have died of the same disease. The patient's urine precipitated on warming to between 40°C and 50°C and cleared at 100°C. Kühne isolated the protein and found that the carbon, hydrogen, and nitrogen levels were similar to those described by Bence Jones. Kühne attributed any differences to the fact that his preparation was more pure than Bence Jones's preparation. He labeled the protein 'albumosurie.'

There was little further interest in the disease until 1889, when Otto Kahler described a striking case involving a 46-year-old physician named Dr Loos.[25] His first symptom, noted in July 1879, consisted of sudden severe pain in the right upper thorax, which was aggravated by taking a deep breath. Six months later, the pain recurred and became localized to the right third rib, which was tender to pressure. During the next 2 years, intermittent pain, aggravated by

exercise, occurred in the ribs, spinal column, left shoulder, upper arm, and right clavicle. Albuminuria was first noticed in September 1881. Skeletal pain, made worse by movement, continued to occur intermittently. Pallor was noted in 1883. In December 1885, Dr Loos was first seen by Kahler, who noted anemia, severe kyphosis, and tenderness of many bones, and albumosuria. The urinary protein had the same characteristics as that described by Bence Jones. When Dr Loos stood, the lower ribs touched the iliac crest. He had recurrent bronchial infections and intermittent hemoptysis. During the following year, kyphosis increased and his height decreased monthly. He became dwarflike. The kyphosis of the upper thoracic spinal column increased, and his chin pressed against the sternum, producing a decubitus ulcer.

On 26 August 1887, Dr Loos died, 8 years after the onset of symptoms. Autopsy revealed hepatosplenomegaly. The ribs were soft and could be broken with minimal effort. Soft gray-reddish masses were noted in the ribs and thoracic vertebrae. Microscopic examination showed large, round cells consistent with myeloma. It is interesting to note that the patient had a high fluid intake and took sodium bicarbonate on a regular basis. This management may have helped prevent renal failure.

OTTO KAHLER

Otto Kahler, born in 1849, was the son of a well-known physician in Prague (Fig. 1.9). He received his MD degree from the University of Prague in 1871 and then worked as an assistant in Professor Halla's clinic. During a sabbatical in Paris, he met two French neurologists, Jean Martin Charcot and G.B.A. Duchenne (Duchenne de Boulogne). Kahler became interested in neurology and particularly in anatomy. He contributed to the pathologic anatomy of the central nervous system and to the anatomy of tabes dorsalis, localization of central oculomotor paralysis, and slow compression of the spinal cord. He became professor of medicine in Prague and, after Halla's resignation, became head of the Second Medical Clinic at the German University of Prague. In 1889, Kahler was invited to succeed the famous physician Heinrich Bamberger as professor at the University of Vienna.[26] In his inaugural address on 13 May 1889, Kahler paid tribute to Professor Bamberger and finished his lecture with one of Bamberger's quotes: 'Ars longa, vita brevis' (the art of medicine is long, life is short).[27] Little did he realize that the tumor on his tongue biopsied in the summer of 1889 was malignant. The carcinoma recurred the following year and a huge tumor developed, causing paralysis of the vagus nerve and

Fig. 1.9 Otto Kahler. (Courtesy of Dr Heinz Ludwig, Vienna.)

compression of the esophagus and main bronchus. Kahler died on 24 January 1893, shortly after his 44th birthday.[28] Kahler was extremely kind to his patients and was an excellent teacher. He emphasized that it was important to cover the entire field of general internal medicine. His obituaries and eulogies made no mention of his famous case report. It is of interest to note that the landmark contributions by Henry Bence Jones and Otto Kahler were not recognized during their lifetimes.

Multiple myeloma in the twentieth century

The first case of multiple myeloma in the United States was reported by Herrick and Hektoen in 1894.[29] The patient, a 40-year-old white woman, had lumbar pain and a nodule on the lower end of the sternum. At autopsy, multiple nodules attached to the sternum, right clavicle, and ribs were found. The sternum was thickened, irregular, and covered with tumor masses but was soft and flexible. Multiple nodules were found on the ribs, which bent readily without cracking. Two of the dorsal vertebral bodies were largely replaced by soft tumor masses. Fungoid masses were seen in the skull. Microscopic examination revealed round, lymphoid cells with large nuclei.

Just 3 years after the discovery of X-rays by Roentgen, Weber reported a case of multiple myeloma and stated that the diagnosis of such cases would be greatly facilitated by the use of X-rays. He concluded that Bence Jones protein was produced by the bone marrow. He also believed that the presence of Bence Jones protein was of 'fatal significance' and nearly always indicated that the patient had multiple myeloma.[30] Weber and Ledingham[31] later suggested that Bence Jones protein came from the cytoplasmic residua of karyolyzed plasma cells.

In 1900, Wright[32] described a 54-year-old man with multiple myeloma and pointed out that the tumor consisted of plasma cells. He emphasized that the neoplasm originated not from red marrow cells collectively but from only one type of cell, the plasma cell. This patient was one of the first in whom roentgenograms revealed changes in the ribs and thus contributed to the diagnosis.

The term 'plasma cell' was coined by Waldeyer in 1875,[33] but his description is not characteristic of plasma cells, and he most likely was describing tissue mast cells. Plasma cells were described accurately by Ramón y Cajal in 1890 during study of syphilitic condylomas; he stated that the unstained perinuclear area (hof) contained the Golgi apparatus.[34] In 1891, Unna[35] used the term 'plasma cell' while describing cells seen in the skin of patients with lupus erythematosus. However, it is not known whether he actually saw plasma cells. In 1895, Marschalkó[36] described the essential characteristics of plasma cells, including blocked chromatin, eccentric position of the nucleus, a perinuclear pale area (hof), and spherical or irregular cytoplasm.

Geschickter and Copeland[37] in 1928 presented an analysis of all 425 cases of multiple myeloma reported since 1848. They called attention to six cardinal features of the disease: back pain, anemia, chronic renal disease, Bence Jones proteinuria, pathologic fractures, and multiple involvement of tumors of the skeletal trunk. Sternal aspiration of the bone marrow, described in 1929 by Arinkin,[38] greatly increased the ante-mortem recognition of multiple myeloma.[39] Bayrd and Heck[40] described eighty-three patients with histologic proof of multiple myeloma seen at the Mayo Clinic through December 1945. The duration of survival ranged from 1 to 84 months (median, 15 months).

Although Jacobson[41] had reported Bence Jones

protein in the serum and urine in 1917, it was not until 1928 that Perlzweig et al.[42] reported hyperproteinemia when they described a patient with multiple myeloma who had 9–11 g of globulin in his serum. The patient also had Bence Jones proteinuria and probably a small amount of Bence Jones protein in the plasma. They also noted that it was almost impossible to obtain serum from the clotted blood because the clot failed to retract, even on prolonged centrifugation. Cryoglobulinemia was recognized by Wintrobe and Buel[43] in 1933 and named 'cryoglobulin' by Lerner and Watson in 1947.[44] In 1938, von Bonsdorff et al.[45] described a patient with cryoglobulinemia in which the globulins crystallized after exposure to the cold for 24 hours.

In 1890, von Behring and Kitasato[46] described a specific neutralizing substance in blood of animals immunized with diphtheria and tetanus toxin. These antibodies were found after the injection of most foreign proteins. In 1937, Tiselius[47] used an electrophoretic technique to separate serum globulins into three components, which he designated α, β, and γ. Interestingly, this article, which led to his Nobel Prize and later to the presidency of the Nobel Foundation, was rejected initially by *Biochemical Journal*.[48] Two years later, Tiselius and Kabat[49] localized antibody activity in the gamma globulin fraction of the plasma proteins. They noted that antibodies to albumin or pneumococcus Type I were found in the area of γ mobility in rabbit serum and antibodies to pneumococcal organisms migrated between β and γ in horse serum. Later, it was recognized that some antibodies migrate in the fast γ region and others in the slow, and some sediment in the ultracentrifuge as 7S and others as 19S molecules; but the concept of a family of proteins with antibody activity was not proposed until late in the 1950s.[50] Before 1960, the term 'gamma globulin' was used for any protein that migrated in the γ mobility region of the electrophoretic pattern. Now these proteins are referred to as immunoglobulins – IgG, IgA, IgM, IgD, and IgE.

In 1939, Longsworth et al.[51] applied electrophoresis to the study of multiple myeloma and demonstrated the tall, narrow-based 'church-spire' peak. This method was cumbersome and difficult; therefore, it was not readily available until the early 1950s, when filter paper was introduced as a supporting medium (zone electrophoresis). Cellulose acetate has since supplanted filter paper.[52] In 1953, Grabar and Williams[53] described immunoelectrophoresis, which has facilitated the diagnosis of multiple myeloma.

Alwall,[54] in 1947, reported that a patient with typical multiple myeloma had a reduction in globulin from 5.9 to 2.2 g/dl, increase in hemoglobin from 60 per cent to 87 per cent, disappearance of proteinuria, and a reduction in bone marrow plasma cells from 33 per cent to 0 per cent when treated with urethane. For almost 20 years, urethane was commonly used for the treatment of myeloma. Holland et al.[55] randomized eighty-three patients with treated or untreated multiple myeloma to receive urethane or a placebo consisting of cherry and cola-flavored syrup. There was no difference in objective improvement nor in survival of the two treatment groups. In fact, the urethane-treated patients died earlier on the average than those treated with placebo. This difference was ascribed to the increased mortality of urethane-treated patients who were azotemic. The patients with poorer prognostic features had a significantly shorter survival with urethane therapy.

In 1958, Blokhin et al.[56] reported benefit in 3 of 6 patients with multiple myeloma who were treated with sarcolysin. Four years later, Bergsagel et al.[57] found significant improvement of 8 in 24 patients with multiple myeloma who were treated with DL-phenylalanine mustard (melphalan, Alkeran). Six other patients obtained improvement in one or more objective factors. Cyclophosphamide-treated patients with myeloma had a median survival of 24.5 months, whereas an ancillary myeloma group had a median survival of 9.5 months.[58] Objective improvement occurred in 81 of 207 patients.

Acknowledgements: Parts of this chapter have been previously published in ref. 3. They are reprinted here by permission of Churchill Livingstone.

References

1. Clamp JR. Some aspects of the first recorded case of multiple myeloma. *Lancet* 1967; **2**: 1354–6.
2. Macintyre W. Case of mollities and fragilitas ossium, accompanied with urine strongly charged with animal matter. *Medico-Chirugical Transactions of London* 1850; **33**: 211–32.
3. Kyle RA. History of multiple myeloma. In: Wiernik PH, Canellos GP, Kyle RA, Schiffer CA eds *Neoplastic diseases of the blood*, 2nd edn. New York: Churchill Livingstone, 1991: 325–32.
4. Bence Jones H. Chemical pathology. *Lancet* 1847; **2**: 88–92.
5. Dalrymple J. On the microscopical character of mollities ossium. *Dublin Quarterly Journal of Medical Science* 1846; **2**: 85–95.
6. Solly S. Remarks on the pathology of mollities ossium. With cases. *Medico-Chirugical Transactions of London* 1844; **27**: 435–61.
7. Bence Jones H. On a new substance occurring in the urine of a patient with mollities ossium. *Philosophical Transactions of the Royal Society of London Biology* 1848: 55–62.
8. Coley NG. Henry Bence-Jones, M.D., F.R.S. (1813–1873). *Notes and Records of the Royal Society of London* 1973; **28**: 31–56.

9. Obituary. Henry Bence Jones, M.D., M.A., F.R.C.P., F.R.S. *Medical Times Gazette* 1873; **1**: 505.

10. Rosenbloom J. An appreciation of Henry Bence Jones, M.D., F.R.S. (1814–1873). *Annals of Medical History* 1919; **2**: 262–4.

11. Bence Jones H. On a deposit of crystallized xanthin in human urine. *Journal of the Chemical Society of London* 1862; **15**: 78–80.

12. Bence Jones H. On intermitting diabetes, and on the diabetes of old age. *Medico-Chirugical Transactions of London* 1853; **36**: 403.

13. Obituary. Dr Henry Bence Jones. *Lancet* 1873; **1**: 614–15.

14. Rosenfeld L. Henry Bence Jones (1813–1873): the best 'chemical doctor' in London. *Clinical Chemistry* 1987; **33**: 1687–92.

15. Heller JF. Die mikroskopisch-chemisch-pathologische Untersuchung. In: von Gaal G ed. *Physikalische Diagnostik und deren Anwendung in der Medicin, Chirurgie, Oculistik, Otiatrik und Geburtshilfe, enthaltend: Inspection, Mensuration, Palpation, Percussion und Auscultation, nebst Einer Kurzen Diagnose der Krankheiten der Athmungs- und Kreislaufsorgane.* Vienna: Braumüller and Seidel, 1846: 576–97.

16. Bradshaw TR. Cited in Bryant T: A case of albumosuria in which the albumose was spontaneously precipitated. *British Medical Journal* 1898; **1**: 1136.

17. Walters W. Bence-Jones proteinuria: a report of three cases with metabolic studies. *Journal of the American Medical Association* 1921; **76**: 641–5.

18. Bayne-Jones S, Wilson DW. Immunological reactions of Bence-Jones proteins. II. Differences between Bence-Jones proteins from various sources. *Bulletin of the Johns Hopkins Hospital* 1922; **33**: 119–25.

19. Korngold L, Lipari R. Multiple-myeloma proteins. III. The antigenic relationship of Bence Jones proteins to normal gamma-globulin and multiple-myeloma serum proteins. *Cancer* 1956; **9**: 262–72.

20. Edelman GM, Galley JA. The nature of Bence-Jones proteins: chemical similarities to polypeptide chains of myeloma globulins and normal γ-globulins. *Journal of Experimental Medicine* 1962; **116**: 207–27.

21. Weber H. Mollities ossium, doubtful whether carcinomatous or syphilitic. *Transactions of the Pathology of the Society London* 1867; **18**: 206–9.

22. Adams W. Mollities ossium. *Transactions of the Patholology Society of London* 1872; **23**: 186–7.

23. von Rustizky J. Multiples myelom. *Deutsche Zeitschrift für Chirurgie (Berlin)* 1873; **3**: 162–72.

24. Kühne W. Ueber Hemialbumose im Harn. *Zeitschrift für Biologie (Munich)* 1883; **19**: 209–27.

25. Kahler O. Zur Symptomatologie des multiplen Myeloms: Beobachtung von Albumosurie. *Prager medizinische Wochenschrift (Prague)* 1889; **14**: 33; 45.

26. Kraus F. Gedächtnisrede auf Otto Kahler. *Wiener klinische Wochenschrift (Wien)* 1894; **27**: 110.

27. Sigmund CL. Zur örtlichen Behandlung: syphilitischer Mund- Nasen- und Rachenaffektionen. *Wiener klinische Wochenschrift (Wien)* 1870; **20**: 781.

28. Nothnagel. Hofrath Otto Kahler. *Wiener klinische Wochenschrift (Wien)* 1893; **6**: 79–80

29. Herrick JB, Hektoen L. Myeloma: report of a case. *Medical News* 1894; **65**: 239–42.

30. Weber FP, Hutchison R, Macleod JJR. Multiple myeloma (myelomatosis), with Bence-Jones proteid in the urine (myelopathic albumosuria of Bradshaw, Kahler's disease). *American Journal of Medical Science* 1903; **126**: 644–65.

31. Weber FP, Ledingham JCG. A note on the histology of a case of myelomatosis (multiple myeloma) with Bence-Jones protein in the urine (myelopathic albumosuria). *Proceedings of the Royal Society of Medicine, London* 1909; **2**: 193–206.

32. Wright JH. A case of multiple myeloma. *Johns Hopkins Hospital Reports* 1900; **9**: 359–66.

33. Waldeyer W. Ueber Bindegewebszellen. *Archiv für mikroskopische Anatomie (Bonn)* 1875; **11**: 176–94.

34. Ramón y Cajal S. Estudios histológicos sobre los tumores epiteliales. *Rev Trimest Microgr* 1896; **1**: 83–111.

35. Unna PG. Über plasmazellen, insbesondere beim Lupus. *Monatsschrift für praktische Dermatologie (Hamburg)* 1891; **12**: 296.

36. Marschalkó T. Ueber die sogenannten Plasmazellen, ein Beitrag zur Kenntniss der Herkunft der entzündlichen Infiltrationszellen. *Archiv für Dermatologie und Syphilis (Wien)* 1895; **30**: 241.

37. Geschickter CF, Copeland MM. Multiple myeloma. *Archives of Surgery* 1928; **16**: 807–63.

38. Arinkin MI. Die intravitale Untersuchungsmethodik des Knochenmarks. *Folia haematologica (Leipzig)* 1929; **38**: 233–40.

39. Rosenthal N, Vogel P. Value of the sternal puncture in the diagnosis of multiple myeloma. *Mt Sinai Journal of Medicine* 1938; **4**: 1001–19.

40. Bayrd ED, Heck FJ. Multiple myeloma: a review of eighty-three proved cases. *Journal of the American Medical Association* 1947; **133**: 147–57.

41. Jacobson VC. A case of multiple myelomata with chronic nephritis showing Bence-Jones protein in urine and blood serum. *Journal of Urology* 1917; **1**: 167–78.

42. Perlzweig WA, Delrue G, Geschickter C. Hyperproteinemia associated with multiple myelomas: report of an unusual case. *Journal of the American Medical Association* 1928; **90**: 755–7.

43. Wintrobe MM, Buell MV. Hyperproteinemia associated with multiple myeloma: with report of a case in which an extraordinary hyperproteinemia was associated with thrombosis of the retinal veins and symptoms suggesting Raynaud's disease. *Bulletin of Johns Hopkins Hospital* 1933; **52**: 156–65.

44. Lerner AB, Watson CJ. Studies of cryoglobulins. I. Unusual purpura associated with the presence of a high concentration of cryoglobulin (cold precipitable serum globulin). *American Journal of Medical Science* 1947; **214**: 410–15.

45. von Bonsdorff B, Groth H, Packalén T. On the presence of a high-molecular crystallizable protein

in the blood serum in myeloma. *Folia haematologica (Leipzig)* 1938; **59**: 184–208.

46. von Behring S, Kitasato T. I. Aus dem hygienischen Institut des Herrn Geheimerath Koch in Berlin. Ueber das Zustandekommen der Diphtherie-Immunität und der Tetanus-Immunität bei Thieren. *Deutsche medizinische Wochenschrift (Leipzig)* 1890; **49**: 1113–45.

47. Tiselius A. Electrophoresis of serum globulin. II. Electrophoretic analysis of normal and immune sera. *Biochemical Journal* 1937; **31**: 1464–77.

48. Putnam FW. From the first to the last of the immunoglobulins: perspectives and prospects. *Clinical Physiology and Biochemistry* 1983; **1**: 63–91.

49. Tiselius A, Kabat EA. An electrophoretic study of immune sera and purified antibody preparations. *Journal of Experimental Medicine* 1939; **69**: 119–31.

50. Heremans JF. Immunochemical studies on protein pathology: the immunoglobulin concept. *Clinica Chimica Acta* 1959; **4**: 639–46.

51. Longsworth LG, Shedlovsky T, MacInnes DA. Electrophoretic patterns of normal and pathological human blood serum and plasma. *Journal of Experimental Medicine* 1939; **70**: 399–413.

52. Kohn J. A cellulose acetate supporting medium for zone electrophoresis. *Clinica Chimica Acta* 1957; **2**: 297–303.

53. Grabar P, Williams CA. Méthode permettant l'étude conjuguée des propriéteés électrophorétiques et immunochimiques d'un mélange de protéines. Application au sérum sanguin. *Biochemica Biophysica Acta* 1953; **10**: 193–4.

54. Alwall N. Urethane and stilbamidine in multiple myeloma: report on two cases. *Lancet* 1947; **2**: 388–9.

55. Holland JF, Hosley H, Scharlau C et al. A controlled trial of urethane treatment in multiple myeloma. *Blood* 1966; **27**: 328–42.

56. Blokhin N, Larionov L, Perevodchikova N et al. Clinical experiences with sarcolysin in neoplastic diseases. *Annals of the New York Academy of Sciences* 1958; **68**: 1128–32.

57. Bergsagel DE, Sprague CC, Austin C, Griffith KM. Evaluation of new chemotherapeutic agents in the treatment of multiple myeloma IV. L-phenylalanine mustard (NSC–8806). *Cancer Chemotherapy Reports* 1962; **21**: 87–99.

58. Korst DR, Clifford GO, Fowler WM et al. Multiple myeloma: II. Analysis of cyclophosphamide therapy in 165 patients. *Journal of the American Medical Association* 1964; **189**: 758–62.

59. Kyle RA. History of multiple myeloma. In: Wiernik PH, Canellos GP, Kyle RA, Schiffer CA eds *Neoplastic diseases of the blood*, 3rd edn. New York: Churchill Livingstone, 1996.

CHAPTER 2

Epidemiology and incidence of plasma cell neoplasms

DANIEL E BERGSAGEL

Introduction	12	The search for other causes of plasma	
The frequency of MGUS and MM	12	cell neoplasms	15
Radiation exposure	15	Genetic factors	18
		References	19

Introduction

Plasma cell neoplasms are distinguished by an idiotypic rearrangement of the immunoglobulin gene, which occurs prior to the malignant transformation of an early B-cell precursor. The clone which develops must increase to about 5×10^9 cells before it produces enough of the idiotypic immunoglobulin to be recognized as a monoclonal 'spike' (M-protein) in a serum electrophoresis pattern. Most subjects with a serum M-protein are asymptomatic; if other causes of an M-protein can be ruled out, they are labeled as monoclonal gammopathies of undetermined significance (MGUS). By definition, the monoclone in MGUS is stable, and the serum M-protein concentration remains level for many years. However, prolonged follow-up of a large group of MGUS subjects at the Mayo Clinic has shown that about 2 per cent of these patients progress per year to develop symptomatic multiple myeloma (MM), macroglobulinemia, malignant lymphoma, chronic lymphocytic leukemia, or amyloidosis.[1] The criteria for distinguishing MGUS and MM are described elsewhere (*see* Chapter 9). MGUS is considered to be a premalignant lesion because the clone does not grow progressively, but is stable and asymptomatic. An additional neoplastic change is required to convert this large, stable clone into MM, a progressively expanding tumor with malignant characteristics.

The frequency of MGUS and MM

The prevalence of an M-protein (i.e. the number of cases in a defined population at a certain time) in people over the age of 25 was determined in the spring of 1964 in the Varmland district of Sweden by doing paper electrophoresis on 6995 consecutive serum samples. The sera of 64 subjects in this sample (0.9 per cent) contained an M-protein.[2] Only one of this group was found to have MM; 63 were classified as MGUS. Among the 6931 subjects without serum M-components in 1964, two were diagnosed as MM (one had light-chain myeloma) within 3 years.[3] The population of Varmland in 1964 was about 10 000. In the 70 per cent of the adult population of this district who had a serum electrophoresis done in the spring of 1964, the prevalence of MGUS was 901/100 000, and of MM 43/100 000. MGUS is encountered much more frequently than MM because these subjects are well, and in contrast to MM patients, who have a median survival of only 32 months from the onset of treatment, MGUS cases tend to survive for prolonged periods, and thus accumulate in the population.

The incidence of MGUS (i.e. the number of new cases developing in a defined population over a defined period of time) has not been determined. In the USA, MM incidence data have been provided by the Surveillance, Epidemiology and End Results (SEER) program since 1973. For the 1973–7 period, the average, annual, age-adjusted (1970 standard) incidence rates per 100 000 for all races in the USA was 3.9 for both sexes, 4.7 for males, and 3.3 for females.[4] In a disease like MM, mortality rates are very similar to the incidence rates because almost all of the people who develop the disease die with it. Mortality rates for MM, however, are usually somewhat lower than incidence rates.

The frequency of MM in a population is strongly influenced by age, race, and the availability of good medical care. MM is slightly more common in men. The incidence of MM appeared to increase markedly in many parts of the world between 1950 and 1980. Persons with IgG and IgA myeloma did not differ in the distribution of age or sex, but a slightly higher proportion of light-chain cases were women (58 versus 45 per cent of all other cases).[5]

AGE

The age-specific incidence rates for MM in Malmö, Sweden, 1970–9, are shown in Fig. 2.1. Myeloma was not detected in patients under the age of 40 in this sample. The incidence increases progressively with age, to reach $64.5/10^5$ in males, and $36.6/10^5$ in females over the age of 80 years.[6] The median age of MM patients eligible for admission to large clinical group studies in North America is about 62 years,[7] but when the total population of myeloma patients in an area is determined, with no exclusions, as was done in the Health Care Region of Western Sweden, the median age at diagnosis was 72 years, and the male/female ratio 1.1.[8]

The frequency of MGUS also increases with age. Axelsson et al. detected an M-protein in 0.2 per cent of subjects aged 30–49, in 1.4 per cent between the ages of 50 and 69, and in 4.0 per cent of subjects aged 70–89.[2] Two surveys of the serum electrophoresis patterns of nonagenarians both reported that a surprising 19 per cent of these sera contain M-proteins.[9,10]

RACE

The age-adjusted incidence (world standard population) of MM from selected population-based cancer registries

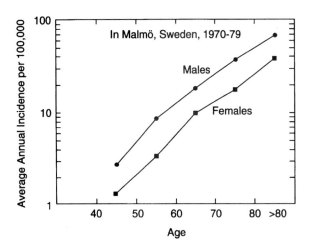

Fig. 2.1 Average annual age-specific incidence rates of multiple myeloma in Malmö, Sweden, 1970–9. Age-adjusted rates in the 1950 US (total) population, and the European standard population (values in brackets) were: males 4.6 (6.2); females 2.2 (3.0); and both sexes 3.2 (4.3).[6]

around the world range from 0.5 in Hawaiian Japanese males to 8.2 in US Bay area black males per 100 000 person-years.[11] The incidence of MM is distinctly lower ($1.4/10^5$, or less, world standard population) for the Chinese of Shanghai and Singapore, and the Japanese of Osaka and Hawaii, than in the Caucasian populations of North America and Europe. Comparisons between incidence rates from cancer registries in different countries must be made cautiously, since the populations served may vary in terms of the availability of medical care and diagnostic services, the completeness of case ascertainment, and the calculation of age-adjusted incidence rates. Still, there do appear to be real differences in the incidence of MM in different races. The observation that the incidence of MM, age-adjusted to the 1970 US standard, in male Chinese ($2.3/10^5$) and Japanese ($1.7/10^5$) living in San Francisco-Oakland and Hawaii, are distinctly lower than for white males ($4.6/10^5$) in the same regions, support this view.[12] Surprisingly, the incidence in Filipino males ($4.7/10^5$) is almost the same as that of white males in these regions.

The average annual age-adjusted (1970 US standard) incidence rates for blacks in America are more than double the rates in whites, at $10.8/10^5$ for black males, and $7.2/10^5$ for black females.[13] The increased incidence of MM in blacks makes this disease the most common hematologic malignancy in the black population of the US (Fig. 2.2). The proportion of patients who develop leukemia is similar in the two groups, while MM accounts for 33 per cent of hematologic malignancies in blacks, and only 13 per

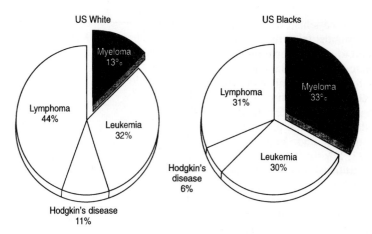

Fig. 2.2 Relative percentages of hematologic malignancies in US whites and blacks, 1982–6. (Reproduced with permission from ref. 11.)

cent in whites. Hodgkin's disease and lymphomas are reduced to 37 per cent of the hematologic malignancies in blacks versus 55 per cent in whites.[11] Socioeconomic status and poverty, as defined by the proportion of people with incomes below the poverty level, do not explain the increased incidence of MM in American blacks, since the rates remain twice those in whites after adjustment.[14] Comprehensive MM incidence data are not available for African populations, but do appear to be high among Jamaicans.[15] Valuable insights into the reasons for the increased incidence of MM in blacks in America might be gained if variations were to be found among black populations around the world.

A comparison of the prevalence of MGUS in subjects over the age of 60 years in retirement communities in Japan and the US revealed an M-protein in 4/146 Japanese (2.7 per cent), versus 11/111 American (10 per cent).[16] The reduced prevalence of MGUS in Japan is in keeping with the lower incidence of MM in this country. Measurements of normal immunoglobulins in these subjects by laser nephelometry provided the surprising observation that for unknown reasons IgG and IgA levels were significantly higher in the elderly Japanese. The frequency of MGUS in sera submitted for electrophoresis at the Houston Veterans Administration Hospital in Texas, was found to be much higher in blacks than in whites, especially in the group over the age of 79,[17] again in agreement with the increased incidence of MM in blacks.

INCREASING MORTALITY RATES FOR MM

Trends in the MM mortality rates for four age groups in the US between 1968 and 1986 are shown in Fig. 2.3. A striking increase occurred in these rates during this

period, especially in the older age groups. Similar increases were noted in the reported myeloma death rates from the United Kingdom, France, Germany, Japan, and Italy.[18] The rate of increase is largest in the group over the age of 85 years, and falls continuously with decreasing age. The rate of increase is substantial by age 70, more than doubling in both sexes in all countries between 1968 and 1986. There is much debate about whether this increasing mortality rate is real, or the result of steadily improving case ascertainment.

Myeloma mortality rates did not increase in Omstead county, Minnesota, during the 33-year period from 1945 to 1977.[19] Most of the medical care for Omstead county is provided by the Mayo Clinic, where there has been a continuing interest in detecting and diagnosing all plasma cell neoplasms accurately. Linos et al. suggest that all of the cases of MM

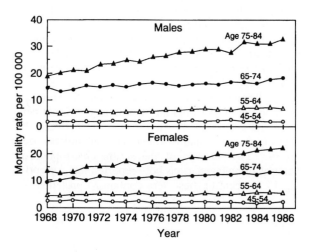

Fig. 2.3 Myeloma mortality rates per 100 000 for four age groups in the US, 1968–86.[79]

have been detected and reported in Omstead county since 1945, and since an increasing incidence has not been observed there, the reported increase in other parts of the US is probably due to improving case ascertainment, and more apparent than real. Other regions with a high standard of medical care and an interest in myeloma, such as Malmö, Sweden, and the canton of Vaud, Switzerland, also have not observed such a marked increase in the incidence of myeloma.[6,20] The authors of these reports agree with the Mayo Clinic interpretation that they represent the asymptote of myeloma detection that may be matched in other areas as their case ascertainment becomes complete. In Denmark there was a two- to three-fold increase in the incidence of MM in both men and women between 1943 and 1962, but since then the incidence has been stable.[21]

Caution must be exercised in the interpretation of the worldwide increase in the incidence and mortality rates for MM, for the increases may be real! Similar patterns of increase in both men and women suggest a ubiquitous, possibly environmental factor, rather than occupational exposure. Cuzick notes that the increasing trends appear to have slowed since 1980, and are even more confined to the oldest age groups. If this increase is due to an environmental factor, then its introduction is now largely complete, suggesting that any such factor was introduced at least 20 years ago.[22]

Radiation exposure

In 1979, Ichimaru et al.[23] concluded that there was a statistically significant increased incidence of MM between 1950 and 1976 among the survivors exposed to radiation dose estimates of more than 1 Gy after the atomic bombs at Hiroshima and Nagasaki. The increased incidence of MM became apparent about 20 years after radiation exposure. The Life Span Study of Atomic Bomb Survivors by the Radiation Effects Research Foundation has added 12 years of follow-up for the occurrence of MM. A recent re-analysis of the data,[24] using Dosimetry System 1986 dose estimates, re-evaluation of all the cases included in previous studies and an estimation of the excess absolute risk (EAR), has changed the conclusions about the effect of radiation on the incidence of MM. The increased EAR for acute lymphocytic leukemia, acute myelogenous leukemia, and chronic myelogenous leukemia was confirmed. There was some evidence of an increased risk of lymphoma in males, but not in females. The risk of developing chronic lymphocytic leukemia and MM was not increased.

Also, the frequency of a monoclonal gammopathy does not appear to have increased in the survivors of the atomic bombs.[25] Cellulose acetate electrophoresis was done on an aliquot of serum from all persons examined in the Adult Health Study in Hiroshima and Nagasaki between October 1979 and September 1981, and in a second survey between June 1985 and May 1987. In the first study, an M-protein was found in 31 of 8796, and in the second, in 68 of 7350 subjects. The relative risk of having an M-protein detected in the two surveys was not significantly increased with increasing radiation dose.

An increased risk of MM has also been reported among workers at nuclear processing plants at Hanford, Washington, and Winscale, UK, and in radium-dial painters.[26] The significance of these findings is weakened by the failure to detect an increased incidence of MM among the atomic bomb survivors.[24]

The search for other causes of plasma cell neoplasms

When a sleuth begins to hunt for a criminal, he looks first for clues associated with the crime, which he tries to organize into a pattern that associates one, or more, of the suspects with the felony. When enough useful clues have been collected, a testable hypothesis is formulated for the identification of the guilty person, and the search moves on. The problem becomes much more complicated when there is more than one criminal. The most blatant, but not necessarily the major culprit, is usually caught first, leaving the defenders of society with the daunting task of tracking down the more devious collaborators.

The induction of plasma cell neoplasms is probably a multistep process. Genetic factors probably play a role in making a person susceptible to a change which results in the proliferation of an early B-cell precursor, to form a stable clone of plasma cells producing a monoclonal protein, as in MGUS. The conversion of the controlled, stable monoclone of MGUS into the uncontrolled, progressive, malignant tumor of MM probably requires one, or more, additional changes.

Since we have no promising leads, we are still in the tiresome phase of looking for clues in the dark.

The major tools employed by epidemiologic detectives in looking for the causes of cancer are cohort studies, and case-control studies.[27]

COHORT STUDIES

A cohort study is often the first, broad-ranging attempt to detect an exposure that may be responsible for causing a cancer. Suspicious exposures are then investigated in more detail, perhaps with the aid of a hypothesis about the biologic mode of action of the agent in question. Cohort studies are not very sensitive, and they are rarely specific enough to identify the causative agent.

In a cohort study, defined populations exposed to factors suspected of increasing, or decreasing the incidence of a cancer, and a control group, not so exposed, are followed for an interval long enough for a number of cancers to develop. The incidence of cancer in the two groups is then compared. Of course, the exposed and unexposed groups must be as alike as possible. In an experimental trial, the exposure under investigation can be randomly assigned, and the subjects followed for the development of cancer according to a predetermined schedule and examination routine. This type of experimental trial is rarely possible in the investigation of the causes of cancer, and so an observational cohort study is substituted, in which the allocation of the exposure is not under the control of the investigators.

Cohort studies compare the cancer incidence in the exposed and unexposed groups. The contribution of the exposure to cancer risk is examined by calculating the ratio of the risks in the two groups, i.e. the relative risk. These studies require very large numbers of subjects, and prolonged observation, in order to have a reasonable chance of detecting a change in the incidence of cancer following exposure.

CASE-CONTROL STUDIES

Case-control studies begin with a group of patients with a cancer, e.g. myeloma, and compare the exposure to various agents of these cases with the exposure of a control group, selected to match the cases in terms of important confounding variables such as age, race, sex, smoking history, etc. The exposure in question is assessed by a questionnaire administered to both the cases and controls; this is one of the problems with case-control studies, for exposure requires recall, which may differ in cases and controls.

The influence of the exposure being investigated on the occurrence of myeloma is assessed with the odds ratio:

Odds ratio =

Number of cases exposed × Number of controls not exposed

Number of controls exposed × Number of cases not exposed

Case-control studies can be done quickly and require fewer subjects to detect a given level of risk. However, they are dependent on the ability of the cases and the controls to remember their exposures, and there are many potential biases in the selection of subjects and the measurement of exposure. Despite these disadvantages, case-control studies may be the best that can be done to investigate the validity of new theories about cancer risk in humans.

SOCIOECONOMIC STATUS

In earlier times myeloma was reported to cause increased mortality rates in members of the higher levels of society.[28] This association probably occurred because the disease was under-ascertained in the poor and uneducated. The reduced detection of myeloma in lower socioeconomic groups has gradually decreased with time, until the 1970s, when it disappeared.[29] A case-control study at Duke University in North Carolina failed to demonstrate any association of myeloma with family income, education, occupation, dwelling size or an index of crowding in the home.[30]

Socioeconomic factors such as household size and family income do not account for the higher incidence of myeloma in blacks in the US. The incidence in blacks remains twice that of whites after adjustment for incomes below and above the poverty level.[14] No occupational or environmental factors have been discovered to explain the marked difference in the incidence of myeloma in blacks and whites.

SMOKING

A prospective cohort study of 34 000 US Seventh Day Adventists followed for 6 years detected nine cases of myeloma in ex-smokers, and two in current smokers. The risk of developing myeloma was three-fold greater in subjects who had ever smoked than in those who had never smoked; a dose–response trend was also noted.[31] This group will have to be followed longer because this positive result differs from the negative association observed in most other studies. In the largest cohort study ever done on the use of tobacco and the occurrence of myeloma, Heineman et al. followed 250 000 US veterans for 26 years.[32] Myeloma deaths occurred in 582 patients in this study, but the relative risk of dying of myeloma was the same for both the smokers and those who had never smoked. Similarly, a smaller prospective cohort study of 17 633 male holders of Lutheran Brotherhood insurance, who completed a self-administered questionnaire about their use of tobacco in 1966, did not observe a significantly increased risk of death from myeloma among those who were smokers. Strangely, and in contrast, the risk of death from non-Hodgkin's lymphoma in this study was increased among men who had ever smoked, and was almost four-fold greater among the heaviest smokers.[33]

OCCUPATIONS

Riedel et al.[11] have reviewed reports of positive associations between specific occupations/industries and myeloma reported in 53 case-control, prospective, and proportionate incidence/mortality studies.

Employment in *agriculture* (predominantly farming) is the occupation most frequently associated with myeloma. Most studies have detected this association, but a few have not.[34–36] These studies have not been able to identify the aspect of agricultural work, be it contact with animals, grains, dust, fertilizers, pesticides or engines, which increases the risk of farmers for myeloma.

Statistical associations have been made between employment as *metal workers* and an increased risk of myeloma. Small numbers of cases and a lack of information on actual exposures make it difficult to determine whether particular metal dusts or fumes, or other occupational exposures, are responsible for the elevated risks observed.

Five reports of increased myeloma mortality in workers employed in *rubber manufacturing* are included in table 23–1 by Riedel et al.,[11] but only one of these studies reported a statistically significant increase, and that was in white male union rubber reclaim workers in the US.[37] A recent evaluation of British rubber reclaim workers did not find an increased risk of myeloma mortality.[38]

There have been several case reports of myeloma in subjects exposed to *benzene*.[39] A cohort analysis of workers (predominantly white males) employed in the manufacture of rubber hydrochloride in pliofilm plants, by a process which includes the dissolution of natural rubber in benzene, detected four deaths from myeloma. The expected number of myeloma deaths in an unexposed normal population of the same age, sex, and race, was one. However, three of the workers who developed myeloma in this study had minimal exposure to benzene (one had worked at the plant for only 4 days), and there was no trend toward an increasing myeloma incidence with greater exposure.[40] An update of this study has added 7 years of follow-up.[41] No new cases of myeloma have developed, and the standardized mortality ratio for this disease is no longer elevated significantly.

Workers in *petroleum refining* and *petroleum production,* and those exposed to *fuel combustion products* (e.g. truck drivers) may also be exposed to benzene, among other petroleum products. A metaanalysis of close to 100 published and unpublished reports of epidemiologic surveys of petroleum industry employees found that the risk of developing myeloma was the same as the risk in the general population.[42] Two recent large studies of exposure to exhaust fumes have reached different conclusions. In Denmark a historical cohort study compared the cancer-specific mortality of 14 225 exposed truck drivers with an unexposed group of 43 024 unskilled labourers, both followed for 10 years.[43] There were five deaths from myeloma among the truck drivers, where 1.25 was expected. This was an unexpected finding, and the author cautions that although it is statistically significant, it may still be due to chance. In contrast, a study of the cancer risk of 160 230 members of a prepaid health care plan who reported their exposure to engine exhaust during a routine health examination, failed to detect an increased risk of myeloma among exposed workers.[44]

An association of an increased risk of myeloma with workers in the *wood, leather* and *textile industries* has been found in some studies, but the results are inconsistent, for the association has not been detected by others.[45]

An increased risk of myeloma has been reported among *painters*.[46–50] The exposure of painters to chemical compounds is complex, for there are various dyes, pigments, and solvents, which are known to be mutagenic, in their environments. The specific agent(s), if there are any, associated with the increased risk of myeloma in painters have not been identified.

The use of *hair dye* has been associated with an increased risk of myeloma in many,[51–54] but not all[55] studies. Hair coloring agents are known to contain constituents, including aromatic, nitro, and amino compounds, that are carcinogenic or mutagenic in animal or laboratory tests. Again, suspicion has not focused on any specific component of hair dyes.

Exposure to *asbestos* has been linked to an increased risk of myeloma in at least two case-control studies.[48,56] In contrast, other case-control studies have not detected this association;[57–59] it will take further work to determine whether asbestos plays a role in the etiology of myeloma.

CHRONIC ANTIGENIC STIMULATION (CAS)

Clinicians have long wondered whether CAS plays a role in the pathogenesis of myeloma.[60] The hypothesis is that CAS stimulates a proliferative response in the immune system, and that myeloma develops in one of the responding cells. There have been several attempts to correlate the occurrence of myeloma with a past history of exposure to viral or bacterial infections, immunizations, allergies, allergy desensitization therapy, and autoimmune disease, but the results are inconsistent.[45] There does, however, appear to be an association between *rheumatoid arthritis* and myeloma. At least two follow-up studies of patients with rheumatoid arthritis have detected a subsequent increased incidence of myeloma,[61–63] and an excess of rheumatoid arthritis has been detected in case-control studies of myeloma in New Zealand and northern Sweden.[64,65] An examination of the frequency of *autoimmune diseases* among first-degree relatives of myeloma patients (cases) discovered a significantly increased risk of rheumatoid arthritis, as compared to the incidence in first-degree relatives of the controls.[66] If this finding is confirmed in larger studies it will suggest that genetic factors underlie the association.

A case-control study of patients with Gaucher's disease in Israel found that these cases have an increased risk of developing hematologic malignancies, including myeloma.[67] A possible mechanism is that the accumulating glucocerebroside acts as a chronic antigenic stimulant of the immune system, as was first suggested by Shoenfeld et al.[68] These authors found that patients with Gaucher's disease had elevated levels of serum immunoglobulins, and that these levels increased with age. In addition, Marti et al.[69] detected diffuse polyclonal hyperglobulinemia in 10, oligoclonal hyperglobulinemia in 6, and monoclonal gammopathy (MGUS) in 2 of 23 patients with Gaucher's disease. The sequence could be that CAS leads to a polyclonal, followed by an oligoclonal lymphocytosis and then a monoclonal proliferation resulting in MGUS, myeloma, a lymphoma, or leukemia.

The discovery of an HIV–1-seropositive patient with myeloma, whose IgG/kappa M-protein specifically recognized the HIV–1 p24 gag antigen, suggests that the antigen-driven response to the viral infection did play a role in the pathogenesis of myeloma in this patient.[70]

Genetic factors

Striking differences in the incidence of plasmacytomas in inbred strains of mice, racial differences in the incidence of plasma cell neoplasms, the association of an increased risk of developing myeloma with certain human leukocyte antigens (HLA), and the occurrence of familial myeloma, all suggest that genetic factors play an important role in the pathogenesis of these neoplasms.

MOUSE PLASMACYTOMAS

There are striking differences in the incidence of monoclonal gammopathy in different inbred strains of mice. About 60 per cent of C57BL/Ka mice develop one or more M-proteins (usually IgG) by 24 months of age.[71] In C3H and NZB mice 40 per cent develop M-proteins of the IgM type, while the incidence in BALB/c and CBA/Rij mice is low.[71] The specific genes responsible for the high incidence of monoclonal gammopathy in these mice have not been identified. Few mice progress to develop a malignant plasma cell neoplasm. There is no correlation between the development of spontaneous gammopathy and susceptibility to the induction of plasmacytomas (usually IgA) by the intraperitoneal injection of mineral oil. BALB/c is the strain that is most susceptible to oil-induced plasmacytomas, while C57BL/Ka are relatively resistant.[72]

RACIAL DIFFERENCES IN THE INCIDENCE OF MM AND MGUS

As mentioned earlier, there appear to be real differences in the incidence of MM and MGUS among different races. The incidence of MM is lowest among the Japanese of Hawaii and Osaka, and the Chinese of Shanghai and Singapore, and highest in the blacks of the Bay area and Connecticut in the USA.[11] The low incidence of MM in Japanese and Chinese populations has moved with them to Hawaii and the USA,[12] suggesting that the incidence of the disease in these populations is determined more by genetic rather than by environmental factors. The prevalence of MGUS in persons over the age of 60 years is also distinctly lower in Japan (2.7 per cent) than in America (10 per cent).[16]

AN INCREASED RISK OF MYELOMA IS ASSOCIATED WITH CERTAIN HLA TYPES

A large population-based study of 46 black male MM cases (with 88 black male controls), and 85 white male MM cases (with 122 white male controls) has been done to determine whether there is any association of human leukocyte antigens of Class I (HLA-A, -B, -C) and Class II (HLA-DR, HLA-DQ) with the disease.[73] Black cases had significantly higher gene frequencies than their controls for *Bw65*, *Cw2*, and *DRw14*, while white cases had higher gene frequencies than controls for *A3* and *Cw2*, and blanks at the *DR* and *DQ* loci. The frequency of *Cw2* in the black and white controls was similar. These findings suggest that the *Cw2* allele, or a gene close to the *C* loci, confers susceptibility to the development of MM, but does not explain the higher risk among blacks. The authors also suggest that undefined Class II antigens may play an etiologic role. New molecular techniques, using genomic DNA, may lead to the identification of these alleles.

FAMILIAL PLASMA CELL NEOPLASIA

Several families have been reported with multiple cases of MM and MGUS (of the IgG, IgA, light-chain and IgD variety). To the forty-one families reviewed by Loth et al.[74] can be added two brothers who developed IgG/κ myeloma,[75] and another family in which a female patient with IgG MM was found to have two sisters and two brothers with IgG MGUS.[76] In these 43 families, MM or MGUS was detected in 7 first-degree relations (parent and child), was most frequent in 32 second-degree relatives (siblings), and was much less frequent in more distant relatives, with occurrences in only one third-degree (aunt–niece), and three fourth-degree (first cousins) relations.

Among the 25 families with IgM M-proteins (Waldenström's macroglobulinemia, WM, or asymptomatic monoclonal IgM) reviewed by Renier et al.,[77] 9 were first-, and 16 second-degree relatives.

It is of interest that the M-protein may have the same heavy and light chain in two relatives with MM or MGUS, or the light chains may differ, or the relative's M-protein may be of the IgG, IgA, or IgD variety, but no relative with an IgM M-protein has been reported. Similarly, relatives of patients with WM have been found to have an IgM M-protein, but not one with any of the other heavy chains.

Although the heavy and light chains of the M-proteins in the relatives may be the same, the proteins in some of these patients have been clearly shown to be different. A rabbit antiserum to a purified urinary lamba light chain of a patient with IgA/λ myeloma did not react with the M-protein of her sister's IgA/λ M-protein.[74] In another family, idiotypic rabbit antisera prepared against each of the IgM M-proteins of four brothers with WM, failed to cross-react, showing that they do not share idiotypes.[77]

The occurrence of multiple cases of a malignancy in a family, without a clear Mendelian pattern of inheritance, suggests that the family members may be exposed to the same environmental hazard, but no hazardous factor has been recognized in any of the MM and WM families that have been investigated.

On the other hand, the discovery that several affected family members have inherited identical HLA haplotypes,[74,75,78] suggests that the tendency to develop these B-cell neoplasms may be inherited. A lod score (the log of the ratio of likelihood of linkage to no linkage) was calculated by the LIPED Program, to test if the occurrence of Waldenström's macroglobulinemia and autoimmune manifestations in a family was segregating with the HLA region on chromosome 6. The lod score was 4.86; this favors chromosomal linkage of a postulated susceptibility gene to the HLA complex.[78]

References

1. Kyle RA, Lust JA. Monoclonal gammopathies of undetermined significance. In: Wiernik PH, Canellos GP, Kyle RA, Schiffer CA eds *Neoplastic diseases of the blood*, vol 1, 2nd edn. New York, Edinburgh, London, Melbourne, Tokyo: Churchill Livingstone, 1991: 571–96.

2. Axelsson U, Bachmann R, Hällén J. Frequency of pathological proteins (M-components) in 6,995 sera from an adult population. *Acta Medica Scandinavica* 1966; **179**: 235–47.

3. Axelsson U, Hällén J. A population study on monoclonal gammopathy: Follow-up after 5 1/2 years on 64 subjects detected by electrophoresis of 6995 sera. *Acta Medica Scandinavica* 1972; **191**: 111–13.

4. Young JL Jr., Percy CL, Asire AJ (eds). *Surveillance, epidemiology, and end results: Incidence and mortality data, 1973–77.* Bethesda: National Institutes of Health, publication 81–2330, 1981:1082. National Cancer Institute Monograph 57.

5. Herrinton LJ, Demers PA, Koepsell TD et al. Epidemiology of the M-component immunoglobulin types of multiple myeloma. *Cancer Causes Control* 1993; **4**(2): 83–92.

6. Turesson I, Zettervall O, Cuzik J et al. Comparison of trends in the incidence of multiple myeloma in Malmö, Sweden, and other countries, 1950–79. *New England Journal of Medicine* 1984; **310**: 421–4.

7. Belch A, Shelley W, Bergsagel D et al. A randomized trial of maintenance versus no maintenance melphalan and prednisone in responding multiple myeloma patients. *British Journal of Cancer* 1988; **57**(1): 94–9.

8. Hjorth M, Holmberg E, Rödjer S, Westin J. Impact of active and passive exclusions on the results of a clinical trial in multiple myeloma. The Myeloma Group of Western Sweden. *British Journal of Haematology* 1992; **80**: 55–61.

9. Englisová M, Englis M, Kyral V, Kourílek K, Dvorák K. Changes in immunoglobulin synthesis in old people. *Experimental Gerontology* 1968; **3**: 125–7.

10. Radl J, Sepers JM, Skvaril F et al. Immunoglobulin patterns in humans over 95 years of age. *Clinical and Experimental Immunology* 1975; **22**: 84–90.

11. Riedel DA, Pottern LM, Blattner WA. Epidemiology of multiple myeloma. In: Wiernik PH, Canellos GP, Kyle RA, Schiffer CA eds *Neoplastic diseases of the blood*, vol 1, 2nd edn. New York, Edinburgh, London, Melbourne, Tokyo: Churchill Livingstone, 1991: 347–72.

12. Devesa SS. Descriptive epidemiology of multiple myeloma. In: Obrams GI, Potter M eds *Epidemiology and biology of multiple myeloma*. Berlin: Springer-Verlag, 1991: 3–12.

13. Sondik EJ. *Cancer statistics review 1973–1986*. Washington, D.C.: US Government Printing Office, 1989. DHHS publication no. NIH 89–2789.

14. McWhorter WP, Schatzkin AG, Horm JW, Brown CC. Contribution of socioeconomic status to black/

white differences in cancer incidence. *Cancer* 1989; **63**: 982–7.

15. Talerman A. Clinico-pathological study of multiple myeloma in Jamaica. *British Journal of Cancer* 1969; **23**(2): 285–93.

16. Bowden M, Crawford J, Cohen HJ, Noyama O. A comparative study of monoclonal gammopathies and immunoglobulin levels in Japanese and United States elderly. *Journal of the American Geriatrical Society* 1993; **41**(1): 11–14.

17. Singh J, Dudley AJ, Kulig KA. Increased incidence of monoclonal gammopathy of undetermined significance in blacks and its age-related differences with whites on the basis of a study of 397 men and one woman in a hospital setting. *Journal of Laboratory and Clinical Medicine* 1990; **116**(6): 785–9.

18. Schwartz J. Multinational trends in multiple myeloma. *Annals New York Academy of Sciences* 1990; **609**(205): 215–24.

19. Linos A, Kyle RA, O'Fallon WM, Kurland LT. Incidence and secular trend of multiple myeloma in Omstead county, Minnesota: 1965–77. *Journal of the National Cancer Institute* 1981; **66**: 17–20.

20. Levi F, La VC. Trends in multiple myeloma [letter]. *International Journal of Cancer* 1990; **46**(4): 755–6.

21. Hansen NE, Karie H, Olsen JH. Trends in the incidence of multiple myeloma in Denmark 1943–1982: A study of 5500 patients. *European Journal of Haematology* 1989; **42**: 72–6.

22. Cuzick J. International time trends for multiple myeloma. *Annals New York Academy of Sciences* 1990; **609**(205): 205–14.

23. Ishimaru M, Ishimaru T, Mickami M et al. Multiple myeloma among atomic bomb survivors at Hiroshima and Nagasaki, 1950–76: *New England Journal of Medicine* 1979; **301**: 439–40.

24. Preston DL, Kusumi S, Tomonaga M et al. Cancer incidence in atomic bomb survivors. Part III. Leukemia, lymphoma and multiple myeloma, 1950–1987. *Radiation Research* 1994; **137**(2 Suppl.):S68–97

25. Neriishi K, Yoshimoto Y, Carter RL et al. Monoclonal gammopathy in atomic bomb survivors. *Radiation Research* 1993; **133**(3): 351–9.

26. Cuzik J. Radiation-induced myelomatosis. *New England Journal of Medicine* 1981; **304**: 204.

27. Boyd NF. The epidemiology of cancer: Principles and methods. In: Tannock IF, Hill RP eds *The basic science of oncology*, 2nd edn. Toronto: Pergamon Press, 1992: 7–22.

28. MacMahon B. Epidemiology of Hodgkin's disease. *Cancer Research* 1966; **26**: 1189–200.

29. Velez R, Beral V, Cuzik J. Increasing trends of multiple myeloma mortality in England and Wales; 1950–1979: Are the changes real? *Journal of the National Cancer Institute* 1982; **69**: 387.

30. Johnston JM, Grufferman S, Bourguet CC et al. Socioeconomic status and risk of multiple myeloma. *Journal of Epidemiology and Community Health* 1985; **39**: 175–8.

31. Mills PK, Newell GR, Beeson WL et al. History of cigarette smoking and risk of leukemia and

myeloma: Results from the Adventist health study. *Journal of the National Cancer Institute* 1990; **82**: 1832–6.

32. Heineman EF, Zahm SH, McLaughlin JK et al. A prospective study of tobacco use and multiple myeloma: evidence against an association [see comments]. *Cancer Causes Control* 1992; **3**(1): 31–6.

33. Linet MS, McLaughlin JK, Hsing AW et al. Is cigarette smoking a risk factor for non-Hodgkin's lymphoma or multiple myeloma? Results from the Lutheran Brotherhood Cohort Study. *Leukemia Research* 1992; **16**(6–7):621–4.

34. Tollerud DJ, Brinton LA, Stone BJ et al. Mortality from multiple myeloma among North Carolina furniture workers. *Journal of the National Cancer Institute* 1985; **74**(4): 799–801.

35. Brownson RC, Reif JS. A cancer registry-based study of occupational risk for lymphoma, multiple myeloma and leukemia. *International Journal of Epidemiology* 1988; **17**: 27–32.

36. Reif JS, Pearce NE, Fraser J. Cancer risks among New Zealand meat workers. *Scandinavian Journal of Work Environment Health* 1989; **15**: 24.

37. Delzell E, Monson RR. Mortality among rubber workers. X. Reclaim workers. *American Journal of Industrial Medicine* 1985; **7**: 307.

38. Sorahan T, Parkes HG, Veys CA et al. Mortality in the British rubber industry 1946–85. *British Journal of Industrial Medicine* 1989; **46**(1): 1–10.

39. Aksoy M, Erdem S, Dincol G et al. Some etiologic factors in the etiology of multiple myeloma. A study in 7 patients. *Acta Haematologia* 1984; **71**: 116–20.

40. Rinsky RA, Smith AB, Hornung R et al. Benzene and leukemia: An epidemiologic risk assessment. *New England Journal of Medicine* 1987; **316**: 1044–50.

41. Paxton MB, Chinchilli VM, Brett SM, Rodricks JV. Leukemia risk associated with the pliofilm cohort: I. Mortality update and exposure distribution. *Journal of Risk Analysis* 1994; **14**: 147–54.

42. Wong O, Raabe GK. Critical review of cancer epidemiology in petroleum industry employees, with a quantitative meta-analysis by cancer site. *American Journal of Industrial Medicine* 1989; **15**: 283–310.

43. Hansen ES. A follow-up study on the mortality of truck drivers. *American Journal of Industrial Medicine* 1993; **23**(5): 811–21.

44. Vandeneeden SK, Friedman GD. Exposure to engine exhaust and risk of subsequent cancer. *Journal of Occupational Medicine* 1993; **35**: 307–11.

45. Riedel DA, Pottern LM. The epidemiology of multiple myeloma. *Hematology and Oncology Clinics of North America* 1992; **6**(2):225–47.

46. Friedman GD. Multiple myeloma: Relation to propoxyphene and other drugs, radiation and occupation. *International Journal of Epidemiology* 1986; **15**: 424–6.

47. Lundberg I. Mortality and cancer incidence among Swedish paint industry workers with long term exposure to organic solvents. *Scandinavian Journal of Work Environment Health* 1986; **12**: 108–13.

48. Cuzick J, De Stavola B. Multiple myeloma – a case-

control study. *British Journal of Cancer* 1988; **57**: 516–20.

49. Bethwaite PB, Pearce N, Fraser J. Cancer risks in painters: study based on the New Zealand Cancer Registry. *British Journal of Industrial Medicine* 1990; **47**(11):742–6.

50. Demers PA, Vaughan TL, Koepsell TD. A case-control study of multiple myeloma and occupation. *American Journal of Industrial Medicine* 1993; **23**(4):629–39.

51. Guidotti S, Wright WE, Peter JM. Multiple myeloma in cosmetologists. *American Journal of Industrial Medicine* 1982; **3**: 169–71.

52. Spinelli JJ, Gallagher RP, Band PR, Threlfall WJ. Multiple myeloma, leukemia and cancer of the ovary in cosmetologists and hairdressers. *American Journal of Industrial Medicine* 1984; **6**: 97–102.

53. Brown LM, Everett GD, Burmeister LF, Blair A. Hair dye use and multiple myeloma in white men. *American Journal of Public Health* 1992; **82**(12):1673–4.

54. Zahm SH, Weisenburger DD, Babbitt PA et al. Use of hair coloring products and the risk of lymphoma, multiple myeloma, and chronic lymphocytic leukemia [see comments]. *American Journal of Public Health* 1992; **82**(7):990–7.

55. Teta MJ, Walrath J, Meigs JM, Flannery JT. Cancer incidence among cosmetologists. *Journal of the National Cancer Institute* 1984; **72**: 1051–7.

56. Linet MS, Harlow SD, McLaughlin JK. A case-control study of multiple myeloma in whites: Chronic antigenic stimulation, occupation and drug use. *Cancer Research* 1987; **47**: 2978–81.

57. Schwartz DA, Vaughan TL, Heyer NJ et al. B cell neoplasms and occupational asbestos exposure. *American Journal of Industrial Medicine* 1988; **14**: 661–71.

58. Boffetta P, Stellman SD, Garfinkel L. A case-control study of multiple myeloma nested in the American cancer society prospective study. *International Journal of Cancer* 1989; **43**: 554–9.

59. Eriksson M, Karlsson M. Occupational and other environmental factors and multiple myeloma: a population based case-control study [see comments]. *British Journal of Industrial Medicine* 1992; **49**(2): 95–103.

60. Isobe T, Osserman EF. Pathologic conditions associated with plasma cell dyscrasia: A study of 806 cases. *Annals of the New York Academy of Sciences* 1971; **90**: 507–18.

61. Isomaki HA, Hakulinen T, Joutsenlahti U. Excess risk of lymphomas, leukemia and myeloma in patients with rheumatoid arthritis. *Journal of Chronic Diseases* 1978; **31**: 691–6.

62. Katusic S, Beard CM, Kurland LT et al. Occurrence of neoplasms in the Rochester, Minnesota, rheumatoid arthritis cohort. *American Journal of Medicine* 1985; **78**(1A): 50–5.

63. Hakulinen T, Isomaki H, Knekt P. Rheumatoid arthritis and cancer studies based on linking nationwide registries in Finland. *American Journal of Medicine* 1985; **78**(1A): 29–32.

64. Pearce NE, Smith AH, Howard JK et al. Case-control study of multiple myeloma and farming. *British Journal of Cancer* 1986; **54**: 493–500.

65. Eriksson M. Rheumatoid arthritis as a risk-factor for multiple myeloma: A case-control study. *European Journal of Cancer* 1993; **29A**: 259–63.

66. Linet MS, McLaughlin JK, Harlow SD, Fraumeni JF. Family history of autoimmune disorders and cancer in multiple myeloma. *International Journal of Epidemiology* 1988; **17**: 512–3.

67. Shiran A, Brenner B, Laor A, Tatarsky H. Increased risk of cancer in patients with Gaucher's disease. *Cancer* 1993; **72**: 219–24.

68. Shoenfeld Y, Gallant LA, Shaklai M et al. Gaucher's disease: A disease with chronic stimulation of the immune system. *Archives of Pathology and Laboratory Medicine* 1982; **106**: 388–91.

69. Marti GE, Ryan ET, Papadopoulas NM et al. Polyclonal B-cell lymphocytosis and hyperglobulinemia in patients with Gaucher disease. *American Journal of Hematology* 1988; **29**: 189–94.

70. Konrad RJ, Kricka LJ, Goodman DRP et al. Brief report: Myeloma-associated paraprotein directed against the HIV–1 p24 antigen in an HIV–1-seropositive patient. *New England Journal of Medicine* 1993; **328**: 1817–9.

71. Radl J, Hollander CF, Van den Berg P, de Glopper E. Idiopathic paraproteinemia. I. Studies in an animal model – the ageing C57BL/KaLwRij mouse. *Journal of Experimental Immunology* 1978; **33**: 395–402.

72. Potter M, Pumphrey JC, Bailey DW. Genetics of susceptibility of plasmacytoma induction. I. BALB/cAnN (C), C57BL/6N (B6), C57BL/Ka (BK), (C x B6)F1 (C x BK)F1 and C x B recombinant inbred strains. *Journal of the National Cancer Institute* 1975; **54**: 1413–17.

73. Pottern LM, Gart JJ, Nam JM et al. HLA and multiple myeloma among black and white men: evidence of a genetic association. *Cancer Epidemiology, Biomarkers and Prevention* 1992; **1**(3): 177–82.

74. Loth TS, Perrotta AL, Lima J et al. Genetic aspects of familial multiple myeloma. *Millitary Medicine* 1991; **156**(8): 430–3.

75. Grosbois F, Gueguen M, Fauchet R et al. Multiple myeloma in two brothers. An immunochemical and immunogenetic familial study. *Cancer* 1986; **58**(11): 2417–21.

76. Bizzaro N, Pasini P. Familial occurrence of multiple myeloma and monoclonal gammopathy of undetermined significance in 5 siblings. *Haematologica* 1990; **75**(1): 58–63.

77. Renier G, Ifrah N, Chevailler A et al. Four brothers with Waldenstrom's macroglobulinemia. *Cancer* 1989; **64**: 1554–9.

78. Blattner WA, Garber JE, Mann DL et al. Waldenström's macroglobulinemia and autoimmune disease in a family. *Annals of Internal Medicine* 1980; **93**: 830–2.

79. Davis DL, Hoel D, Fox J, Lopez AD. International trends in cancer mortality in France, West Germany, Italy, Japan, England and Wales, and the United States. *Annals New York Academy of Sciences* 1990; **609**(5): 5–48.

CHAPTER 3

Basic concepts – the plasma cell in multiple myeloma

FEDERICO CALIGARIS-CAPPIO, MARIA GRAZIA GREGORETTI

Introduction	22	Growth factor requirements of MM plasma cells	28
Plasma cells in the life history of B-lineage cells	23	Cytokines produced by MM plasma cells and their	
Normal plasma cell populations are heterogeneous	24	role in disease progression	29
Plasma cell properties vary according to the type		Plasma cells and bone marrow microenvironment:	
of immune response	24	cellular and molecular interactions	29
MM plasma cell precursors	25	Conclusions	31
Circulating plasma cell precursors in MM	26	References	32
Phenotype of MM plasma cells	27		

Introduction

Plasma cells are the end-effector cells of the B-lymphocyte lineage that produce and secrete antigen (Ag)-specific antibodies (Ab). They originate from Ag-specific B cells after a number of activation, proliferation, and differentiation steps. These processes may occur in several different anatomical sites as witnessed by the existence of multiple plasma cell-rich areas: the lamina propria of the intestine, the medullary cords of lymph nodes, the white pulp and peri-arterioral sheaths of the spleen, the submucosa of the upper airways, and the bone marrow.[1,2]

The prototype plasma cell malignancy is multiple myeloma (MM), a neoplasm characterized by the accumulation of monoclonal plasma cells which show three major features.[3,4] First, the secreted monoclonal immunoglobulins are usually of the IgG or IgA isotype. Second, in contrast with the distribution of normal plasma cells, MM plasma cells localize uniquely within the bone marrow. Even if the lamina propria of the intestine contains more immunoglobin-producing cells than all other tissues in the body, it is never a site where MM develops, not even the IgA1- and IgA2-producing myelomas.[5] The involvement of the spleen and/or lymph nodes, though typical of Waldenström's macroglobulinemia, is likewise very unusual in MM. The third relevant feature of MM plasma cells is their ability to produce a number of cytokines, some formally identified as osteoclast activating factors (OAF).[3] This property has been taken to explain why in myeloma the accumulation of plasma cells within the bone marrow leads to the typical punched out osteolytic lesions.

On these bases, it is the purpose of this chapter to review a number of issues that may provide a link between experimental findings and clinical observations:

1. The place of plasma cells in the life history of B-lineage cells.
2. The heterogeneity of normal plasma cell populations and their properties.

3. The phenotype and the growth factor requirements of MM plasma cells.
4. The cytokines produced by MM plasma cells.
5. The mechanisms that allow the growth of MM plasma cells only within the bone marrow.

Plasma cells in the life history of B-lineage cells

The life history of B lymphocytes can be operationally divided into two main phases: an Ag-independent and an Ag-dependent phase (reviewed in refs 6–8). The Ag-independent phase involves the generation of B-lineage cells from stem cells and their subsequent differentiation into mature virgin B lymphocytes. In adults, these different steps occur within the bone marrow microenvironment, a properly organized meshwork of cells which are collectively named 'stromal cells' and are loosely distributed around bone marrow sinusoids. Stromal cells include fibroblasts, myofibroblasts, macrophages, adipocytes, endothelial cells, reticular cells, and adventitial cells and produce both extracellular matrix proteins (ECMP) and a large array of cytokines.[9–11] The differentiation of B-cell precursors into mature B cells is marked by two events. First, a series of rearrangements of the variable, diversity, and joining segments (VDJ) at the immunoglobulin loci to form the exons encoding the heavy and light chains of Ab molecules. Second, the orderly acquisition and loss of surface markers that pinpoint the subsequent differentiation steps.[12] Virgin B cells express surface (s)IgM and sIgD, are CD19+, CD20+ and leave the bone marrow to migrate into peripheral lymphoid organs, mainly the spleen. In the absence of the specific Ag, the virgin B-lymphocyte lifespan is very short and most of them die *in situ* after few days.

The specific Ag triggers the Ag-dependent phase of B lymphocytes in peripheral lymphoid organs (*see* ref. 6 and Fig. 3.1). The subsequent steps of B-cell activation, proliferation, and differentiation are regulated by a number of cytokines, adhesion structures and surface molecules and lead to the generation of B memory and plasma cells. Specialized Ag-presenting cells and T lymphocytes are central to these events.[13,14] The Ag-dependent phase is marked by two highly specialized genetic processes, the isotype switch and the somatic hypermutation. These processes allow the B cells that are able to produce the Ab with the best fit for the Ag to be selected for expansion and survival. The unselected cells die *in situ* by apoptosis and are disposed of by macrophages. The selected cells develop either into plasmablasts → plasma cells or into B memory cells. The precise signals that teach an Ag-selected B cell to terminally differentiate and become an Ab-secreting plasma cell or to mature into a memory cell are presently undefined. Recent experimental findings suggest that the stimulation of CD40 will drive the progeny of Ag-selected B cell into the memory pathway, while signals provided by CD23 and IL-1α will favor the development into plasmablast stage.[15] Memory B cells enrich the mantle of secondary follicles, recirculate and may be recruited by their specific Ag for another round of Ag-dependent activation, proliferation and differentiation (Fig. 3.1). At present, no known phenotypic markers can safely distinguish memory B cells from virgin B cells.

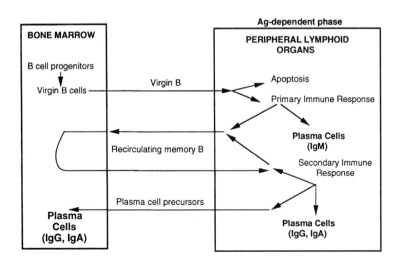

Fig. 3.1 Plasma cell generation in different microenvironments according to the type of immune response.

Normal plasma cell populations are heterogeneous

The widespread anatomical distribution of plasma cells suggests that microenvironmental influences may be important in determining the fate of B cells and implies that normal B cells can follow multiple pathways to give rise to immunoglobulin-producing cells *in vivo*.[14–19] The question is therefore raised whether plasma cell populations are heterogeneous and whether their properties permit dissection of this heterogeneity.

Plasma cells can develop within two different antigenic scenarios: (a) T-cell-dependent versus T-cell-independent immune response; (b) primary versus secondary immune response (i.e. virgin B cells versus memory B cells). The different immune responses that lead to plasma cell development occur within different microenviroments. These include: (a) macrophage-rich areas, such as the red pulp of the spleen, where T-cell-independent responses take place; (b) areas of extrafollicular secondary lymphoid tissues rich in T lymphocytes and interdigitating cells, and (c) germinal centers (GC) of secondary lymphoid follicles where follicular dendritic cells (FDC) and T lymphocytes are present. It is generally accepted that virgin B cells undergo primary antigenic stimulation in the extrafollicular areas of peripheral lymphoid organs, while memory B cells are recruited by secondary Ag response in lymphoid follicles. During secondary Ag response, primary follicles, i.e. B-cell agglomerates devoid of germinal centers, transform into secondary follicles characterized by the progressive development of prominent germinal centers. Anatomically, germinal centers can be divided into two main compartments: the dark and the light zone. The dark zone is packed with proliferating centroblasts; the light zone is mainly occupied by non-dividing centrocytes which are in close contact with the intertwining processes of follicular dendritic cells that hold Ag in an immunogenic form on their surface and provide a long-standing stimulation. Several experimental findings indicate that centroblasts undergo the somatic hypermutations which lead to the generation of high-affinity Ab and subsequently differentiate into centrocytes (reviewed in ref. 14). The centrocytes that express immunoglobulin with a high affinity for the Ag presented by follicular dendritic cells give rise to memory B cells or to long-lived plasma cells.

B-lineage cells have recently been re-classified as B1 or B2 B lymphocytes according to the presence (B1) or the absence (B2) of the surface CD5 molecule.[20] This dichotomy is better defined in mouse than in man and no data are available to properly discriminate B1-derived from B2-derived plasma cells.

Plasma cell properties vary according to the type of immune response

The properties of plasma cells, including the morphology, immunoglobulin isotype, anatomical distribution and lifespan, vary according to the type of immune response that has led to their development.

On morphological grounds, two major types of plasma cells can be identified:[21] the Marschalko type (or reticular plasma cell) and the lymphatic (or lymphoplasmacytoid) plasma cell. It is widely held that the Marshalko type plasma cells develop from germinal centers of secondary follicles with the cooperation of T cells and are mainly responsible for the production of IgG and IgA, while the lymphatic plasma cells originate outside the germinal centers and are mainly responsible for the production of low-affinity IgM. Immunoglobulin-producing tumors support to a certain extent this morphological dicotomy, as IgG- and IgA-producing multiple myelomas are composed of reticular plasma cells while IgM-producing lymphomas contain mainly lymphoplasmacytoid cells.[22] Interestingly, a proportion of IgM-producing lymphomas are also CD5+ (B1 lymphocytes),[22] while we have been unable to detect CD5 mRNA in MM plasma cells. These findings suggest an affiliation of MM plasma cells to the conventional B2 lineage.

Plasma cells generated in macrophage-rich areas after T-cell-independent Ag stimulation secrete IgM and are short-lived. Primary Ag challenge in the extrafollicular areas gives rise to short-lived plasma cells which remain in the lymph node extramedullary regions. Secondary T-cell-dependent Ab responses lead to the production of IgG- or IgA-secreting plasma cells with a lifespan of few weeks (reviewed in ref. 1).

Even if the steps of Ag processing and presentation that lead to the generation of IgG and IgA plasma cells occur only in secondary lymphoid follicles, the bone marrow is a major site where IgG and IgA are produced in T-cell-dependent secondary immune responses (Fig. 3.1). Plasma cell precursors (plasmablasts) with specific traffic commitments originate from secondary lymphoid organs and migrate to the bone marrow a few days after the antigenic challenge (Fig. 3.1). Only plasma cell precursors generated in the follicles of the spleen or peripheral lymph nodes migrate to the bone marrow, while those derived from

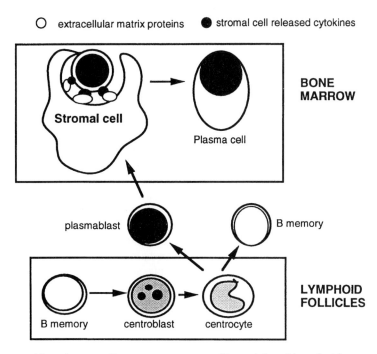

○ extracellular matrix proteins   ● stromal cell released cytokines

Fig. 3.2   Bone marrow-seeking plasma cell precursors generated in peripheral lymphoid organs differentiate in contact with bone marrow stromal cells.

the follicles of Peyer's patches or mesenteric lymph nodes migrate to the lamina propria of the gut.[1] It is assumed that marrow-seeking plasma cell precursors receive a differentiation signal when they come in contact with bone marrow stromal microenvironment (Fig. 3.2).

These observations lead to the conclusion that different plasma cell populations exist with different properties. Therefore it may be asked which normal plasma cell population can be more easily equated to MM malignant plasma cells. As plasma cells are terminal cells which are bound to die in few days or few weeks, the question may be rephrased by asking which are MM plasma cell precursors.

MM plasma cell precursors

Are MM plasma cell precursors early bone marrow stem cells[23] or are they late peripheral B cells?[24] Stated in these terms, the question is misleading. The terms MM stem cell and MM plasma cell precursors have been used almost interchangeably and the cell population whose original transformation has ultimately generated the plasma cell progeny that we call multiple myeloma has been equated to the B-cell population that allows the disease dissemination throughout the axial skeleton. We feel that, in order

to clarify the issue, it is more appropriate to keep the two populations conceptually separated (Fig. 3.3).

The identity of MM stem cells is unknown, i.e. we ignore both the cellular target of the primary transforming event and where the unknown cellular target was hit by the transforming event. More information is available on the B-cell population that feeds the downstream compartment of plasma cells and disseminates the disease. The expression of several hematopoietic lineage-associated markers (myeloid, erythroid, megakaryocytic) by MM plasma cells[23] as well as the incidence of secondary myeloid abnormalities like acute myeloblastic leukemia in MM patients[25] have been taken as evidence that MM plasma cell precursors may be early hematopoietic cells. However, a number of simple, though basic, observations favor the view that late peripheral B cells may be a more appropriate candidate for this role:

1. A number of reports[26–28] present evidence that paraproteins may be directed against a wide variery of infectious agents, suggesting that MM and Ag stimulation may be causally related.
2. The immunoglobulin isotype of MM plasma cells is generally IgG or IgA; IgM-producing MM are exceptional.[3] This suggests that the predominant phenotype of MM tumor cells is post-switch.
3. There is no evidence that MM immunoglobulin V_H genes have ongoing somatic hypermutations,[29]

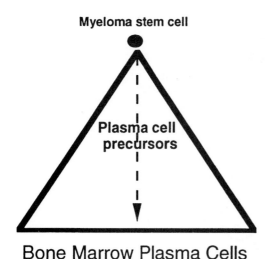

Fig. 3.3 Schematic representation of the three possible cell populations involved in multiple myeloma: myeloma stem cells, plasma cell precursors, and plasma cells.

indicating that the clonal proliferation occurs in a cell type that has a post-switch phenotype and has already passed through the stage of somatic hypermutation.

4. MM is a neoplasm of bone marrow plasma cells.[3]

The most plausible explanation for these findings is that the precursors of MM plasma cells have been generated in peripheral lymphoid organs during secondary T-cell-dependent Ab response, are programmed to home to the bone marrow and are committed to differentiate once they come into close association with the bone marrow microenvironment (Fig. 3.2 and ref. 24). It is difficult to imagine that a pool of early hematopoietic cells originating in the bone marrow would continuously replenish the MM plasma cell compartment by migrating in lymph nodes and/or spleen to undergo always the same isotype switch and then returning to the bone marrow for final maturation. Such a cell would need to have the capacity to move consistently through both the Ag-independent and the Ag-dependent phases of maturation to expand only at the terminal differentiation level, and, at the same time, it would need to be refractory to every possible interference in between such a long and repetitive journey. The most likely candidate to the physiological B-lymphocyte equivalent of MM plasma cell precursor is a memory B cell. Interestingly, it has recently been demonstrated in mice that class-switched memory B cells can be generated both in the presence and absence of intentional immunization, presumably in response to environmental Ag.[30]

Bringing all these observations together, it appears

not unreasonable to conclude that MM is a disease of memory B cells whose progeny circulate and home to the bone marrow where they interact with elements of the bone marrow microenvironment (Fig. 3.2).

Circulating plasma cell precursors in MM

A number of experimental data favor the existence of circulating plasma cell precursors. First of all, the disease is widespread throughout the axial skeleton from the earliest recognizable stage, when very few plasma cells are seen in the peripheral blood.[3] This observation has suggested that human MM, like murine plasmacytoma,[31] may be disseminated by clonogenic cells circulating in the peripheral blood. Their existence is favored by different lines of evidence, including the presence of circulating B lymphocytes that express the MM paraprotein idiotype,[32] the detection of a monoclonal immunoglobulin gene rearrangement within peripheral blood mononuclear cells (PBMC),[33] the identification of DNA-aneuploid cells in peripheral blood samples,[3] the growth of plasma cell colonies from peripheral blood[34] and the induction of patients' PBMC to plasma cell differentiation by cytokines.[35]

The existence of circulating precursors is more a functional than a cytological concept, as their phenotype, notwithstanding several experimental approaches (*see* ref. 36), is still undefined. More specifically, it is as yet unknown whether circulating plasma cell precursors in MM are plasmablasts or more mature B cells. Similarly, their precise traffic relationship with the plasma cells compartment is also not clear. Two non-mutually exclusive possibilities exist, either that circulating precursor cells are 'coming' from the bone marrow, spilling over from an overcrowded environment, or that they are 'going' to the bone marrow after having been primed in peripheral lymphoid organs. Recent data obtained in human MM by amplifying the complementarity-determining region (CDR) 3 have shown that the incidence of circulating monoclonal tumor-related cells is independent of both tumor burden and stage of disease.[37] These findings match the classic phenotypic analysis of MM bone marrow samples that reveals the coexistence of plasma cells together with monoclonal B cells with the morphology of activated lymphocytes (ref. 38 and Fig. 3.4). By means of the CDR 3 amplification of rearranged heavy chain alleles the existence has been shown, within the BM of MM patients, of pre-switch B cells which are clonally related to malignant plasma cells.[39,40] This population has an extremely tiny size and is not seen

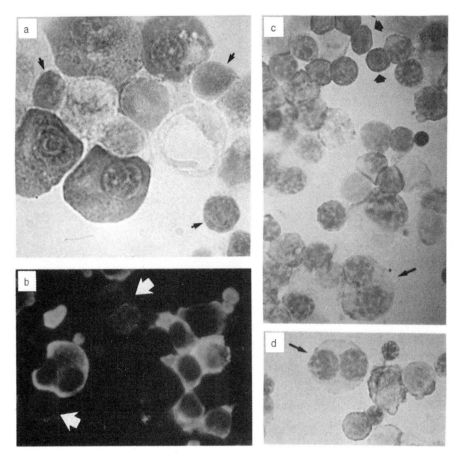

Fig. 3.4 The phenotypic analysis of cytospinned MM bone marrow samples with either immunohistochemical (alkalkine phosphatase–anti-alkaline phosphatase: a,c,d) or immunofluorescence (b) reveals the coexistence of plasma cells together with monoclonal B cells with the morphology of activated lymphocytes (a,c,d). In a proportion of MM patients both bone marrow plasma cells (small arrows) and some monoclonal B cells (large arrowheads) are CD10+ (b-d).

in PB where only post-switch B cells, with the tumor VDJ sequences linked to Cα or Cγ, have been detected.[39] The precise place of BM pre-switch cells in the evolution of MM clone is presently unknown. The possibility that they might belong to the MM stem cell pool cannot be easily dismissed.[41] Still, for the reasons detailed before, it is easier to think that they might represent either a 'blind alley' in the natural history of the malignant clone or clonal memory B cells that have not yet completed their full differentiation and are still awaiting for the switch process to occur.

Phenotype of MM plasma cells

The differentiation of B-lineage cells to plasma cells is accompanied by the orderly acquisition and loss of surface markers. The precise phenotype of normal plasma cells is not fully established due to their scanty number in normal bone marrow samples and, in general, their phenotypic features have been mutuated by the phenotype of MM plasma cells. Likewise, besides the classical PCA-1,[42] very few plasma cell-specific monoclonal antibodies (mAbs) have been produced. By using the cytoplasmic (cy) expression of immunoglobulin with very little or no surface immunoglobulin as a reference marker to scan the expression of surface molecules, it appears that the phenotype of plasma cells has two interesting properties. While conventional B cell surface markers are almost absent at the plasma cell level, as most B-specific and B-associated surface markers are lost during the transition to the plasma cell stage, a very large array of adhesion structures are acquired (Table 3.1).

The best surface marker of plasma cells is CD38,

Table 3.1 Surface markers and adhesion molecules expressed by normal and malignant plasma cells

	Normal plasma cells	Malignant plasma cells
CD10	unknown	+[a]
CD19	+	−
CD38	+	+
CD18 (LFA-1)	−	+[b]
CD44 (H-CAM)	+	+
CD54 (I-CAM-1)	+	+
CD56 (N-CAM)	−	+
CD58 (LFA-3)	−	+
VLA-4	+	+
VLA-5	+	+

[a] Irregular and variable expression.
[b] *See* ref. 52.

Table 3.2 Osteoclast-activating factors (OAF) in multiple myeloma bone marrow microenvironment

Plasma-cell-secreted:	IL-1β, TNF-β, M-CSF
Stromal-cell-secreted:	IL-6
T-cell-secreted:	IL-3

whose expression is shared by both normal and MM plasma cells.[43,44] CD38 is not a marker restricted to the B lineage nor is a stage-specific marker within the B-cell lineage, as it is expressed both by plasma cells and by B blasts proliferating in the germinal center.[43] CD38 is related to an enzyme (ADP-ribosyl cyclase) involved in a recently discovered pathway of intracellular Ca^{2+} mobilization that is distinct from the inositol 1,4,5-triphosphate pathway.[45,46] Normal plasma cells have been reported to be CD19+, while MM plasma cells are usually CD19−.[44] In a proportion of MM patients both bone marrow plasma cells and some monoclonal B cells circulating in peripheral blood are CD10+.[38,47] As CD10 is expressed by early B-lineage committed cells in the bone marrow,[12] this phenotypic feature has reinforced the speculation that MM might be a pre-B-cell (pre-switch) malignancy. However, this possibility is hindered by the observation that CD10 is also expressed by germinal center B blasts as well as by activated B lymphocytes,[13,48] again bringing the attention back to the possible peripheral B cell nature of MM plasma cell precursors.

MM plasma cells (Fig. 3.4) express adhesion molecules on their surface like H-CAM (CD44), I-CAM-1 (CD54), N-CAM (CD56), LFA-3 (CD58), the proteoglycan syndecan, a receptor for hyaluronan-mediated motility and frequently also CD11/CD18.[49–57] It is conceivable that MM surface adhesion structures, by interacting with their homologous ligands in the bone marrow microenvironment, may allow malignant B cells to be entrapped within the bone marrow stromal cell web, where they would be exposed to the cytokines, like IL-6, that promote B-cell proliferation and differentiation (Fig. 3.4 and ref. 59). The most interesting surface adhesion molecule appears to be CD56, as normal plasma cells are CD19+, CD56−, while MM plasma cells are almost invariably CD56+, thus

suggesting a role for this molecule in the pathophysiology of the disease.[57,58] CD56 is expressed by bone marrow macrophages and osteoclasts, not by bone marrow fibroblasts that represent the bulk of bone marrow stromal cells (Table 3.2). This would suggest that CD56 expression is more likely to be implicated in the homotypic adhesions of tumor cells which lead to the formation of MM nodules and clusters scattered throughout the marrow. It has been reported that the loss of CD56 is associated with a more aggressive course of the disease and a tendency to disseminate plasma cells in peripheral blood.[58] CD56 binds also specifically to heparan sulfate, a member of the extracellular matrix protein (ECMP) family, and it might thus be involved in tumor cell–ECMP interactions.[57–59]

Growth factor requirements of MM plasma cells

Growth factor of MM plasma cells will only be briefly reviewed here (*see* Chapter 8). *In vitro*, the growth of human MM cell lines can be improved by IL-6 or is dependent on IL-6 producing feeder cells.[60,61] IL-6 is a potent growth factor for murine hybridomas and plasmacytomas and, at present, is the most relevant growth factor, known not only for human MM cell lines, but also for fresh MM samples.[62–65] Other cytokines, like granulocyte macrophage (GM)-CSF, IL-1, IL-3 and IL-5, frequently in synergy with IL-6, may increase the [^{3}H]-Tdr incorporation of purified MM plasma cells cultured *in vitro*.[35,66,67] Finally, though IL-11 has been shown to have an IL-6-like activity on plasmacytomas,[68] it appears to be ineffective in promoting the growth of human MM plasma cells.[69]

It is generally agreed that the cytokines active *in vitro* play a role *in vivo* as well. This is demonstrated by the presence of high levels of IL-6 in the sera of patients with aggressive or progressive MM[70,71] and by the therapeutic effect of the infusion of anti-IL-6 antibodies in patients with plasma cell leukemia.[72]

Activated T cells, fibroblasts and accessory cells are able to produce several cytokines potentially relevant to the growth of MM (reviewed in ref. 73). The corollary of these findings is that a number of cells, including accessory cells and T lymphocytes, are potentially involved in the growth of the malignant cell clone through mutually interacting paracrine loops. Studies of the possible role of T lymphocytes in MM have been focused on two problems. First, the possibility that T cells might be (at least partially) responsible for the depressed primary and secondary Ab response that leads to low levels of normal polyclonal immunoglobulin. Second, the possibility that T-cell (sub)populations might be involved in a series of immunoregulatory circuits controling the disease progression. A number of observations (reviewed in refs 31 and 74) have shown the existence of T-cell-driven immunoregulatory circuits in murine plasmacytomas. In human MM, the data (reviewed in ref. 74) document that a deterioration of cell-mediated immunity is associated with tumor progression. The recently[75] described hyperreactivity of MM T cells suggests that T lymphocytes may be implicated in the expansion of the malignant B-cell clone.

Cytokines produced by MM plasma cells and their role in disease progression

MM malignant plasma cells are not inert vehicles of monoclonal immunoglobulin. They also produce a number of cytokines endowed with the capacity of influencing different cell populations which may have a role in the pathophysiology of MM. The cytokines produced by MM plasma cells include interleukin (IL)-1β, tumor necrosis factor (TNF)-β and a functionally active truncated version of monocyte–macrophage colony stimulating factor (M-CSF).[76–79] All these cytokines activate stromal and accessory cells and also have a significant osteoclast-activating factor (OAF) activity (Table 3.2). A minority of human MM cell lines autonomously produce small amounts of IL-6, but it is debated whether fresh MM plasma cells may also produce IL-6.[60–63,80] IL-6 promotes B-cell proliferation and differentiation and has been recently shown to have an important OAF activity (Table 3.2).

Plasma cells and bone marrow microenvironment: cellular and molecular interactions

The influence of bone marrow stromal cells in the process of cell differentiation occurs through multiple molecular contacts which are mediated by adhesion molecules and EMCP.[11] The result is the local production of both stimulatory and inhibitory

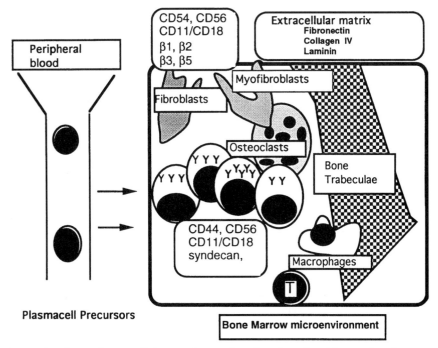

Fig. 3.5 Adhesion molecules and extracellular matrix proteins allow functional interactions between the malignant B-cell clone and bone marrow microenvironment.

Table 3.3   Adhesion molecules of multiple myeloma bone marrow stromal cells

	Fibroblasts	Macrophages	Osteoclasts
β1 integrin	+[a]	+[a]	+[a]
β2 integrin (CD18)	−	+[c]	−
β3 integrin	+[b]	−	+[c]
β5 integrin	+[b]	−	−
CD11a (α L)	−	+	−
CD11b (α M)	−	+	−
CD11c (α X)	−	+	−
CD54 (I-CAM-1)	+	+	+
CD56 (N-CAM)	−	+	+

[a] Diffuse distribution on the cell surface.
[b] Concentrated in adhesion plaques.
[c] Concentrated in podosomes.

cytokines that are smeared on to ECMP (Fig. 3.2 and ref. 81). Growth factors are functionally active when bound to ECMP components like proteoglycans and laminin.[11] A strict interplay exists between cytokines and adhesion molecules: cytokines may regulate cell adhesion which, in return, may modify the cellular response to cytokines.[81,82]

Microenvironmental stromal cells play an essential role in the growth of plasma cell tumors. In mice, the survival and expansion of plasmacytoma requires a stromal cell layer.[83] In patients with MM, stromal cells can be cultured *in vitro* from the bone marrow[59,84] and form a confluent meshwork of intertwining elements that may be used as a feeder layer for autologous peripheral blood mononuclear cells (PBMC) (Fig. 3.5). With this experimental approach a monoclonal B lineage cell population has been shown to develop and grow tightly adherent to the stromal cell layer in a number of MM cases.[59] Four major populations can be identified in MM bone marrow stromal cell cultures: fibroblasts, myofibroblasts, macrophages, and osteoclasts (Fig. 3.5); these are fully equipped with adhesion molecules (Table 3.3 and ref. 85). Fibroblast populations react strongly with antibodies to CD54 (I-CAM-1), integrin β1, β3, β5 and some associated α chains (Table 3.3). CD14+ macrophages are CD11a+ (αL), CD11b+ (αM), CD11c+ (αX), CD54+, CD56+ (N-CAM), β1 and β2 (CD18) integrin positive (Table 3.3). MM osteoclasts show a weak diffuse staining with CD54 and CD56 Abs (Table 3.3).

As MM bone marrow stromal cells *in vitro* support the expansion of putative circulating monoclonal plasma cell precursors[59] and *in vivo* represent the microenvironment that allows the growth of the malignant clone,[3] it may be asked whether there is a difference between the stromal cells that can be cultured from the marrow of patients with MM, those with MGUS and normal donors. The difference exists

and can be summarized by saying that bone marrow stromal cells from patients with MM are 'activated'. This activated state is revealed by at least two properties. First, MM bone marrow stromal cells are actively growing as compared to age-matched controls.[59] Second, only MM marrow stromal cells release significant amounts IL-6 and IL-8 in the culture supernatant.[71] A further difference involves a lower deposition and simpler organization of the ECMP (fibronectin, laminin, collagen Type IV) produced by MM fibroblasts.[85] The stromal cells growing from MGUS marrow are rare, do not support the growth of autologous B cells and produce very low amounts of IL-6 in the culture supernatants.[59]

These findings lead to the conclusion that several environmental cells that normally would be quiescent, above all in a certain age range, are instead activated in the bone marrow of MM. As normal marrow stromal cells produce IL-6 only after activation by inflammatory mediators like IL-1,[86] it is not unreasonable to consider the activated state of stromal cells in MM as a direct consequence of the influence of accessory cell-activating cytokines produced by the expanding monoclonal B cell population (Fig. 3.6). In MGUS patients, the clonal population is below the theshold size that may produce enough cytokines to initiate the activation of marrow stromal cells. Once the threshold size is reached, i.e. MGUS has evolved to MM, marrow stromal cells are activated and trigger a self-perpetuating mechanism of mutual help and recruitment between malignant plasma cells and marrow stromal cells that favors the progressive expansion of the B-cell clone (Fig. 3.6).

Taken together, a number of observations indicate that MM bone marrow stromal cells are well supplied with a large series of adhesion and extracellular matrix molecules (Fig. 3.5 and Table 3.3). Both homotypic and heterotypic interactions are mediated

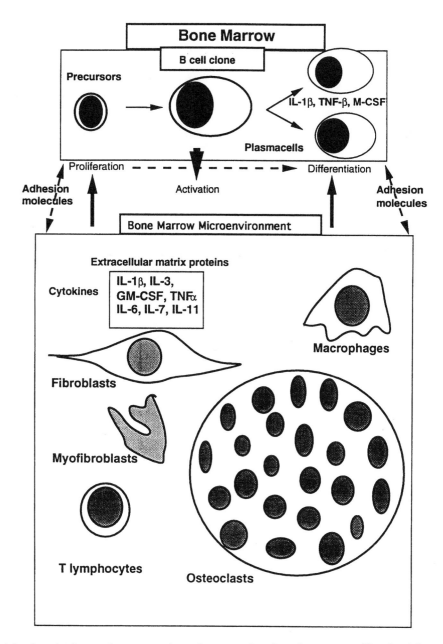

Fig. 3.6 Model of multiple myeloma growth and progression based upon a self-maintaining series of mutual interactions between the B-cell clone and the bone marrow microenvironment.

by cell surface adhesion receptors.[82] Further, ECMP secreted by fibroblast-like cells may provide anchorage sites where cells are selectively exposed to the locally released growth factors.[11,87,88] MM bone marrow stromal cells also produce cytokines known to play a crucial role in the evolution of the disease. IL-6 has a well-defined plasma cell growth promoting activity[62] and IL-8 can upregulate the expression of adhesion molecules.[89] A relevant *in vitro* property of MM bone marrow stromal cells is the ability to sup-

port the growth of peripheral blood-borne monoclonal B-cell precursors:[59] it is reasonable to assume that such an ability mirrors the events occurring *in vivo*.

Conclusions

In conclusion, the bone marrow microenvironment in MM is well organized, in terms of adhesive properties, to entrap marrow-seeking circulating precursors and is able to produce the cytokines that offer them the optimal conditions for local growth and final differentiation.[73] Bone marrow stromal cell features go hand in hand with the malignant B-cell clone properties, as MM plasma cells are equipped with adhesion molecules that act as ligands for stromal cell adhesion structures and may establish close contacts with bone marrow stromal cells and stromal cell-produced ECMP (reviewed in ref. 73). Further, plasma cells release a number of cytokines, including IL-1β and M-CSGF, that activate bone marrow stromal cells to produce cytokines, like IL-6, which, in return, are able to stimulate the proliferation of plasma cells.[60–65,73]

These experimental findings, linked to clinical observation, lead to the attractive hypothesis (Fig. 3.6) that a self-maintaining series of mutual interactions between the malignant B-cell clone and the bone marrow microenvironment may explain the progression of MM.[73] The expansion of the malignant B-cell clone would cause the production of ever-increasing amounts of cytokines capable of recruiting and activating several microenvironmental cells, including osteoclasts and accessory cells. Activated accessory cells, in turn, would produce IL-6 and thus further enhance the size of the malignant B-cell clone, triggering a vicious spiral that causes the progressive growth and dissemination of the disease. This model also explains why the expansion of the B cell clone goes alongside with the activation and numerical increase of the osteoclast population and is associated with osteolytic lesions. Most cytokines produced within MM bone marrow both by the malignant plasma cells and by several microenvironmental elements, including IL-1β, TNF-β, M-CSF, IL-3, and IL-6, have OAF properties (refs 90–94 and Table 3.2). It can therefore be reasonably stated that the mutual interactions between the B-cell clone and the bone marrow microenvironment are also responsible for osteoclast activation.

Acknowledgements: This work was supported by A.I.R.C., Milano and by C.N.R. Progetto Finalizzato Applicazioni Cliniche della Ricerca Oncologica, grant No.92.02152.PF39. M.G.G. is recipient of a fellowship of F.I.R.C.

References

1. MacLennan ICM, Chan EYT. The origin of bone marrow plasma cells. In: Obrams GI, Potter M eds *Epidemiology and biology of multiple myeloma.* Berlin: Springer-Verlag, 1991: 129–35.
2. Benner R, Hijmans W, Haajman JJ. The bone marrow: the major source of serum immunoglobulins, but still a neglected site of antibody formation. *Clinical and Experimental Immunology* 1981; **46**: 1–8.
3. Barlogie B, Epstein J, Selvanayagam P, Alexanian R. Plasma cell myeloma: new biological insights and advances in therapy. *Blood* 1989; **73**: 865–79.
4. Salmon SE, Seligmann M. B-cell neoplasia in man. *Lancet* 1974; **2**: 1230–5.
5. Leonard RFL, Maclennan ICM, Sart Y et al. Light chain isotype-associated suppression of normal plasma cell numbers in patients with myelomatosis. *International Journal of Cancer* 1979; **24**: 385–92.
6. MacLennan ICM, Gray D. Antigen-driven selection of virgin and memory B cells. *Immunological Reviews* 1986; **91**: 61–85.
7. Cooper MD. B lymphocytes. Normal development and function. *New England Journal of Medicine* 1987; **317**: 1452–6.
8. Rolink A, Melchers F. Generation and regeneration of cells of the B-lymphocyte lineage. *Current Opinions in Immunology* 1993; **5**: 207–17.
9. Kinkade P, Lee G, Pietrangeli CE et al. Cells and molecules that regulate B lymphopoiesis in bone marrow. *Annual Reviews in Immunology* 1989; **7**: 111–43.
10. Dorshkind K. Regulation of hemopoiesis by bone marrow stromal cells and their products. *Annual Reviews in Immunology* 1990; **8**: 111–37.
11. Juliano RL, Haskill S. Signal transduction from the extracellular matrix. *Journal of Cell Biology* 1993; **120**: 577–85.
12. Greaves MF. Differentiation-linked leukemogenesis in lymphocytes. *Science* 1986; **234**: 697–704.
13. Stein H, Gerdes J, Mason DY. The normal and malignant germinal centre. *Clinics in Haematology* 1982; **11**: 531–59.
14. Liu YJ, Johnson GD, Gordon J, MacLennan ICM. Germinal centres in T-cell dependent antibody responses. *Immunology Today* 1992; **13**: 17–21.
15. Liu YJ, Mason DY, Johnson GD et al. Germinal center cells express bcl-2-protein after activation by signals which prevent their entry to apoptosis. *European Journal of Immunology* 1991; **21**: 1905–10.
16. Ho F, Lortan J, Khan M, Maclennan ICM. Distinct short-lived and long-lived antibody-producing cell populations. *European Journal of Immunology* 1986; **16**: 1297–301.
17. DiLosa RM, Maeda K, Masuda A et al. Germinal center B cells and antibody production in the bone marrow. *Journal of Immunology* 1991; **146**: 4071–7.
18. Benner R, Hijmans W, Haaijman JJ. The bone marrow: the major source of serum immunoglobulins, but still a neglected site of antibody formation. *Clinical and Experimental Immunology* 1981; **46**: 1–8.
19. Hall JG, Hopkins J, Orlans E. Studies on lymphoblasts in the sheep III: the destination of lymphborne immunoblasts according to their tissue

origin. *European Journal of Immunology* 1977; **7**: 30–9.

20. Kantor AB. The development and repertoire of B-1 cells (CD5 B cells). *Immunology Today* 1991; **12**: 389–91.

21. Lennert K, Mohri N, Stein H et al. *Malignant lymphomas*. Berlin, Heidelberg, New York: Springer-Verlag, 1978.

22. Stein H, Lennert K, Feller AC, Mason DY. Immunohistological analysis of human lymphoma: correlation of histological and immunological categories. *Advances in Cancer Research* 1984; **42**: 67–147.

23. Epstein J, Xiao H, He XY. Markers of multiple hematopoietic-cell lineages in multiple myeloma. *New England Journal of Medicine* 1990; **322**: 664–9.

24. MacLennan ICM. In which cells does neoplastic transformation occur in myelomatosis? *Current Topics in Microbiology and Immunology* 1992; **182**: 209–13.

25. Bergsagel DE. Chemotherapy of myeloma: drug combinations versus single agents, an overview, and comments on acute leukemia in myeloma. *Hematology and Oncology* 1988; **6**: 159–64.

26. Potter M. Myeloma proteins (M-components) with antibody-like activity. *New England Journal of Medicine* 1971; **284**: 831–8.

27. Seligmann M, Brouet JC. Antibody activity of human myeloma globulins. *Seminars in Hematology* 1973; **10**: 163–77.

28. Konrad RJ, Kricka LJ, Goodman DBP et al. Myeloma-associated paraprotein directed against the HIV-1 p24 antigen in an HIV-1-seropositive patient. *New England Journal of Medicine* 1993; **328**: 1817–19.

29. Bakkus MHC, Heirman C, Van Riet I et al. Evidence that multiple myeloma Ig heavy chain VDJ genes contain somatic mutations but show no intraclonal variation. *Blood* 1992; **80**: 2326–35.

30. Schittek B, Rajewsky K. Natural occurrence and origin of somatically mutated memory B cells in mice. *Journal of Experimental Medicine* 1992; **176**: 427–38.

31. Lynch RG, Rohrer JM, Odermatt B et al. Immunoregulation of murine myeloma cell growth and differentiation: a monoclonal model of B cell differentiation. *Immunology Reviews* 1977; **48**: 45–73.

32. Osterborg A, Steinitz M, Lewin N et al. Establishment of idiotype bearing B-lymphocyte clones from a patient with monoclonal gammopathy. *Blood* 1991; **78**: 2642–9.

33. Berenson J, Wong R, Kim K et al. Evidence of peripheral blood B lymphocyte but not T lymphocyte involvement in multiple myeloma. *Blood* 1987; **70**: 1550–3.

34. Hamburger AW, Salmon SE. Primary bioassay of human myeloma stem cells. *Journal of Clinical Investigation* 1977; **60**: 846–54.

35. Bergui L, Schena M, Gaidano G et al. Interleukin 3 and interleukin 6 synergistically promote the proliferation and differentiation of malignant plasma cell

precursors in multiple myeloma. *Journal of Experimental Medicine* 1989; **170**: 613–18.

36. Jensen GS, Belch AR, Mant MJ et al. Expression of multiple adhesion molecules on circulating monoclonal B cells in myeloma. *Current Topics in Microbiology and Immunology* 1992; **182**: 187–93.

37. Billadeau D, Quam L, Thomas W et al. Detection and quantitation of malignant cells in the peripheral blood of multiple myeloma patients. *Blood* 1992; **80**: 1818–24.

38. Caligaris-Cappio F, Bergui L, Tesio L et al. Identification of malignant plasma cell precursors in the bone marrow of multiple myeloma. *Journal of Clinical Investigation* 1985; **76**: 1243–51.

39. Corradini P, Boccadoro M, Voena C, Pileri A. Evidence for a bone marrow B cell transcribing malignant plasma cell VDJ joined to Cμ sequence in IgG and IgA secreting multiple myelomas. *Journal of Experimental Medicine* 1993; **178**: 1091–6.

40. Billadeau D, Ahmann G, Greipp P, Van Ness B. The bone marrow of multiple myeloma patients contains B cell populations at different stages of differentiation that are clonally related to the malignant plasma cell. *Journal of Experimental Medicine* 1993; **178**: 1023–31.

41. Kubagawa H, Vogler CB, Capra JD et al. Studies on the clonal origin of multiple myeloma. *Journal of Experimental Medicine* 1979; **150**: 792–807.

42. Anderson KC, Park EK, Bates MP et al. Antigens on human plasma cells identified by monoclonal antibodies. *Journal of Immunology* 1992; **130**: 1132–8.

43. Malavasi F, Caligaris-Cappio F, Milanese C et al. Characterization of a murine monoclonal antibody specific for human early lymphohemopoietic cells. *Human Immunology* 1984; **9**: 9–20.

44. Harada H, Kawano MM, Huang N et al. Phenotypic difference of normal plasma cells from mature myeloma cells. *Blood* 1993; **81**: 2658–63.

45. States DJ, Walseth TF, Lee HC. Similarities in amino acid sequences of Aplysia ADP-ribosyl cyclase and human lymhocyte antigen CD38. *Trends in Biochemical Sciences* 1992; **17**: 495–9.

46. Galione A. Cyclic ADP-ribose: a new way to control calcium. *Science* 1993; **259**: 325–6.

47. Ruiz-Arguelles GJ, Katzmann JA, Greipp PR et al. Multiple myeloma: circulating lymphocytes that express plasmacell antigens. *Blood* 1984; **64**: 352–6.

48. Warburton, Joshua DE, Gibson J, Brown RD. CD10-(CALLA)-positive lymphocytes in myeloma: evidence that they are a malignant precursor population and are of germinal center origin. *Leukemia Lymphoma* 1989; **1**: 11–20.

49. Van Camp B, Durie BGM, Spier C et al. Plasma cells in multiple myeloma express a natural killer cell-associated antigen: CD56 (NKH-1; Leu-19). *Blood* 1990, **76**: 377–82.

50. Lewinsohn DM, Nagler A, Ginzton N et al. Hematopoietic progenitor cells expression of the H-CAM (CD44) homing-associated adhesion molecule. *Blood* 1990; **76**: 589–94.

51. Drach J, Gattringer C, Huber H. Expression of the

neural cell adhesion molecule (CD56) by human myeloma cells. *Clinical and Experimental Immunology* 1991; **83**: 418–22.

52. Ashmann EJM, Lokhorst HM, Dekker AW, Bloem AC. Lymphocyte function-associated antigen-1 expression on plasma cells correlates with tumor growth in multiple myeloma. *Blood* 1992; **79**: 2068–75.

53. Uchiyama H, Barut BA, Chauhan D et al. Characterization of adhesion molecules on human myeloma cell lines. *Blood* 1992; **80**: 2306–14.

54. Barker HF, Hamilton MS, Ball J et al. Expression of adhesion molecules LFA-3 and N-CAM on normal and malignant human plasma cells. *British Journal of Haematology* 1992; **81**: 331–5.

55. Turley EA, Belch AJ, Poppema S, Pilarski L. Expression and function of a receptor for hyaluronan-mediated motility on normal and malignant B lymphocytes. *Blood* 1993; **81**: 446–53.

56. Ridley RC, Xiao H, Hata H et al. Expression of syndecan regulates human myeloma plasma cell adhesion to type I collagen. *Blood* 1993; **81**: 767–74.

57. Van Riet I, Van Camp B. The involvement of adhesion molecules in the biology of multiple myeloma. *Leukemia Lymphoma* 1993; **9**: 441–52.

58. Barker HF, Ball J, Drew M et al. The role of adhesion molecules in multiple myeloma. *Leukemia Lymphoma* 1993; **8**: 189–96.

59. Caligaris-Cappio F, Bergui L, Gregoretti MG et al. Role of bone marrow stromal cells in the growth of human multiple myeloma. *Blood* 1991; **77**: 2688–93.

60. Klein B, Zhang XG, Jourdan M et al. Paracrine rather than autocrine regulation of myeloma-cell growth and differentiation by interleukin-6. *Blood* 1989; **73**: 517–26.

61. Jernberg H, Pettersson M, Kishimoto T, Nilsson K. Heterogeneity in response to interleukin 6 (IL-6), expression of IL-6 and IL-6 receptor mRNA in a panel of established human multiple myeloma cell lines. *Leukemia* 1991; **5**: 255–65.

62. Kishimoto T. The biology of Interleukin-6. *Blood* 1989; **74**: 1–10.

63. Nilsson K, Jernberg H, Pettersson M. IL-6 as a growth factor for human multiple myeloma cells – a short overview. In: Melchers F, Potter M eds *Current Topics of Microbiology and Immunology*, vol. 166. Berlin: Springer-Verlag, 1990: 3–8.

64. Nordan RP, Potter M. A macrophage derived factor required by plasmacytomas for survival and proliferation *in vitro*. *Science* 1986; **233**: 566–9.

65. Zhang XCG, Klein B, Bataille R. Interleukin-6 is a potent myeloma-cell growth factor in patients with aggressive multiple myeloma. *Blood* 1989; **74**: 11–13.

66. Anderson KC, Jones RM, Morimoto C et al. Response patterns of purified myeloma cells to hematopoietic growth factors. *Blood* 1989; **73**: 1915–24.

67. Zhang XCG, Bataille R, Jourdan M et al. GM-CSF synergizes with interleukin-6 in supporting the proliferation of human myeloma cells. *Blood* 1990; **76**: 2599–605.

68. Paul SR, Bennett F, Calvetti JA et al. Molecular cloning of a cDNA encoding interleukin 11, a stromal cell-derived lymphopoietic and hematopoietic cytokine. *Proceedings of the National Academy of Sciences* 1990; **87**: 7512–16.

69. Paul SR, Barut BA, Bennett F et al. Lack of a role of interleukin 11 in the growth of multiple myeloma. *Leukemia Research* 1992; **16**: 247–52.

70. Bataille R, Jourdan M, Zhang XG, Klein B. Serum levels of interleukin-6, a potent myeloma cell growth factor, as a reflection of disease severity in plasma cell dyscrasias. *Journal of Clinical Investigation* 1989; **84**: 2008–11.

71. Merico F, Bergui L, Gregoretti MG et al. Cytokines involved in the progression of multiple myeloma. *Clinical and Experimental Immunology* 1993; **92**: 27–31.

72. Klein B, Wijdenes J, Zhang XG et al. Murine anti-interleukin-6 monoclonal antibody therapy for a patient with plasma cell leukemia. *Blood* 1991; **78**: 1198–204.

73. Caligaris-Cappio F, Gregoretti MG, Ghia, Bergui L. *In vitro* growth of human multiple myeloma: implications for biology and therapy. *Hematology / Oncology Clinics of North America* 1992; **6**: 257–71.

74. Hoover RG, Kornbluth J. Immunoregulation of murine and human myeloma. *Hematology / Oncology Clinics of North America* 1992; **6:** 407–24.

75. Massaia M, Attisano C, Peola S et al. Rapid generation of antiplasma cell activity in the bone marrow of myeloma patients by CD3-activated T cells. *Blood* 1993; **82**: 1787–97.

76. Cozzolino F, Torcia M, Aldinucci D et al. Production of interleukin-1 by bone marrow myeloma cells. *Blood* 1989; **74**: 380–7.

77. Mundy GR, Raisz LG, Cooper RA et al. Evidence for the secretion of of an osteoclast stimulating factor in myeloma. *New England Journal of Medicine* 1974; **291**: 1041–6.

78. Ross Garrett I, Durie BGM, Nedwin GE et al. Production of lymphotoxin, a bone-resorbing cytokine, by cultured human myeloma cells. *New England Journal of Medicine* 1987; **317**: 526–32.

79. Nakamura M, Merchav S, Carter A et al. Expression of a novel 3.5-kb macrophage colony-stimulating factor transcript in human myeloma cells. *Journal of Immunology* 1989; **143**: 3543–7.

80. Kawano M, Hirano T, Matsuda T et al. Autocrine generation and requirement of BSF-2/IL-6 for human multiple myeloma. *Nature* 1988; **332**: 83–6.

81. Thiery JP, Boyer B. The junction between cytokines and cell adhesion. *Current Opinions in Cell Biology* 1992; **4**: 782–92.

82. Hynes RO. Integrins: versatility, modulation and signalling in cell adhesion. *Cell* 1992; **69**: 11–25.

83. Degrassi A, Hilbert DM, Rudikoff S et al. In vitro culture of primary plasmacytomas requires stromal cell feeder layers. *Proceedings of the National Academy of Sciences* 1993; **90**: 2060–4.

84. Bernerman ZN, Chen ZZ, Ramael M et al. Human long-term bone marrow cultures (HLTBMCs) in myelomatous disorders. *Leukemia* 1989; **3**: 151–4.

85. Gregoretti MG, Gottardi D, Ghia P et al. Characterization of bone marrow stromal cells from multiple myeloma. *Leukemia Research* 1994; **18**: 675–82.

86. Nemunaitis J, Andrews F, Mochizuki D et al. Human marrow stromal cells: response to interleukin-6 (IL-6) and control of Il-6 expression. *Blood* 1989; **74**: 1929–36.

87. Williams DA, Rios M, Stephens C, Patel VP. Fibronectin and VLA-4 in haematopoietic stem cell-microenvironment interactions. *Nature* 1991; **352**: 438–41.

88. Roberts R, Gallagher J, Spooncer E et al. Heparan sulphate bound growth factors: a mechanism for stromal cell mediated haemopoiesis. *Nature* 1988; **332**: 376–8.

89. Detmers PA, Lo SK, Olsen-Egbert E et al. Neutrophil-activating protein 1 / interleukin 8 stimulates the binding activity of the leukocyte adhesion receptor CD11b/CD18 on human neutrophils. *Journal of Experimental Medicine* 1990; **171**: 1155–62.

90. Bertolini DR, Nedwin GE, Bringman TS et al. Stimulation of bone resorption and inhibition of bone formation in vitro by human tumor necrosis factor. *Nature* 1986; **319**: 516–18.

91. Kodama H, Nose M, Niida S, Yamasaki A. Essential role of macrophage colony-stimulating factor in the osteoclast differentiation supported by stromal cells. *Journal of Experimental Medicine* 1991; **173**: 1291–4.

92. Kurihara N, Bertolini D, Suda T et al. IL-6 stimulates osteoclast-like multinucleated cell formation in long term human marrow cultures by inducing IL-1 release. *Journal of Immunology* 1990; **144**: 4226–30.

93. Barton BE, Mayer R. IL-3 induces differentiation of bone marrow precursor cells to osteoclast-like cells. *Journal of Immunology* 1989; **143**: 3211–15.

94. Stashenko P, Dewhirst FE, Rooney ML et al. Interleukin1β is a potent inhibitor of bone formation *in vitro*. *Bone Mineral Research* 1987; **2**: 559–64.

CHAPTER 4

Immunoglobulins in multiple myeloma

ANDERS ÖSTERBORG, LARS-OLOF HANSSON, HÅKAN MELLSTEDT

Introduction	36	Idiotype-bearing B cells in multiple myeloma	40
Structure of the M-component	36	Idiotypic immune network regulation	42
Normal immunoglobulin gene rearrangement	37	Detection and quantification of M-components	44
M-Component isotypes in monoclonal gammopathies	39	Can an idiotype-specific T-cell immunity be induced/enhanced *in vivo*?	46
Biclonal gammopathy	40	References	47
Antigen-reactivity of M-components	40		

Introduction

The term 'monoclonal gammopathy' refers to a group of related B-cell disorders which have in common the production of a specific and, for each patient, unique monoclonal immunoglobulin, usually referred to as the 'M-component'.[1] Monoclonal gammopathies include multiple myeloma, Waldenström's macroglobulinemia, monoclonal gammopathy of undetermined significance (MGUS), and heavy chain disease. The term MGUS denotes the presence of an M-component in serum and/or urine from patients without evidence of a B-cell malignancy at the time of detection of the monoclonal immunoglobulin. The majority (64 per cent) of patients with a newly diagnosed M-component have MGUS and 16 per cent have multiple myeloma; 2 per cent have Waldenström's macroglobulinemia and in the remaining patients the M-component is associated with malignant lymphoma or primary systemic amyloidosis.[2]

Structure of the M-component

Each immunoglobulin molecule of the M-component is a bilaterally symmetrical glycoprotein composed of two heavy polypeptide chains of the same class (γ_{1-4}, α_{1-2}, μ, δ or ε) and two light chains of the same type (κ or λ), held together by non-covalent bonds and interchain disulfide bridges (Fig. 4.1). The molecular weight ranges from 150 000 kDa (IgG) to 190 000 kDa (IgE). Soluble IgM is usually present as a pentamer with a total molecular mass of about 900 000 kDa. The heavy and light chains are divided into globular domains of approximately 100 amino acids residues. The most N-terminal domain of the heavy and light chain consists of the variable region and the remaining domains form the constant region. The variable domains of the heavy and light chains form together the antigen-binding site, where unique amino acid sequences constitute the idiotope containing three complementary determined regions (CDR) (also termed hypervariable (HV) regions) interspaced by framework structures. These structures form the antigenic specificity of the immunoglobulin molecule.

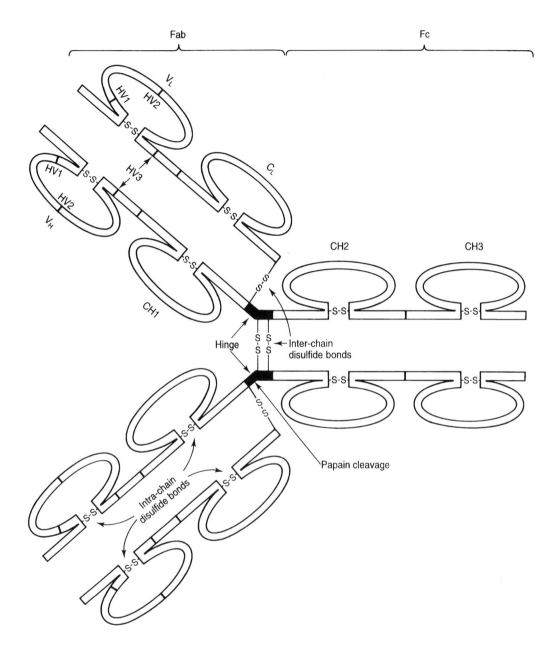

Fig. 4.1 Schematic presentation of an IgG molecule. V_L, V_H, variable domains of the light and heavy chains; C_L, constant region of the light chain; CH1, CH2, CH3, constant domains of the heavy chain; HV1, HV2, HV3, hypervariable or complementarity-determining regions; SS-, disulfide bonds; Fab, papain-derived antigen-binding fragment; Fc, papain-derived crystallizable fragment. (Adapted from ref. 113.)

Normal immunoglobulin gene rearrangement

In the 1970s, Tonegawa and co-workers demonstrated that the immunoglobulin molecule was encoded by multiple gene segments and the segments had to be combined to form a functional immunoglobulin gene.[3] Immunoglobulin molecules are encoded by three gene families located on three separate chromosomes: the heavy-chain genes on chromosome 14, the κ light-chain gene on chromosome 2 and λ light-chain gene on chromosome 22.[4–6] In the germline configuration, the heavy-chain genes consist of discontinuous gene segments (exons) coding for the variable (V), diversity (D), joining (J) and constant (C)

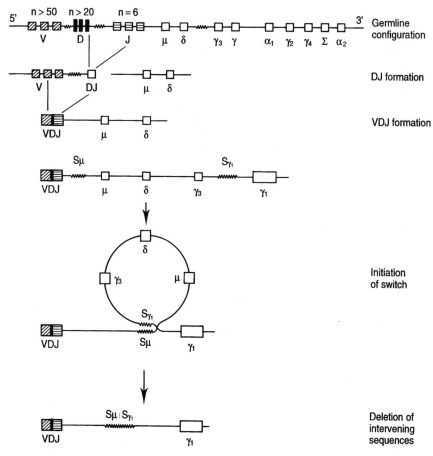

Fig. 4.2   Immunglobulin heavy-chain gene rearrangement and isotype switching. S, switch region. (Adapted from ref. 16.)

regions of the immunoglobulin molecule[7] (Fig. 4.2). These regions are separated by non-coding sequences (introns). Light-chain genes consist of V, J and C (but not D) regions.[3] There are at least 50 different V genes, at least 20 D genes but only 6 J genes.[8]

The first event to take place during heavy chain rearrangement is transposition of one specific D gene to one of the J genes followed by deletion of the intervening sequences of DNA. This process occurs on both chromosomes. Then one of the V genes combines with the newly formed DJ segment, giving a complete VDJ gene segment (Fig. 4.2). This complex codes for the variable region of the immunoglobulin heavy chain and occurs primarily only on one of the two chromosomes. Transcription is started, facilitated by enhancer elements located between the J and C regions.[9] Formation of mRNA occurs by splicing and is followed by synthesis of complete μ heavy chains.[10] The second chromosome will rearrange only if a non-productive rearrangement takes place on the first chromosome (allelic exclusion).[11]

Thereafter, rearrangement takes place of the κ

light-chain genes. This is carried out in a similar way, resulting in formation of a VJ gene complex. If a non-productive rearrangement occurs, rearrangement of the κ gene on the other chromosome will take place. If this rearrangement also fails to produce a functional gene, the cell will rearrange the λ genes on chromosome 22.

Thus, rearrangements of the light-chain genes take place in an ordered sequence $\kappa^a \rightarrow \kappa^b \rightarrow \lambda^a \rightarrow \lambda^b$.[12,13] As a consequence, the majority of normal B cells express κ chains rather than λ chains.[14] A complete IgM molecule is formed by binding of the light chains to the μ heavy chains in the endoplasmic reticulum. The immunoglobulin molecule is then transported to and expressed on the cell surface of the immature B cell.

ISOTYPE SWITCHING

Due to the close location of μ and δ genes, a long primary transcript including the VDJ gene segment, μ and δ genes is produced in mature B cells. After splicing into separate VDJ-μ and VDJ-δ transcripts, separate mRNA and protein molecules are synthesized, which, after combination with the immunoglobulin light chain in the endoplasmic reticulum, are expressed simultaneously as surface IgM and IgD molecules.[14]

During differentiation, an irreversible switch mechanism is utilized by activated B cells. This involves cutting off DNA sequences in a switch region located 5′ to the μ gene. After combining with a complementary switch region (located 5′ to another heavy chain gene), the looping-out intervening sequences are deleted[15,16] (Fig. 4.2). The transient expression of three different isotypes at the same time is explained by the fact that expression of the μ and δ genes decreases gradually.[17]

The mechanisms of immunoglobulin rearrangement summarized above have theoretical and functional implications. They enable a B cell to maintain antigenic and idiotypic specificity during maturation. Thus, specific reactivity and clonal restriction are conserved, whilst isotype switching results in functionally different antibodies. However, if a pre-B cell undergoes proliferation after rearrangement of the heavy-chain genes but before rearrangement of the light-chain genes, clones may emerge which display different light-chain gene rearrangements, with various specificities or exhibiting different light-chain isotypes despite the fact that they have a common origin.[18,19]

Exceptions from the strict hierarchial rearrangement pathway have been reported. Kleinfield et al.[20] showed replacement of a functional rearranged heavy-chain gene complex by another V gene segment with a change in specificity. A similar replacement has been described for VJ segments of the light-chain genes.[21] Hardy et al.[22] found expression of λ chains on the cell surface of lymphoma cells which originally expressed κ chains. However, the λ-positive cells continued to produce κ chain mRNA. It has also been shown that a B cell may replace the original rearranged light chain by rearranging and expressing a new light-chain gene from the other allele.[23]

GENERATION OF ANTIBODY DIVERSITY

The extreme diversity of the normal antibody repertoire of a healthy individual has been developed to achieve the ability to mount a specific immune response against practically every existing foreign antigen. Several mechanisms contribute to the generation of the diversity. First, a random combination of V, D and J segments takes place for the heavy chain, as well as for V and J segments for the light chain. If there are 50 V genes, 20 D genes and 6 J genes, the maximal number of VDJ combinations will be 6000. Moreover, the exact site of V–J, V–D and D–J joining may vary by up to ten nucleotides, with a further increase in diversity. Insertion of extra nucleotides might also occur.[24] Finally, B cells use somatic hypermutation to increase diversity.[24] Since this mechanism takes place predominantly during an immune response it is probably not a random process[25] and selection for mutant B cells producing antibodies of high affinity may also occur. This might also enable malignant B cells to change idiotype specificity, which might during certain conditions favor an escape of recognition by the immune system (*see below*).

M-Component isotypes in monoclonal gammopathies

In a large sample of unselected patients with monoclonal gammopathy, IgG accounted for 61 per cent of the cases, followed by IgM (18 per cent), IgA (11 per cent) and IgD (0.5 per cent). Six per cent of the patients had Bence Jones light chains only.[26] A biclonal gammopathy was found in 3.5 per cent of the cases.

Table 4.1 M-Component immunoglobulin isotypes (per cent) in relation to diagnosis

	IgG	IgA	IgM	IgD	IgE	Bence Jones light chains only
MGUS	75–80	15–20	5–10	a	a	b
Multiple myeloma	55–60	20–25	–	1–2	c	15–20
Waldenström's macroglobulinemia	–	–	100	–	–	–

[a] One case has been reported.
[b] The exact percentage of cases unknown.
[c] Eleven cases reported.

The distribution of the M-component isotypes in relation to diagnosis is shown in Table 4.1, from which it can be seen that IgG is the most common isotype in MGUS and in multiple myeloma.

The M-component type may have prognostic implications. The median survival has usually been considered to be shorter for IgA than for IgG myelomas.[27,28] Myeloma patients producing only λ light chains seem to have a poorer prognosis compared to patients with IgG, IgA or Bence Jones κ myeloma.[29] In addition, patients with Bence Jones κ myeloma responded better to chemotherapy than patients who produced λ light chains.[30] The reasons for these differences are not yet understood.

Biclonal gammopathy

Biclonal gammopathies are characterized by the simultaneous appearance of two different M-components. The incidence is about 1 per cent of all monoclonal gammopathies.[31] The most common combination is IgG + IgA, which occurs in 33 per cent of cases, followed by IgM + IgG with a reported frequency of 24 per cent. Two different IgG M-components were present in 17 per cent of the patients, whereas other combinations occurred in lower frequency. Triclonal gammopathy has also been described in rare cases.[32]

Biclonal gammopathy is associated with multiple myeloma, gammopathy of undetermined significance, Waldenström's macroglobulinemia and other lymphoproliferative disorders e.g. non-Hodgkin's lymphoma, chronic lymphocytic leukemia.[33] The distribution between various clinical entities, the clinical features of the patients as well as the prognosis seems to be similar to that of monoclonal gammopathies.[33]

From a theoretical point of view, the main question is whether the two M-components are of true biclonal nature, i.e. originate from two unrelated clones, or if they are derived from a single parenteral clone in which an isotype switch has occurred. In a statistical analysis, identical light chain isotypes were found more often than would be expected if only random selection had occurred.[31] This finding indicates that at least some of these apparently biclonal gammopathies have a common clonal origin. In an individual case, different light chains on the two M-components have been considered to be a strong evidence of a true biclonal origin.[34,35] However, if the clonal precursor cell is a pre-B cell where light chain gene rearrangement has not occurred, both κ and λ production might be encountered. Using anti-idiotypic antisera, serological identity of the two M-components strongly favours a common origin.[36] However, lack of idiotypic conformity[36,37] does not exclude a common origin

as B lymphocytes may use somatic hypermutation to increase antigenic diversity. Moreover, Southern blot analysis has demonstrated identical immunoglobulin heavy chain rearrangement patterns in two B cell clones that expressed different light-chain isotypes.[38] Using amino acid sequencing, the finding of identical sequences of the variable region in combination with different constant regions of the heavy chain strongly indicated a clonal origin.[39] Different sequences in the hypervariable region of the N-terminal region do not exclude a common origin. Immunofluorescence studies of plasma cells have shown that both immunoglobulin isotypes could be synthesized by the same cell[40,41] but also that the M-components could be synthesized by different cells.[42,43] Furthermore, the serum concentrations of the two M-components followed the activity of the disease in most cases.

Thus, most data speak in favour of the notion that, in the majority of biclonal gammopathies, the two M-components originate from the same clone.

Antigen-reactivity of M-components

In most cases, it is difficult to shown an antigenic reactivity of M-component from patients with monoclonal gammopathies, probably due to difficulties in identifying the proper antigen. Thus, specific reactivity has only occasionally been detected. M-Components with rheumatoid factor activity as well as IgG M-components with antistreptolysin activity have been described.[44,45] IgM M-components may have cold-agglutinin activity.[44] IgA M-components may interact with serum lipoproteins and give elevated cholesterol levels.[44] Recent observations indicate that M-components may be preferentially directed against auto-antibodies within an idiotypic regulatory network. This hypothesis is consistent with the frequent occurrence of cross-reactive idiotypes.[46] However, as the M-components are normal immunoglobulin molecules it is anticipated that all M-components possess antigen-binding specificity.

Idiotype-bearing B cells in multiple myeloma

Multiple myeloma has previously been regarded to be a malignancy originating from mature plasma cells. This opinion has mainly been based on the morphology of the bone marrow. However, increased numbers of peripheral blood lymphocytes with atypical morphology have been found.[47] Using anti-idiotypic antisera, it has been demonstrated that a proportion of the blood B lymphocytes carry the same idiotypic immunoglobulin surface structure as that of the serum M component and the cytoplasmic immunoglobulin in myeloma plasma cells.[48,49] In this study, the idiotypic surface immunoglobulin could be removed by trypsin and reappeared at the cell surface after *in vitro* incubation, indicating an endogenous production. The presence of circulating monoclonal B cells has been confirmed by cytogenetic studies[50] including Southern blot analysis, showing identical rearrangements of heavy and light chain genes in peripheral blood B cells and in bone marrow plasma cells.[51] In analogy to the normal B-cell development, it might be assumed that idiotype-bearing B lymphocytes are progenitors to idiotype-secreting plasma cells, and thus multiple myeloma should be regarded as a differentiating B-cell malignancy (Fig. 4.3). Up to 40 per cent of idiotype-bearing monoclonal blood B cells may be found in advanced multiple myeloma,[52] but usually less than 5 per cent of idiotypic B cells are present.[53] In MGUS, the numbers of clonal blood B cells are low.[53,54] Using a polymerase chain reaction (PCR)-based method, circulating myeloma precursor B cells with clonal immunoglobulin gene rearrangements could be detected in practically all myeloma patients.[55] These cells are hyperploid and seem to be chemotherapy resistant.[56] The presence of such cells might have several therapeutic implications. If these cells are the feeder pool then efforts should be made to eliminate these cells and not only the plasma cells

(*see below*). Furthermore, using peripheral blood stem cell support in combination with intensive ablative therapy the existence of these precursor cells in the transplant might be a source of relapse.

The presence of idiotype-bearing B cells in peripheral blood should be regarded as a sign of dissemination of the disease, which might be due to an overload of tumor cells in the bone marrow with subsequent leakage of cells into the circulation. Alternatively, idiotype-bearing blood B cells might comprise a certain phenotype of cells prone to circulate and programmed for migration across vascular endothelium to the bone marrow where they undergo terminal differentiation to myeloma plasma cells. This hypothesis is strengthened by an increased myeloma B-cell expression of adhesion molecules participating in the cellular traffic over the vessel wall.[57,58]

The detection of individually unique idiotypic surface structures is a laborious method since individually produced anti-idiotypic antibodies are required. Thus, only a limited number of patients can be studied with this technique. A simple immunofluorescence method, which enables the study of a large number of patients, is based on the fact that every individual B lymphocyte expresses surface immunoglobulin of either κ or λ light chain type but not both. Thus, the normal B lymphocyte population contains κ- as well as λ-expressing cells. Using fluorochrome-coupled murine antibodies against human κ and λ light chains respectively, the percentages of κ- and λ-expressing lymphocytes in the peripheral blood can be determined. In healthy controls, the ratio of κ-positive to λ-positive lymphocytes ranged between 0.9 and 3.5.[59] A κ/λ ratio higher than the upper normal range (>3.5) (in multiple myeloma of κ-type) or lower than the normal range (<0.9) (in multiple myeloma of λ-type) might thus indicate the presence of monoclonal B cells, 'clonal B cell excess' (CBE). About 50 per cent of untreated multiple myeloma patients and 15 per cent of MGUS patients had CBE.[54,59] The presence of blood CBE in multiple myeloma patients was associated with poor prognosis.[59] In patients with MGUS, blood CBE might predict a subsequent malignant transformation. Fifty-three per cent of CBE-positive MGUS patients compared to only 17 per cent of CBE-negative patients developed a malignant disease during follow-up (to be published).

Highly sensitive PCR-based methods enable the detection of very small numbers ($1/10^5$–$1/10^4$) of circulating monoclonal B cells.[60,61] Using immunoglobulin gene fingerprinting, circulating monoclonal cells can be detected in the majority of patients with active myeloma but not in patients with complete or good partial remission.[62] Moreover, this method might be useful to detect minimal residual disease after bone marrow transplantation.[60] The clinical

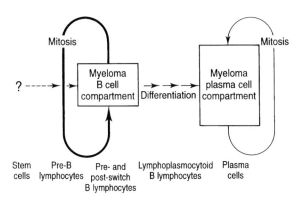

Fig. 4.3 A schematic model of multiple myeloma as a differentiating B-cell malignancy. A small compartment of rapidly dividing myeloma precursor B lymphocytes differentiates towards mature myeloma plasma cells. Most of the neoplastic cells are confined to the plasma cell compartment where the mitotic activity is low. It is still uncertain whether the myeloma cell clone includes pre-B cells or bone marrow stem cells.

significance of minimal residual disease (PCR positivity) in patients with long-term complete remission is not yet known. Using allele-specific PCR for the CDR3 region of the rearranged immunoglobulin heavy chain allele, Reece et al. detected a persistent clonal B cell population in the bone marrow from a patient treated with allogeneic bone marrow transplantation. A complete clinical remission was, however, still present 43 months later.[61]

It is a matter of debate which is the earliest B cell belonging to the myeloma cell clone. Phenotypic and morphological analyses have demonstrated that the majority of clonal cells are late B lymphocytes and plasma cells.[63,64] However, a substantial fraction are B cells of an earlier ontogeny. Some studies indicate that the neoplastic event occurs in a post-switch memory B cell, since the isotype of the surface idiotypic immunoglobulin was found to be identical to that of the immunoglobulin secreted by the myeloma clone.[65] However, in IgG myeloma idiotypic structures have not only been detected on lymphocytes expressing surface IgG, but also on cells expressing surface IgM and/or IgD.[66] Van't Veer et al.[67] found a dual expression of IgM and IgG in the cytoplasm of plasma cells from an MGUS patient with an IgG serum M-component, indicating a pre-switch origin. This suggestion is also supported by cytogenetic analyses, showing that one-third of the patients had a clonal immunoglobulin gene rearrangement using the *Hin*dIII restriction enzyme only but not with other restriction enzymes.[68] As the 3' ends of the *Hin*dIII restriction fragments are located upstream from the μ switch region, their size does not vary during isotype switching, indicating that the clonogenic fraction originated from a pre-switch B cell. Moreover, the coexpression of κ and λ cytoplasmic light-chain genes as well as the expression of germline immunoglobulin genes, further suggests involvement of an early B-cell progenitor.[69] Using PCR-based techniques it has been confirmed that pre-switch B lymphocytes in multiple myeloma belongs to the tumor clone[70] (Van Ness, personal communication). It is, however, not yet clear whether pre-B cells belong to the myeloma tumor clone. Kubagawa et al.[71] described B lymphocytes with cytoplasmic idiotypic structures together with μ chains only, i.e. a pre-B cell.

Using purified and isolated subsets of B lymphocytes in various ontogenetic stages, Berenson and coworkers could not identify clonal immunoglobulin gene rearrangements in CD34-enriched stem-cell fractions nor in pre-B cells (J. Berenson, personal communication).

Antigens as well as megakaryocytic or erythroid cell surface markers have been found on multiple myeloma plasma cells.[72,73] Clonal rearrangement of the β chain of the T-cell receptor has also been found in myeloma plasma cells.[74] These findings have been suggested to reflect a stem-cell origin of multiple myeloma but may also just reflect an aberrant expression due to the malignant transformation. A stem-cell origin has also been suggested from epidemiological studies demonstrating familial aggregations of hematological malignancies.[75]

Taken together, the results provide convincing evidence that the tumor clone in myeloma originates from a pre-switch B lymphocyte. Whether stem cells are involved or not is not clear (Fig. 4.3).

Idiotypic immune network regulation

The majority of the bone marrow myeloma clonal cells are plasma cells. However, a fraction of B lymphocytes also belong to the malignant cell clone.[52,53] The proliferative activity is higher in the precursor monoclonal B-lymphocyte fraction than in the myeloma plasma cell pool[76,77] (Fig. 4.3). This might have therapeutic implications. Therapy might preferentially be directed against the myeloma progenitor B lymphocytes. However, idiotype-bearing monoclonal B lymphocytes seem to be more resistant to conventional chemotherapy than myeloma plasma cells.[78] On the other hand, myeloma B lymphocytes express idiotypic immunoglobulin molecules on the cell surface, which may render the cells (myeloma B

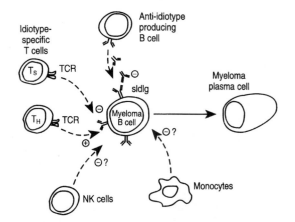

Fig. 4.4 A hypothetical schematic model of an immune network regulatory circuit in myeloma. Idiotype-specific regulation may be operating selectively on the myeloma B-lymphocyte precursors as myeloma plasma cells do not express surface idiotypic immunoglobulin molecules. T_S, suppressor/cytotoxic T cell; T_H, helper T cell; TCR, T cell receptor; sIdIg, surface idiotypic immunoglobulin; NK, natural killer cell.

lymphocytes but not plasma cells) sensitive to regulatory signals provided by cellular and humoral components of the idiotypic immune network system.[79] According to the immune network theory,[79] antibodies interact through complementary idiotypic and anti-idiotypic structures, to maintain a balance within the immune system. The network theory also postulates that T lymphocytes participate through the variable region of the α/β chains of the T-cell receptor (TCR). Within this network every T- and B-cell receptor is recognized by a complementary anti-idiotype which is also an idiotype for another lymphocyte clone. A clonal expansion of idiotypic B cells may accordingly lead to induction and expansion of complementary idiotypes as well as anti-idiotypic B and T cell clones. Both anti-idiotypic antibodies and idiotype-specific T cells might be operating to regulate the proliferation and differentiation of the myeloma cell clone (Fig. 4.4).

A humoral anti-idiotypic immunity with regulatory impact on the myeloma clone was described in a murine myeloma system in 1971 by Sirisinha and Eisen.[80] BALB/c mice produced antibodies specific for idiotypic immunoglobulin determinants after immunization with the idiotype coupled to an adjuvant. Immunized mice were resistant to challenge with otherwise lethal numbers of myeloma cells. The resistance was found to be idiotype-specific. The idiotypic immunoglobulin structures produced by the myeloma cells may thus be regarded as a tumor-specific antigen.[81] The role of anti-idiotypic antibodies in an idiotype-specific transplantation rejection was, however, questionable. Later studies suggested a minor role for a humoral anti-idiotypic immunity but a major role for T cells recognizing idiotypic structures. Nevertheless, the administration of anti-idiotypic antibodies to mice inhibited the myeloma tumor cell growth.[82] The presence of anti-idiotypic antibodies has also been demonstrated in human multiple myeloma.[83] A high production of anti-idiotypic antibodies was noted in patients with a low tumor burden, whereas the production was low in patients with advanced disease.[84]

A T-cell-mediated response against the tumor clone might be of even greater significance than a humoral response. Daley et al.[85] demonstrated in mice that the specific myeloma transplantation resistance was eliminated by post-immunization thymectomy, providing convincing evidence that the resistance was mediated by idiotype-specific T cells. Similarly, idiotype-specific T cells were found to inhibit myeloma cell growth *in vitro* in a murine system.[86] Furthermore, idiotype-specific T cells could inhibit differentiation and immunoglobulin secretion of myeloma B cells.[87]

Idiotype-specific T cells have also been demonstrated in human myeloma. Dianzani et al.[88]

described idiotype-reactive T cells with an activated phenotype (CD8+/HLA−DR+) in peripheral blood of myeloma patients. However, it is not clear whether the cells were truly idiotype-specific or bound to the constant regions of the immunoglobulin molecule. Isotype-specific CD8+ T cells have been described in a murine myeloma system as well as in humans, which cells suppressed myeloma growth and immunoglobulin secretion by production of a soluble receptor for the Fc part of the immunoglobulin molecule.[89,90]

Most T cells recognize processed antigenic peptides together with major histocompatibility complex (MHC) molecules. The generally accepted paradigm is that peptides from endogenous protein are presented on Class I MHC molecules while peptides derived from exogenous proteins are presented on Class II molecules.[91] Thus, malignant B cells may display processed idiotypes together with MHC Class I molecules in addition to unprocessed idiotypic immunoglobulin molecules on their cell surface (Fig. 4.5). The CD8 structure binds to Class I MHC molecules. CD8+ T cells have cytolytic function, and thus may be important effector cells in an idiotype- or isotype-specific regulation of growth of the myeloma B-cell clone.[88–90]

Malignant B cells may also express the idiotype of the myeloma immunoglobulin together with MHC Class II molecules. In this case the idiotype may be exported externally from the B cell and then taken up through the endocytic pathway (recycling), followed by proteolytic processing and presentation on the cell

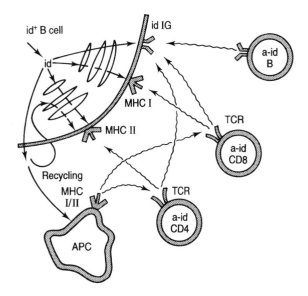

Fig. 4.5 Interplay between the idiotype, anti-idiotypic antibodies, MHC molecules and idiotype-specific T cells in myeloma.

surface in association with MHC Class II molecules.[92] However, the idiotype might also be processed and presented with MHC Class II molecules without having been outside the cell before processing.[93] Idiotype-specific CD4+ T cells, which bind to MHC Class II processed idiotypic peptides in the groove, may exert cytolytic activity and play a significant role in an immunological surveillance of idiotypic B-cell tumors. Both CD4/Th$_1$ and CD4/Th$_2$ clones might be active.[94] Malignant plasma cells may, however, lack MHC Class II molecules. Nevertheless Lauritzsen and Bogen[95] recently demonstrated that idiotype-specific, MHC Class II restricted T cells exhibited protection *in vivo* against the MHC Class II molecule negative plasmocytoma cell line MOPC-315. Moreover, T cells inhibited the growth of MOPC-315 *in vitro* provided that antigen-presenting cells (APC) and surplus of idiotypic immunoglobulin were present. It was proposed that high concentrations of locally produced idiotypic immunoglobulin enabled neighboring APCs to process and present the idiotype to the idiotype-specific Class II restricted T cells. Such activated T cells may secrete locally large amounts of cytokines which may affect the growth of the myeloma cells.[94] Alternatively, monocytes activated by locally produced cytokines might exert cytotoxic activity against the myeloma cells. These results provide evidence for a role of monocytes in the regulation of the myeloma tumor cell clone.

Idiotype-specific T cells may also recognize unprocessed idiotypes.[96–98] T cells of either CD4+ or CD8+ phenotypes that bound fluorescein-labelled F(ab′)$_2$ fragments of the autologous M-component, but not an isotype-matched non-relevant M-component, were identified in the blood of myeloma patients.[99] The number of idiotype-reactive T cells was low (8–15 per cent of the specific subset), but in some patients the cells comprised 25 per cent of the total T cells. The idiotype bound specifically to cells stained with an anti-TCRII (T$_{\alpha/\beta}$) mAb and to cells binding a specific TCR mAb (against one V$_\alpha$ or V$_\beta$ gene segment membrane product). Some patients, however, had a fraction of idiotype-specific T cells that were not stained by the panel of TCR mAbs. The reason for this discordance might be ascribed to the fact that the TCR mAb panel used in that study only covered 25 per cent of the normal TCR repertoire. Other patients had a predominant expansion of a T-cell clone exhibiting one TCR V$_{\alpha/\beta}$ gene segment which did not bind the idiotype.[99] The specificity of the latter T-cell population remains to be established. Probably myeloma patients, apart from having T cells which bind unprocessed immunoglobulin, also have T cells recognizing the idiotype in a processed form (*see above*) and thus may not be detected by the technique used in that study. This might have functional implications, as B cells may undergo somatic hypermutation. Even a single mutation may result in a dramatic change of the three-dimensional structure of the antigen-binding site of the immunglobulin molecule, resulting in an inability of idiotype-specific T cells to recognize unprocessed idiotypic immunoglobulin but not processed amino acid sequences exhibiting a linear structure. Somatic hypermutation of malignant B cells seems, however, to be a rare phenomenon in patients with myeloma.[100] As MHC-restricted idiotype-specific T cells are probably functionally more important than antibodies, the problem of somatic hypermutation as well as blocking of the TCR by circulating idiotypic immunoglobulin might be of minor importance.

To further elucidate the existence of idiotype-specific T cells, peripheral blood T lymphocytes were stimulated *in vitro* with the autologous M-component. DNA synthesis as well as IFN-λ and IL-2 production were assessed. *In vivo* delayed type hypersensitivity (DTH) reaction against the idiotype was determined as the DTH reaction is considered to reflect a key function in tumor rejection. In 7 out of 9 patients at least one of these four tests was positive indicating the presence of idiotype-reactive T cells. In four of the patients three or more tests were positive.[101]

Idiotype-reactive T cells were also analyzed at the clonal level. From three patients, idiotype-specific T-cell clones were generated which showed a significant proliferative response against the idiotype. The majority of the clones expressed the CD8 structure, but CD4+ clones were also obtained.[102] The high frequency of idiotype-specific CD8+ clones is of interest as T-cell clones generated against conventional protein antigens usually exhibit the CD4+ phenotype. The functional significance of the idiotype-specific T-cell clones remains to be established.

Detection and quantification of M-components

The most common signal to a doctor indicating the presence of a monoclonal immunoglobulin (M-component) is an accidentally detected strong increase of the erythrocyte sedimentation rate (ESR) in a patient without clinical symptoms. However, the ESR test has both very low sensitivity as well as specificity and cannot be used as a screening method for M-components. The only reliable technique to detect monoclonal immunoglobulins is high-resolving electrophoresis of the patient's plasma (serum) and urine samples in combination with immunochemical quantification of the immunoglobulins (IgA, IgG, and IgM).

The first laboratory test for the presence of an increase in a pathological protein in urine was the urine heat-stability test first described in 1846 by Sir Henry Bence Jones.[103] Indeed, this was the very first description of a tumor-specific marker. The test can detect monoclonal free light chains in urine (Bence Jones proteinuria) when the concentration is > 300 mg/l. Today this test is mostly replaced by electrophoresis of the patient's urine sample or by immunological quantification of kappa and lambda light chains. Bence Jones proteinuria are not detected by the various dipsticks used to detect proteinuria; these tests almost exclusively detect albumin only.

Since the introduction of high-resolving agarose-gel electrophoresis during the 1970s this has been the method of choice for the visualization of an M-component in both plasma and urine samples, even though cellulose–acetate electrophoretic methods are common and still dominate in many countries. The resolving power of cellulose–acetate electrophoresis is low compared to agarose–gel electrophoresis. High-quality ready-made agarose gels are commercially available. These gels can be used both for the primary electrophoretic investigation of plasma and urine samples and for the follow-up of M-components by immunofixation. Agarose-gel electrophoresis can detect M-components in a concentration of < 0.5 g/l of plasma samples stained with amido black. Using a sample volume of 20 µl and a Paragon violet stain (Beckman Instruments, Inc., CA, USA) the agarose-gel electrophoresis method can detect monoclonal free light chains with a concentration of < 20 mg/l in unconcentrated urine samples.

The electrophoretic pattern of the immunoglobulins in plasma samples from healthy subjects appears as an evenly stained film representing polyclonal immunoglobulins. During an inflammatory process, the immunoglobulin pattern may appear as a general polyclonal increase of immunoglobulins but at the same time also an oligoclonal immunoglobulin pattern may appear, with two or more immunoglobulin bands, representing a stimulation of a restricted number of lymphocyte clones. In lymphoproliferative diseases the electrophoretic immunoglobulin pattern of plasma samples may vary from a depressed pattern ('hypogammaglobulinemia') to a mono-, bi-, or even triclonal immunoglobulin pattern and to a oligo- or polyclonal increase of the immunoglobulins.

The most common electrophoretic immunoglobulin picture in multiple myeloma, Waldenström's macroglobulinemia, and MGUS is a single monoclonal immunoglobulin band (M-component) combined with a normal or depressed polyclonal immunoglobulin background. The position of the M-component in the electrophoretic field is stable in each patient but may vary considerably from patient to patient depending on the type and charge of the M-component at the actual pH of the buffer used in the electrophoretic system. Small IgA M-components (< 1–2 g/l) might be difficult to detect if the M-component co-migrates with one of the normal protein bands (e.g. transferrin, complement factor 3 (C3) and fibrinogen). A careful inspection of the gel by an experienced investigator will, however, reveal all relevant M-components by comparing the actual electrophoretic lane with adjacent lanes on the agarose plate. The examiner should have access to quantitative results from the determinations of the plasma IgG, IgA, and IgM when interpreting the electrophoretic pattern.

The first time an M-component is identified in a patient's plasma the M-component has to be classified preferentially by immunofixation[104,105] using monospecific antisera against heavy and light immunoglobulin chains. The immunoelectrophoretic method developed by Grabar and Williams[106] is difficult to interpret in the case of a small M-component and with normal or increased background immunoglobulin levels. Normally there is no need to perform repeated typing of the M-component.

Information on the clinical condition of a patient is also obtained by investigation of the patient's urine, preferably after an overnight or 24-hour sampling time. Minimum analyses of the urine should include high-resolution electrophoresis and quantification of at least albumin and preferably also of IgG and free kappa and lambda light chains. Total protein measurement gives no further information compared to these methods. There is little need to classify the M-component of the patient's urine by immunofixation, as long as high-resolution electrophoresis of the urine and quantification of, at least, urinary albumin is performed. These analyses will always disclose if there is an M-component in the urine and if the M-component is composed of complete immunoglobulin molecules or only free light chains.

Quantification of the immunoglobulin content in an M-component has for a long time been of major concern. The methods used until now have been either densitometric scanning or visual evaluation of the electrophoretic pattern. Both methods are, however, often unreliable due to the extreme high concentration of protein in the M-component. A newly developed automatic capillary electrophoretic method with a resolving power comparable to agarose-gel electrophoresis has recently been introduced. This technique will probably solve the problem of quantification of the M-component, as separated proteins will pass a photometric detector cell where continuous quantification of the protein concentration in the different peaks (bands) is carried out.

Immunochemical quantification of IgA, IgG, and

IgM can either be performed manually by the gel techniques, single radial immunodiffusion[107] or electroimmunoassay,[108] by automated immuno-turbidimetric methods using common clinical chemistry analysers or by immuno-nefelometric methods with dedicated instruments for specific immunochemical protein quantification. As long as internationally certified calibrators are used, all four principally different methods will give acceptable results in terms of accuracy and imprecision. The choice of method depends on the workload, manual and instrumental competence of the laboratory. The manual methods are robust but time-consuming with results available within 1–3 days compared to within 1–2 hours using automated methods. In normal routine work the imprecision (CV) is 2–5 per cent when using automated methods.

Immunochemical quantification of monoclonal IgM might generate incorrect results with all of the above methods. This is partly because of the presence of IgM molecules of mono-, di-, tri-, tetra- or pentameric forms and partly because of the extreme tendency of the IgM molecule to aggregate. These problems may be circumvented by degrading the immunoglobulin to heavy and light chains followed by quantification of the free μ chains. The calibrator has to be treated in the same way as the sample. If a patient has cryoglobulins it is of utmost importance to draw the blood sample at 37°C and then transport, centrifuge and store the sample at 37°C until assayed to avoid loss of a large proportion of the monoclonal immunoglobulin.

Quantification of light chains, kappa and lambda, in the urine may be done by measuring the total amount of the respective light chain by turbidimetric or nephelometric techniques irrespective of whether the light chains are free or bound to heavy chains. An alternative method is to measure the free light chains after separation of free light chains from the complete immunoglobulins by precipitation of the urine sample using polyethylene glycol 6000 (PEG 6000) at a final concentration of 20 per cent followed by centrifugation. The separated free light chains of the supernatant are then measured by single radial immunodiffusion.

Can an idiotype-specific T-cell immunity be induced/ enhanced in vivo?

If idiotype-reactive T cells have the potential to regulate growth of the myeloma tumor cell clone, a therapeutic intervention leading to expansion of cells with suppressive or cytotoxic effects on the tumor clone might be a rewarding approach. Immunizing with the autologous M-component may induce a specific immunity. Active immunization with idiotypic immunoglobulin molecules has produced resistance to tumor cell challenges in several animal systems including a myeloma model.[80,85,109,110] R. Levy's group at Stanford[111] (personal communication) have immunized twenty-six patients with non-Hodgkin lymphoma of the B-cell type with idiotypic immunoglobulin conjugated with KLH (keyhole limpet hemocyanin) precipitated in aluminium phosphate (alum). Twelve patients developed a humoral anti-idiotypic response and 9 patients a specific T-cell response. In 17 (65 per cent) of the patients both a humoral and cellular response were induced. Two patients with a detectable disease at start of immunization entered a complete remission. Eight out of 9 evaluable patients who did not develop anti-idiotype immunity progressed during the observation period, while only 2 out of 17 patients exhibiting a detectable anti-idiotype response had progressive disease.

We have recently initiated an active immunization protocol in myeloma patients with untreated asymptomatic disease. Prior to immunization there was a low but significant DNA synthesis after stimulation with the autologous idiotype. After the second immunization, the proliferation response to the idiotype increased 15–30-fold. Similar results were obtained using an assay analysing for idiotype-specific IL-2 and IFN-γ production at the single cell level. No humoral response to the autologous idiotype could be detected.[112] These preliminary results suggest that it might be possible to augment an idiotypic-specific T-cell response *in vivo* in patients with myeloma. Clinical studies are urgently required to explore the clinical significance of an immunization protocol using the idiotypic immunoglobulin. This treatment approach should be used in remission when the tumor load is minimal but still a sufficient number of chemotherapy-resistant myeloma B lymphocytes are present which might be responsible for the relapse.

In conclusion, *in vitro* and *in vivo* data allow the presentation of a plausible scenario in multiple myeloma considered to be a differentiating B-cell malignancy, where the clonal growth is up- and downregulated by signals delivered by different components of the immune system. A better understanding of these regulatory mechanisms might be a prerequisite for a major breakthrough in the treatment of human myeloma, in which disease the idiotypic immunoglobulin, the hallmark of the disease, may be regarded as one of the few tumor-specific antigens so far identified in humans.

References

1. Gutman AB. The plasma proteins in disease. *Advances in Protein Chemistry* 1948; **4**: 155.
2. Kyle RA. Monoclonal gammopathy of undetermined significance and smoldering multiple myeloma. *European Journal of Haematology* 1989; **43** (suppl. 51): 70–5.
3. Hozumi N, Tonegawa S. Evidence for somatic rearrangement of immunoglobulin genes coding for variable and constant regions. *Proceedings of the National Academy of Sciences* 1976; **73**: 3628–32.
4. Croce CM, Shander M, Martinis J et al. Chromosomal location of the genes for human immunoglobulin heavy chains. *Proceedings of the National Academy of Sciences* 1979; **76**: 3416–19.
5. Malcolm S, Barton P, Murphy C et al. Localization of human immunoglobulin kappa light chain variable region genes to the short arm of chromosome 2 by *in situ* hybridization. *Proceedings of the National Academy of Sciences* 1982; **79**: 4957–61.
6. Erikson J, Martinis J, Croce CM. Assignment of the genes for human λ immunoglobulin chains to chromosome 22. *Nature* 1981; **294**: 173–5.
7. Sakano H, Maki R, Kurosawa Y et al. Two types of somatic recombination are necessary for the generation of complete immunoglobulin heavy-chain genes. *Nature* 1980; **286**: 676–83.
8. Ravetch JV, Siebenlist U, Korsmeyer S et al. Structure of the human immunoglobulin μ locus: Characterization of embryonic and rearranged J and D genes. *Cell* 1981; **27**: 583–91.
9. Banerji J, Olson L, Schaffner W. A lymphocyte-specific cellular enhancer is located downstream of the joining region in immunoglobulin heavy chain genes. *Cell* 1983; **33**: 729–40.
10. Korsmeyer SJ, Hieter PA, Ravetch JV et al. Developmental hierarchy of immunoglobulin gene rearrangements in human leukemic pre-B-cells. *Proceedings of the National Academy of Sciences* 1981; **78**: 7096–100.
11. Alt FW. Exclusive immunoglobulin genes. *Nature* 1984; **312**: 502–3.
12. Hieter PA, Korsmeyer SJ, Waldman TA, Leder P. Human immunoglobulin κ light-chain genes are deleted or rearranged in λ-producing B cells. *Nature* 1981; **290**: 368–78.
13. Vogler LB. Bone marrow B cell development. In: *Clinics in haematology*, vol. 11, no. 3. Philadelphia: W.B. Saunders, 1982: 509–29.
14. Gordon J. Molecular aspects of immunoglobulin expression by human B cell leukemias and lymphomas. *Advances in Cancer Research* 1984; **41**: 71–154.
15. Shimuzu A, Honjo T. Immunoglobulin class switching. *Cell* 1984; **36**: 801–3.
16. Cooper MD. B lymphocytes. Normal development and function. *New England Journal of Medicine* 1987; **317**: 1452–6.
17. Cooper MD, Kearney JF, Gathings WE, Lawton AR. Effects of anti-Ig antibodies on the development and differentiation of B cells. *Immunological Reviews* 1980; **52**: 29–53.
18. Riley SC, Brock EJ, Kuehl WM. Induction of light chain expression in a pre-B cell line by fusion to myeloma cells. *Nature* 1981; **289**: 804–6.
19. Caton AJ. A single pre-B cell can give rise to antigen-specific B cells that utilize distinct immunoglobulin gene rearrangements. *Journal of Experimental Medicine* 1990; **172**: 815–25.
20. Kleinfield R, Hardy RR, Tarlinton D et al. Recombination between an expressed immunoglobulin heavy-chain gene and a germline variable gene segment in a Lyl⁺ B cell lymphoma. *Nature* 1986; **322**: 843–6.
21. Feddersen RM, Van Ness BG. Double recombination of a single immunoglobulin κ-chain allele: implications for the mechanism of rearrangement. *Proceedings of the National Academy of Sciences* 1985; **82**: 4793–7.
22. Hardy RR, Dangl JL, Haykawa K et al. Frequent λ light chain gene rearrangement and expression in a Ly−1 B lymphoma with a productive κ chain allele. *Proceedings of the National Academy of Sciences* 1986; **83**: 1438–42.
23. Berinstein N, Levy S, Levy R. Activation of an excluded immunoglobulin allele in a human B lymphoma cell line. *Science* 1989; **244**: 337–9.
24. Tonegawa S. Somatic generation of antibody diversity. *Nature* 1983; **302**: 575–81.
25. O'Brien RL, Brinster RL, Storb U. Somatic hypermutation of an immunoglobulin transgene in κ transgenic mice. *Nature* 1987; **326**: 405–9.
26. Kyle RA. Diagnostic criteria of multiple myeloma. *Hematology/Oncology Clinics of North America* 1992; **6**: 347–58.
27. Hansen OP, Jensen B, Videbaek A. Prognosis of myelomatosis on treatment with prednisone and cytostatics. *Scandinavian Journal of Haematology* 1973; **10**: 282–90.
28. Baldini L, Radaelli F, Chiorboli O et al. No correlation between response and survival in patients with multiple myeloma treated with vincristine, melphalan, cyclophosphamide and prednisone. *Cancer* 1991; **68**: 62–7.
29. Bergsagel DE, Phil D, Bailey AJ et al. The chemotherapy of plasma-cell myeloma and the incidence of acute leukemia. *New England Journal of Medicine* 1979; **301**: 743–8.
30. Bergsagel DE, Migliore PJ, Griffith KM. Myeloma proteins and the clinical response to melphalan therapy. *Science* 1965; **148**: 376–7.
31. Bouvet JP, Feingold J, Oriol R, Liacopoulos P. Statistical study on double paraproteinaemia. Evidence for a common cellular origin of both myeloma globulins. *Biomedicine* 1975; **2 2**: 517–23.
32. Ray RA, Schlotters SB, Jacobs A, Rodgerson DO. Triclonal gammopathy in a patient with plasma cell dyscrasia. *Clinical Chemistry* 1986; **32** 205–6.
33. Kyle RA, Robinson RA, Katzman JA. The clinical aspects of biclonal gammopathies. Review of 57 cases. *American Journal of Medicine* 1981; **71**: 999–1008.

34. Graziani MS, Lippi U. Multiple myeloma with serum IgM kappa and Bence-Jones lambda biclonal gammopathy. *Clinical Chemistry* 1986; **32**: 2220–1.
35. Finco B, Schiavon R. Multiple myeloma with serum IgG kappa and Bence-Jones lambda biclonal gammopathy. *Clinical Chemistry* 1987; **33**: 1305–6.
36. Bast EJEG, Slaper-Cortenbach CM, Verdonck LF et al. Transient expression of second monoclonal component in two forms of biclonal gammopathy. *British Journal of Haematology* 1985; **60**: 91–7.
37. Bonewald L, Virella G, Wang A-C. Evidence for the biclonal nature of a Waldenström's macroglobulinemia. *Clinica Chimica Acta* 1985; **146**: 53–63.
38. Miyamura K, Osada H, Yamauchi T et al. Single clonal origin of neoplastic B-cells with different immunoglobulin light chains in a patient with Richter's syndrome. *Cancer* 1990; **66**: 140–4.
39. Wang A-C, Gergely J, Fudenberg HH. Amino acid sequences at constant and variable regions of heavy chains of monotypic immunoglobulins G and M of a single patient. *Biochemistry* 1973; **12**: 528–34.
40. Rudders RA, Yakulis V, Heller P. Double myeloma. Production of both IgG type lambda and IgA type lambda myeloma proteins by a single plasma cell line. *American Journal of Medicine* 1973; **55**: 215–21.
41. Bouvet JP, Buffe D, Oriol R, Liacopoulos P. Two myeloma globulins $IgG_1\kappa$ and $IgG_1\lambda$ from a single patient. (Im) II. Their common cellular origin revealed by immunofluorescence studies. *Immunology* 1974; **27**: 1095–01.
42. Bjerrum OJ, Weeke B. Two M-components γGK, γML in different cells in the same patients. *Scandinavian Journal of Haematology* 1968; **5**: 215–34.
43. Rosen BJ, Smith TW, Bloch KJ. Multiple myeloma associated with two serum M-components γG type K and γA type L. *New England Journal of Medicine* 1967; **17**: 902–7.
44. Sherwood LM, Parris EE. Myeloma proteins (M-components) with antibody-like activity. *New England Journal of Medicine* 1971; **284**: 831–8.
45. Waldenström J, Winblad S, Hällén J, Ljungman S. The occurrence of serological 'antibody' reagins or similar γ-globulins in conditions with monoclonal hypergammaglobulinemia, such as myeloma, macroglobulinemia etc. *Acta Medica Scandinavica* 1964: **176**: 619–31.
46. Berenson JR, Lichtenstein A, Hart S et al. Expression of shared idiotypes by paraproteins from patients with multiple myeloma and monoclonal gammopathy of undetermined significance. *Blood* 1990; **75**: 2107–11.
47. Mellstedt H, Jondal M, Holm G. *In vitro* studies of lymphocytes from patients with plasma cell myeloma. II. Characterization by cell surface markers. *Clinical and Experimental Immunology* 1973; **15**: 321–30.
48. Mellstedt H, Hammarström S, Holm G. Monoclonal lymphocyte population in human plasma cell myeloma. *Clinical and Experimental Immunology* 1974; **17**: 371–84.
49. Mellstedt H, Holm G, Björkholm M. Multiple myeloma. Waldenström's macroglobulinemia and benign monoclonal gammopathy. Characteristics of the B cell clone, immunoregulatory cell populations and clinical implications. *Advances in Cancer Research* 1984; **41**: 257–89.
50. MacKenzie MR, Lewis JP. Cytogenetic evidence that the malignant event in multiple myeloma occurs in a precursor lymphocyte. *Cancer Genetics and Cytogenetics* 1985; **17**: 13–20.
51. Berenson J, Wong E, Kim K, Lichtenstein A. Evidence for peripheral blood B lymphocyte but not T lymphocyte involvement in multiple myeloma. *Blood* 1987; **70**: 1550–3.
52. Pettersson D, Mellstedt H, Holm G. Monoclonal B lymphocytes in multiple myeloma. *Scandinavian Journal of Immunology* 1980; **12**: 375–82.
53. Bast EJEG, Van Camp B, Reynaert P et al. Idiotypic peripheral blood lymphocytes in monoclonal gammopathy. *Clinical and Experimental Immunology* 1982; **47**: 677–82.
54. Petterson D, Mellstedt H, Holm G, Björkholm M. Monoclonal blood lymphocytes in benign monoclonal gammopathy and multiple myeloma in relation to clinical stage. *Scandinavian Journal of Haematology* 1981; **27**: 287–93.
55. Billadeau D, Quam L, Thomas W et al. Detection and quantitation of malignant cells in peripheral blood of multiple myeloma patients. *Blood* 1992; **80**: 1818–24.
56. Pilarski LM, Belch AR. Circulating monoclonal B cells expressing P glycoprotein may be a reservoir of multidrug-resistant disease in multiple myeloma. *Blood* 1994; **83**: 724–36.
57. De Greef C, Van Riet I, Bakkus M et al. Molecular aspects of N-CAM expression in multiple myeloma. Abstract. Paper presented at the IV International Workshop on Multiple Myeloma. Rochester, MO, USA, 2–5 October 1993, p. 134.
58. Jensen GS, Belch AR, Kherani F et al. Restricted expression of immunoglobulin light chain mRNA and of the adhesion molecule CD11b on circulating monoclonal B lineage cells in peripheral blood of myeloma patients. *Scandinavian Journal of Immunology* 1992; **36**: 843–53.
59. Österborg A, Nilsson B, Björkholm M et al. Blood clonal B cell excess at diagnosis in multiple myeloma. Relation to prognosis. *European Journal of Haematology* 1987; **38**: 173–8.
60. Bird JM, Russell NH, Samson D. Minimal residual disease after bone marrow transplantation for multiple myeloma: evidence for cure in long-term survivors. *Bone Marrow Transplantation* 1993; **12**: 651–4.
61. Reece D, Billadeau D, Van Ness B et al. Intensive therapy (IT) and allogeneic bone marrow transplantation (allo BMT) in multiple myeloma (MM). Preliminary clinical and molecular results. Abstract. Paper presented at the IV International Workshop on Multiple Myeloma, Rochester, MO, USA, 2–5 October, 1993, p. 147.
62. Bird JM, Samson D. Detection of clonally rearranged cells in the peripheral blood of myeloma

patients using immunoglobulin gene fingerprinting. Abstract. Paper presnted at the IV International Workshop on Multiple Myeloma, Rochester, MO, USA, 2–5 October, 1993, p. 130.

63. Jensen GS, Mant MJ, Belch AJ et al. Selective expression of CD45 isoforms defines CALLA⁺ monoclonal B-lineage cells in peripheral blood from myeloma patients as late stage B cells. *Blood* 1991; **78**: 711–19.

64. Österborg A, Steinitz M, Lewin N et al. Establishment of idiotype bearing B-lymphocyte clones from a patient with monoclonal gammopathy. *Blood* 1991; **78**: 2642–9.

65. Van Camp B, Reynaert PH, Broddtaerts L. Studies on the origin of the precursor cells in multiple myeloma, Waldenström's macroglobulinemia and benign monoclonal gammopathy. I. Cytoplasmic isotype and idiotype distribution in peripheral blood and bone marrow. *Clinical and Experimental Immunology* 1981; **44**: 82.

66. Pettersson D, Mellstedt H, Holm G. Immunoglobulin isotypes on monoclonal blood lymphocytes in human plasma cell myeloma. *Journal of Clinical and Laboratory Immunology* 1980; **3**: 93–8.

67. Van't Veer MB, Radl J, Waltman FL et al. Simultaneous expression of cytoplasmic IgM and IgG in plasma cells in a patient with IgG paraproteinaemia. *European Journal of Haematology* 1989; **43**: 178–81.

68. Palumbo A, Battaglio S, Astolfi M et al. Multiple independent immunoglobulin class-switch recombinations occurring within the same clone in myeloma. *British Journal of Haematology* 1992; **82**: 676–80.

69. Hata H, Selvanayagam P, Petrucci MT et al. Germline immunoglobulin genes expression and T-cell receptor-delta rearrangements in myeloma (MM). Abstract. *Annual Meeting of the American Society of Clinical Oncology* 1990; **31**: A215.

70. Bakkus M, Van Riet I, Van Camp B, Thielemans K. Evidence that the clonogenic cell in multiple myeloma originates from a pre-switched but somatically mutated B cell. *British Journal of Haematology* 1994; **87**: 68–74.

71. Kubagawa H, Vogler LB, Capra JD et al. Studies on the clonal origin of multiple myeloma. Use of individualy specific (idiotype) antibodies to trace the oncogenic event to its earliest point of expression in B cell differentiation. *Journal of Experimental Medicine* 1979; **150**: 792–807.

72. Grogan TM, Durie BGM, Spier CM et al. Myelomonocytic antigen positive multiple myeloma. *Blood* 1989; **73**: 763–9.

73. Epstein J, Xiao H, He X-Y. Markers of multiple hematopoetic-cell lineages in multiple myeloma. *New England Journal of Medicine* 1991; **322**: 664–8.

74. Berenson J, Lichtenstein A. Clonal rearrangement of the β-T cell receptor gene in multiple myeloma. *Leukemia* 1989; **3**: 133–6.

75. Shpilberg O, Modan M, Modan B, Ramot B. Multiple hematopoietic-cell lineages in multiple myeloma. *New England Journal of Medicine* 1990; **323**: 277.

76. Mellstedt H, Killander D, Pettersson D. Bone marrow kinetic studies on three patients with myelomatosis. Indications for malignant proliferation within both the plasma cell and lymphoid cell compartments. *Acta Medica Scandinavica* 1977; **202**: 413–17.

77. Chan CS, Wormsley SB, Pierce LE, Schechter GP. B-cell surface phenotypes of proliferating myeloma cells: Target antigens for immunotherapy. *American Journal of Hematology* 1990; **33**: 101–9.

78. Pilarski LM, Jensen GS, Mant MJ, Belch AR. Abnormal monoclonal B lineage cells in blood may comprise the malignant stem cell in myeloma: Expression of β1 and β2 integrins. Abstract. Paper presented at the Third International Workshop on Multiple Myeloma, Torino, Italy, 9–12 April, 1991, pp. 58–60.

79. Jerne NK. Towards a network theory of the immune system. *Annales D Immunologie* 1974; **1256**: 373–89.

80. Sirisinha S, Eisen HN. Autoimmune-like antibodies to the ligand-binding sites of myeloma proteins. *Proceedings of the National Academy of Sciences* 1971; **68**: 3130–5.

81. Lynch RG, Graff RJ, Sirisinha S et al. Myeloma proteins as tumor-specific transplantation antigens. *Proceedings of the National Academy of Sciences* 1972; **69**: 1540–4.

82. Chen Y, Yakulis V, Heller P. Passive immunity to murine plasmacytoma by rabbit anti-idiotypic antibody to myeloma protein. *Proceedings of the Society for Experimental Biology and Medicine* 1976; **151**: 121–5.

83. Pilarski LM, Piotrowska-Krezolak M, Gibney DY et al. Specificity repertoire of lymphocytes from multiple myeloma patients. I. High frequency of B cells specific for idiotypic and F(ab')₂-region determinants on immunoglobulin. *Journal of Clinical Immunology* 1985; **5**: 275–84.

84. Bergenbrant S, Österborg A, Holm G et al. Antiidiotypic antibodies in patients with monoclonal gammopathies: relation to the tumor load. *British Journal of Haematology* 1991; **78**: 66–70.

85. Daley MJ, Gebel HM, Lynch RG. Idiotype-specific transplantation resistance to MOPC-315: Abrogation by post-immunization thymectomy. *Journal of Immunology* 1978; **120**: 1620–4.

86. Flood PM, Phillipps C, Taupier MA, Schreiber H. Regulation of myeloma growth *in vitro* by idiotype-specific T lymphocytes. *Journal of Immunology* 1980; **124**: 424–30.

87. Lynch RG. Immunoglobulin-specific suppressor T cells. *Advances in Immunology* 1987; **40**: 135–51.

88. Dianzani U, Pileri A, Boccadoro M et al. Activated idiotype-reactive cells in suppressor/cytotoxic subpopulations of monoclonal gammopathies: correlation with diagnosis and disease status. *Blood* 1988; **72**: 1064–8.

89. Hoover RG, Hickman S, Gebel HM et al. Expansion of Fc receptor-bearing T lymphocytes in patients

with immunoglobulin G an immunoglobulin A myeloma. *Journal of Clinical Investigation* 1981; **67**: 308–11.

90. Roman S, Moore JS, Darby C et al. Modulation of Ig gene expression by Ig binding factors. Suppression of alpha-H chain and lambda-2-L chain mRNA, accumulation in MOPC-315 by IgA-binding factor. *Journal of Immunology* 1988; **140**: 3622–30.

91. Grey HM, Sette AS, Buus S. How T cells see antigen. *Scientific American* 1989; **261**: 38–46.

92. Bogen B, Weiss S. B lymphoma cells process and present their endogenous Ig: implications for network theory. *Advances in Experimental Medicine and Biology* 1988; **237**: 877–82.

93. Weiss S, Bogen B. MHC class II-restricted presentation of intracellular antigen. *Cell* 1991; **64**: 767–76.

94. Lauritzen GF, Weiss S, Bogen B. Anti-tumor activity of idiotype-specific, MHC-restricted Th1 and Th2 clones *in vitro* and *in vivo*. *Scandinavian Journal of Immunology* 1993; **37**: 77–85.

95. Lauritzen GF, Bogen B. The role of idiotype-specific, CD4$^+$ T cells in tumor resistance against major histocompatibility complex class II molecule negative plasmacytoma cells. *Cellular Immunology* 1993; **148**: 177–88.

96. Bottomly K, Mosier DE. Antigen specific helper T cells required for dominant idiotype expression are not H-2 restricted. *Journal of Experimental Medicine* 1989; **154**: 411–21.

97. Becker-Dunn E, Bottomly K. T15-specific helper T cells: analysis of idiotype-specificity by competetive inhibition analysis. *European Journal of Immunology* 1985; **15**: 728–32.

98. Cerny J, Smith JS, Webb C, Tucker PW. Properties of anti-idiotypic T cell lines propagated with syngeneic B lymphocytes. I. T cells bind intact idiotypes and discriminate between the somatic idiotypic variants in a manner similar to the anti-idiotypic antibodies. *Journal of Immunology* 1988; **141**: 3718–25.

99. Österborg A, Janson CH, Bergenbrant S et al. Peripheral blood T lymphocytes in patients with monoclonal gammopathies: Expanded subsets as depicted by capacity to bind to autologous monoclonal immunoglobulins or reactivity with anti-V gene-restricted antibodies. *European Journal of Haematology* 1991; **47**: 185–91.

100. Bakkus MHC, Heirman C, Van Riet I et al. Evidence that multiple myeloma Ig heavy chain VDJ genes contain somatic mutations but show no intraclonal variation. *Blood* 1992; **80**: 2326–35.

101. Österborg A, Yi Q, Bergenbrant S et al. Idiotype-specific T cells in multiple myeloma stage I: an evaluation by four different functional tests. *British Journal of Haematology* 1995; **89**: 110–16.

102. Österborg A, Masucci M, Bergenbrant S et al. Generation of T cell clones binding F(ab′)$_2$ fragments of the idiotypic immunoglobulin in patients with monoclonal gammopathy. *Cancer Immunology and Immunotherapy* 1991: **34**: 157–62.

103. Bence Jones H. III. On a new substance occurring in the urine of patients with mollities ossium. *Philosophical Transactions of the Royal Society, London* 1848; 55–62.

104. Alper CA, Johnson AM. Immunofixation electrophoresis: a technique for the study of protein polymorphism. *Vox Sanguinis* 1969; **17**: 445–52.

105. Ritchie RF, Smith R. Immunofixation. I. General principles and application to agarose gel electrophoresis. *Clinical Chemistry* 1976; **22**: 497–9.

106. Grabar P, Williams CA. Méthode immunélectrophoretique d'analyse de melanges de substances antigen'niques. *Biochimica Biophysica Acta* 1955: **17**: 67–74.

107. Mancini G, Carbonara AO, Heremans JF. Immunochemical quantitation of antigens by single radial immunodiffusion. *Immunochemistry* 1965; **2**: 235–54.

108. Laurell C-B. Electrophoretic and electroimmunochemical analysis of proteins. *Scandinavian Journal of Clinical Laboratory Investigation* 1972; **29** (suppl. 124): 21–37.

109. Stevenson FK, Gordon J. Immunization with idiotypic immunoglobulin protects against development of B lymphocytic leukemia, but emerging tumor cells can evade antibody attack by modulation. *Journal of Immunology* 1983; **130**: 970–3.

110. Campbell MJ, Esserman L, Byars NE et al. Idiotype vaccination against murine B cell lymphoma: humoral and cellular requirements for the full expression of antitumor immunity. *Journal of Immunology* 1990; **145**: 1029–36.

111. Kwak LW, Campbell MJ, Czerwinski DK et al. Induction of immune responses in patients with B-cell lymphoma against the surface-immunoglobulin idiotype expressed by their tumors. *New England Journal of Medicine* 1992; **327**: 1209–15.

112. Österborg A, Yi Q, Bergenbrant S, Gigliotti D et al. Regulatory T cells populations in multiple myeloma. In: Dammacco F Barlogie B eds *Challenges of modern medicine*, vol. 4. Ares-Serano Symposia Publications, 1993: 71–82.

113. Rutihoff S. Principles of tumor immunity: Biology of antibody-mediated responses. In: De Vita VT, Rosenberg SA eds *Biologic therapy of cancer*. Philadelphia: JB Lippincott, 1991.

CHAPTER 5

Human myeloma-induced bone changes

RÉGIS BATAILLE

Introduction	51	Cytokines and other factors responsible for	
Excessive osteoclastic resorption	51	myeloma-induced bone changes	52
Uncoupled bone remodeling	52	Conclusion	53
		References	53

Introduction

Lytic bone lesions and hypercalcemia are common features in patients with multiple myeloma (MM). In fact, few patients with MM do not develop lytic bone lesions, and occasional patients (< 1 per cent) present with osteosclerotic lesions. In contrast, other B-cell malignancies (except for hairy cell leukemia) are not associated with bone involvement despite bone marrow invasion. It is thus critical to clarify the mechanisms of bone lesions in MM (and those of bone protection observed in a minority of MM patients) and to understand why bone involvement is such a characteristic feature of MM.

There have been three major phases in the understanding of myeloma-induced bone changes. First, the realization that myeloma cells trigger increased osteoclast activity. Second, the crucial involvement of osteoblasts, which are active in early disease but inhibited (uncoupled) as lytic lesions evolve. Third, the investigation of cytokines and other factors which mediate the observed changes. Each phase will be discussed separately.

Excessive osteoclastic resorption

Increased osteoclastic resorption on quantitative bone biopsy is a characteristic feature of patients with active MM. It is observed in the close vicinity of myeloma cells in all patients with active disease regardless of the presence (or not) of lytic bone lesions on radiography and it is related to an increase of both trabecular osteoclast number which is easily quantified in 75 per cent of patients[5] and single osteoclast activity. This excessive osteoclastic resorption is not observed on non-invaded biopsy and/or on biopsy from patients in remission. MM-activated osteoclasts have normal length, in contrast to the large multinucleated osteoclasts encountered in primary hyperparathyroidism and Paget's disease of bone.[6] The increased osteoclastic resorption explains the high sensitivity of myeloma patients to the acute effects of calcitonin, a potent anti-osteoclastic agent[7,8] and supports the potentially beneficial effects of bisphosphonates in the treatment of myeloma-induced bone changes as illustrated by published results[9–12] and in agreement with previous mouse studies.[13] When quantitative bone biopsies were performed in individuals with early (i.e. infra-clinical) MM, an excessive bone resorption was also observed, as marked as that in patients with overt MM.[5] On the other hand, this abnormal remodelling was not found in individuals with either benign mono-clonal gammopathy or smoldering MM.[5] These critical data have shown that the excessive osteoclastic resorption was an early phenomenon in MM which could be observed several years before the first clinical symptoms of the disease and thus is a useful parameter to discriminate between benign monoclonal gammopathy or smoldering MM and early active MM.[5]

Uncoupled bone remodeling

The increased osteoclastic resorption in almost all the patients with MM demonstrates that lytic bone lesions cannot be explained simply by increased bone resorption.[14] At diagnosis, when we compared the histomorphometric features of patients with lytic bone lesions to those of patients lacking such lesions, it was evident that these subsets of patients had opposite bone profiles. An uncoupling bone process (i.e. increased bone resorption with decreased bone formation) was a characteristic feature of patients with lytic bone lesions. Such an uncoupling was not observed in patients lacking lytic bone lesions.[14] Thus, these data have shown that the inhibition of bone formation was as critical in the occurrence of lytic bone lesions as the excessive bone resorption.[14] This concept was further supported by studies we have performed in 10 patients who did not develop lytic bone lesions or who presented sclerotic MM.[15] Eight of these patients had increased parameters of both bone resorption and bone formation, whereas two had a selective increase of parameters of bone formation. Serial biopsies have confirmed the occurrence of an uncoupling process during bone destruction and, conversely, the maintenance of a coupling process in patients lacking lytic bone lesions despite active disease. The critical role of bone formation in the occurrence of lytic bone lesions in patients with MM was further supported by our studies with the bone gla protein (BGP).[16,17] A good correlation was found between BGP serum levels and bone formation rates (evaluated on bone biopsies) in patients with either early or overt MM.[5,15,17] Extensive studies of serum BGP in MM have shown an inverse correlation between BGP serum levels and the lytic potential of the tumor.[16,17] Patients lacking lytic bone lesions had significantly higher BGP serum levels than those with lytic bone lesions.

Cytokines and other factors responsible for myeloma-induced bone changes (Table 5.1)

The generation of new osteoclasts with increased osteoclastic resorption (at the single-cell level) in the close vicinity of myeloma cells is an early and characteristic feature of myeloma bone marrow.[5] With advanced disease, bone formation is suppressed, whereas in early disease osteoblastic activity is maintained or rather stimulated.[5] Another critical observation is the strong natural killer (NK) cell activity which has been found in the bone marrow of patients with MM and not in that of patients with B-cell malignancies other than MM.[18] Thus, accelerated maturation of both osteoclasts and NK cells are the most specific features of MM when compared to other B-cell malignancies and could be related abnormalities.

Since Mundy et al.'s first study,[3] several works have confirmed the presence of a strong bone resorbing activity in the bone marrow of patients with MM.[19–27] In three of these studies, IL–1β rather than TNF-α or -β was shown to be the cytokine supporting the bone-resorbing activity present in the bone marrow of patients with MM.[22,24,27] We and others have shown that interleukin-6 was an essential paracrine rather than autocrine myeloma cell growth factor.[28–33] The recent demonstrations of (1) the bone-resorbing activity of IL–6 *in vitro* and *in vivo* (in mouse), (2) the induction of IL–6 by IL–1β, via prostaglandins E_2 (PGE$_2$), (3) the synergy between IL–1β and IL–6 *in vitro* in terms of bone-resorbing activity, and (4) the IL–6 mediation of IL–1β bone-resorbing activity, show that IL–1β, PGE$_2$ and IL–6, which are overproduced in the microenvironment of MM, are the most critical and final products involved in myeloma-induced bone changes

Table 5.1 Summary of cytokines and other factors involved with production of lytic bone lesions

Factor	Effect
Prostaglandins (E$_2$)	Induction of IL–1β and IL–6
IL–1β INF–α/β IL–6	} Osteoclast-activating factors
M–CSF	Expansion/induction of osteoclasts
Natural killer cells	?Facilitate/mediate osteoblast inhibition/osteoclast activation

and in realizing an avalanche of osteoclast-activating factors in the myeloma intermediate milieu.[34] Intensive research has to be devoted to the nature of known and unknown factors produced by the myeloma cells themselves which are able to both stimulate the production of IL–6 by the tumoral microenvironment and synergize with IL–6 and IL–1β to increase bone resorption and myeloma cell growth. M–CSF could be a critical factor in this process. Indeed, myeloma cell lines do produce M–CSF *in vitro*[35] and M–CSF is overproduced *in vivo* in MM, in relation to disease severity.[36] Taken together, these data suggest that M–CSF, which is overproduced *in vivo*, maybe by the myeloma cells themselves, could be important in the pathogenesis of myeloma-induced bone changes.

Conclusion

Bone involvement, mainly bone destruction, is a characteristic and usual feature of MM. On the other hand, it is exceptional in B-cell malignancies other than MM. Bone destruction is the consequence of an uncoupling process associating an increased osteoclastic resorption with an inhibition of bone formation. Conversely, patients lacking lytic bone lesions or those with sclerotic MM have an increased bone resorption but maintain a normal or have an increased bone formation (coupling process). This excessive osteoclastic resorption is an early phenomenon, as opposed to the inhibition of bone formation. It is observed several months or years before the occurrence of the first clinical symptoms of the disease. Thus, it is an early criterion of malignancy, useful for discriminating between benign monoclonal gammopathy or smoldering MM and early active MM. Several osteoclast-activating factors, produced either by the myeloma cells themselves or the hematopoietic microenvironment, are probably involved in the pathogenesis of such bone lesions. At the present time, IL–6, PGE$_2$, IL–1β appear to be the most critical factors. Indirect arguments suggest that other hematopoietic growth factors (mainly M–CSF) could play a role. Taken together, these data demonstrate a close relationship between myeloma cell growth factors and osteoclast-activating factors.

Although much has been learned about the pathogenesis of myeloma bone disease, many details remain unexplained. For example, why do approximately 20 per cent of patients have diffuse osteopenia rather than focal lytic lesions? Are the myeloma cells less adhesion-dependent? Or perhaps higher levels of cytokines (e.g. IL–1β/IL–6) induce more diffuse osteoclast activation in trabecular bone.

An important practical point is whether or not bone proteins and other substances released from the resorption surfaces are responsible for sustaining, or even accelerating myeloma growth. There is some evidence that bisphosphonates not only inhibit bone resorption, but can retard myeloma progression. With the introduction of more potent bisphosphonates this may become a useful approach to myeloma treatment. In addition, inhibition of PGE$_2$ activity may have an additive effect. It is important to note that PGE$_2$ not only triggers IL–1β and IL–6 but can induce DNA damage in myeloma cells. An aggressive approach to the management of myeloma bone disease can become a more central facet of standard care.

References

1. Bataille R, Chappard D, Alexandre C, Sany J. Importance of quantitative histology of bone changes in monoclonal gammopathy. *British Journal of Cancer* 1986; **53**: 805.
2. Grauer JL, Blanc D, Zagala A et al. L'histomorphométrie osseuse dans les dysglobulinémies monoclonales. *Revue du Rhumatisme* 1986; **53**: 517.
3. Mundy GR, Raisz LG, Cooper RA et al. Evidence for the secretion of an osteoclast stimulating factor in myeloma. *New England Journal of Medicine* 1974; **291**: 1041.
4. Valentin-Opran A, Charmon SA, Meunier PJ et al. Quantitative histology of myeloma-induced bone changes. *British Journal of Haematology* 1982; **52**: 601.
5. Bataille R, Chappard D, Marcelli C et al. The recruitment of new osteoblasts and osteoclasts is the earliest critical event in the pathogenesis of human multiple myeloma. *Journal of Clinical Investigation* 1991: **88**: 62.
6. Chappard D, Rossi JF, Bataille R, Alexandre C. Cytomorphometry of osteoclasts demonstrates an abnormal population in B-cell malignancies but not in multiple myeloma. *Calcified Tissue International* 1991; **48**: 13.
7. Bataille R, Sany J. Clinical evaluation of myeloma osteoclastic bone lesions: II. Induced hypocalcemia test using salmon calcitonin. *Metabolic Bone Disease and Related Research* 1982; **4**: 39.
8. Bataille R, Legendre C, Sany J. Acute effects of salmon calcitonin in multiple myeloma: a valuable method for serial evaluation of osteoclastic lesions and disease activity. A prospective study of 125 patients. *Journal of Clinical Oncology* 1985; **3**: 229.
9. Ascari E, Attardo-Parrinello G, Merlini G. Treatment of painful bone lesions and hypercalcemia. *European Journal of Haematology* 1989; **43**: 135.
10. Delmas PD, Charrhon S, Chapuy MC et al. Long-term effects of dichloromethylene diphosphonate (C12MDP) on skeletal lesions in multiple myeloma. *Metabolic Bone Disease and Related Research* 1982; **4**: 163.

11. Paterson AD, Kanis JA, Cameron EC et al. The use of dichloromethylene diphosphonate for the management of hypercalcemia in multiple myeloma. *British Journal of Haematology* 1983; **54**: 121.

12. Siris ES, Sherman WH, Baquiran DC et al. Effects of dichloromethylene diphosphonate on skeletal mobilization of calcium in multiple myeloma. *New England Journal of Medicine* 1980; **302**: 310.

13. Radl J, Croese JW, Zurcher C et al. Influence of treatment with APD-bisphosphonates on the bone lesions in the mouse 5T2 multiple myeloma. *Cancer* 1985;**55**: 1030.

14. Bataille R, Chappard D, Marcelli C et al. Mechanism of bone destruction in multiple myeloma. The importance of an unbalanced process in determining the severity of lytic bone disease. *Journal of Clinical Oncology* 1989; **7**: 1909.

15. Bataille, R, Chappard D, Marcelli C et al. Osteoblast stimulation in multiple myeloma lacking lytic bone lesions. *British Journal of Haematology* 1990; **76**: 484.

16. Bataille, R, Delmas P, Sany J. Serum bone gla-protein (osteocalcin) in multiple myeloma. *Cancer* 1987; **59**: 329.

17. Bataille, R, Delmas PD, Chappard D, Sany J. Abnormal serum bone gla-protein levels in multiple myeloma: crucial role of bone formation and prognostic implications. *Cancer* 1990; **66**: 167.

18. Uchida A, Yagita M, Sugiyama H et al. Strong natural killer (NK) cell activity in bone marrow of myeloma patients: accelerated maturation of bone marrow NK cells and their interaction with other bone marrow cells. *International Journal of Cancer* 1984; **34**: 375.

19. Durie BGM, Salmon SE, Mundy GR. Relation of osteoclast activating factor production to extent of bone disease in multiple myeloma. *British Journal of Haematology* 1981; **47**: 21.

20. Gailani S, McLimans WF, Mundy GR et al. Controlled environment culture of bone marrow explants from human myeloma. *Cancer Research* 1976; **36**: 1299.

21. Garrett JR, Durie BGM, Nedwin GE et al. Production of lymphotoxin, a bone resorbing cytokine, by cultured human myeloma cells. *New England Journal of Medicine* 1987; **317**: 526.

22. Gozzolino F, Torcia M, Aldinucci DL et al. Production of interleukin-1 by bone marrow myeloma cells. *Blood* 1989; **74**: 380.

23. Josse RG, Murray TM, Mundy GR et al. Observations on the mechanism of bone resorption induced by multiple myeloma marrow culture fluids and partially purified osteoclast-activating factor. *Journal of Clinical Investigation* 1981; **67**: 1472.

24. Kawano M, Yamamoto I, Iwato K et al. Interleukin-I beta rather than lymphotoxin as the major bone resorbing activity in human multiple myeloma. *Blood* 1989; **73**: 1646.

25. Rossi JF, Bataille R. *In vitro* osteolytic activity of human myeloma plasma cells and the clinical evaluation of myeloma osteoclastic bone lesions. *British Journal of Cancer* 1984; **50**: 119.

26. Schecter GP, Wahl LM, Horton JE. *In vitro* bone resorption by human myeloma cells. In: Potter M ed. *Progress in myeloma. Biology of myeloma.* Elsevier North Holland, 1980: 67.

27. Yamamoto I, Kawano M, Sone T et al. Production of interleukin-1β, a potent bone resorbing cytokine, by cultured myeloma cells. *Cancer Research* 1989; **49**: 4242.

28. Bataille R, Jourdan M, Zhang XG, Klein B. Serum levels of interleukin-6, a potent myeloma cell growth factor, as a reflect of disease severity in plasma cell dyscrasias. *Journal of Clinical Investigation* 1989; **84**: 2008.

29. Kawano M, Hirano T, Matsuda T et al. Autocrine generation and essential requirement of BSF/2 IL-6 for human multiple myeloma. *Nature* 1988; **322**: 73.

30. Klein B, Widjenes J, Zhang XG et al. Murine, anti-Interleukin-6 monoclonal antibody therapy for a patient with plasma cell leukemia. *Blood* 1991; **78**: 1198.

31. Klein B, Zhang XG, Jourdan M et al. Paracrine rather than autocrine regulation of myeloma cell growth and differentiation by interleukin-6. *Blood* 1989; **73**: 517.

32. Portier M, Rajzbaum G, Zhang XG et al. *In vivo* paracrine but not autocrine interleukin-6 gene expression in multiple myeloma. *European Journal of Immunology* 1991; **21**: 1759.

33. Zhang XG, Klein B, Bataille R. Interleukin-6 is a potent myeloma-cell growth factor in patients with aggressive multiple myeloma. *Blood* 1989; **74**: 11.

34. Roodman GD. Interleukin-6: an osteotropic factor? *Journal of Bone and Mineral Research* 1992; **7**: 475.

35. Nakamura M, Merchav S, Carter A et al. Expression of a novel 3.5 kb macrophage colony-stimulating factor transcript in human myeloma cells. *Journal of Immunology* 1989; **143**: 3543.

36. Wieczorek AJ, Belch AR, Jacobs A et al. Increased circulating colony-stimulating factor-1 in patients with pre leukemia, leukemia and lymphoid malignancies. *Blood* 1991; **77**: 1796.

CHAPTER 6

Chromosome abnormalities in multiple myeloma

GUNNAR JULIUSSON, GÖSTA GAHRTON

Introduction	55	Clonal evolution	59
Pre-banding studies	56	Clinical association to karyotypes	60
Banding studies	56	Chromosomes and oncogenes	60
Chromosome 1	58	Summary	61
The t(11;14)(q13;q32) translocation	58	References	61

Introduction

It was suggested as early as 1914 that abnormalities in the chromosomes were a characteristic feature of malignant tumors.[1] In 1959, the year before the first observation of the Philadelphia chromosome in chronic myelocytic leukemia,[2] the first study of chromosomes in myeloma was published.[3] However, no clonal abnormalities were found. A few years later abnormal markers were identified,[4] and since then more than 1400 cases have been published. However, in contrast to the advancement in the identification of specific chromosome aberrations with corresponding gene abnormalities in other haematologic malignancies, such as chronic myelocytic leukemia, acute lymphoblastic leukemia, and follicular lymphoma, we have still very little knowledge about relevant genetic changes in myeloma.

This lack of progress has several reasons. One major difficulty in myeloma as compared to acute leukemias is the difficulty in achieving metaphases from the relevant cells. Only a small proportion of the myeloma patients studied have been evaluable cytogenetically, and it seems that those from which metaphases were achieved had advanced disease with highly complex karyotypes. Myeloma cells from patients with early disease do not, in general, proliferate *in vitro*, and karyotypic evolution with the accumulation of additional unspecific chromosome changes seems to be rapid in myeloma. In contrast to the case in chronic lymphocytic leukemia,[5,6] B-cell mitogen activation has not significantly improved the outcome of chromosome analysis, despite the fact that circulating B cells with idiotypic markers identical to the cells with the myeloma phenotype is a well-documented finding,[7] and such B cells ought to be responsive to B-cell activation *in vitro* similar to other types of malignant B cells.

Recent improvements in techniques for studies of genetic changes, such as fluorescence *in situ* hybridization (FISH), molecular genetic techniques, gene expression studies, and analyses of gene products with assays, such as single-strand conformation polymorphism, might well help to identify disease-related specific gene abnormalities in myeloma.

The current review is a summary of chromosome data in myeloma collected over more than 30 years, which might serve as a basis for further attempts at the genetic characterization of the underlying pathophysiologic mechanisms in myeloma.

Pre-banding studies

The first chromosome study in myeloma, reported in 1959 by Baikie et al., failed to identify abnormalities in two patients.[3] In the early 1960s, however, multiple reports on missing chromosomes,[8] and clonal markers[4] were published. These early data[9–13] have been reviewed by Anday et al.[14] and Philip.[15] Large marker chromosomes in multiple myeloma and Waldenström's macroglobulinemia[16] became known as MG markers (from monoclonal gammopathy). As pointed out by Philip[15] and Dewald and Jenkins,[17] these varied in size and shape. Early, it was clear that these markers were not a myeloma equivalent to the Philadelphia chromosome in chronic myeloid leukemia.[2] We now know (*below*) that structural abnormalities in myeloma preferentially involve chromosome 1,[18] which by definition is the largest chromosome, and thereby large marker chromosomes were created.[19]

Banding studies

To date, chromosome studies on more than 1400 patients have been published (Table 6.1). Commonly, patients with all types of monoclonal serum immunoglobulins have been considered together. However, Waldenström's disease,[40] and other immunoglobulin-secreting subgroups of immunocytoma have a genotype, phenotype, clinical presentation, and response to therapy clearly different from myelomatosis, and are thus not discussed here. In contrast, plasma cell leukemia is an aggressive subtype of myeloma. The success rate in the karyotyping of plasma cell leukemia is greater than for other myelomas, but the spectrum of specific abnormalities seems to be similar to that of other myeloma subtypes. We have therefore included plasma cell leukemia in the current analyses. The high proportion of clonal abnormalities is probably due to the high proliferation rate of plasma cell leukemia cells, which is

Table 6.1 Published chromosome studies with banding in myeloma

First author	Year	Technique	Total number	Clonal	#1	t(11;14)	14q+	Other recurrent abnormalities
Travis[20]	1993		325	102		27		13q−
Durie[21]	1991		262	70	35		18	6q−, 7q−
Gould[22]	1988	No culture	140	53		5	13	t(8;14)
Facon[23]	1993		121	52	18	3	10	6q−,7,t(1;16)(p11;p11)
Dermitzaki[24]	1990		71	41	15		12	6q−,22q−
Kokkinou[25]	1990	Blood+bone marrow, B-cell mitogen	62	33	5	0	2	inv(7)(p11p22)
Weh[26]	1993	IL-6, G	110	33		4	10	t(8;14)
Dewald[27]	1985		82	29	12	3	11	
Facon[28]	1993	GM-CSF±IL-6±IL-3	46	19	3+	1	3	
Lisse[29]	1988		88	18	6	3	4	
Chen[30]	1986		26	15	6	2	3	
Philip[31]	1980	No culture	25	12	8	0	3	
Lewis[32]	1984	±culture, phythemagglutinin, pokeweed mitogen	30	10	5	0	3	
Nishida[33]	1989		28	7	4	0	5	t(14;18)
Liang[34]	1979		18	6	5	2	5	del(6)(q25)
Ranni[35]	1987	1–2 days culture	6	6	1	0	0	
Ueshima[36]	1983	Only plasma cell leukemia	6	6	6	1	3	
Ferti[37]	1984	Blood or bone marrow, no culture	10	5	4	1	3	
Kowalczyk[38]	1991	No culture	14	4	0	0	1	
Moscinski[39]	1991		26	4				
Total			1450	502	130	51	106	

Note: Figures might be too low due to incomplete data in reports.

Table 6.2 Extensively published karyotypes in myeloma

Total number of patients	166	
Single abnormalities, all	29	
Single numerical abnormalities		18
Single structural abnormalities		11
Multiple abnormalities, all	137	
Multiple abnormalities, hypodiploidy (< 45 chromosomes)		26
Multiple abnormalities, pseudodiploid (45–50 chromosomes)		72
Multiple abnormalities, hyperdiploid (> 50 chromosomes)		39

important for the analysis of the tumor cells. An improved technique to induce mitosis in relevant myeloma cells might reveal an increased frequency of clonal abnormalities in future studies, since there are indications that the metaphases achieved in cultures with no clonal abnormalities are non-malignant cells.[41]

Some investigators have attempted to increase the cytogenetically evaluable cases by *in vitro* activation of the myeloma cells.[23,25,26,28] This approach was crucial for the identificaton of chromosome abnormalities in chronic lymphocytic leukemia.[5,6] In myeloma, B-cell mitogen stimulation and cytokine combinations have been utilized (Table 6.1). However, no major breakthrough has as yet been achieved with current culture techniques, although some improvement is claimed from the use of myeloma cell cultures with activation through multiple interleukins.[23,28] Another approach to increase the proportion of evaluable cases is the use of fluorescent *in situ* hybridization (FISH),[42] or other molecular techniques.

Among cases of myelomatosis with karyotype analysis performed, slightly more than one-third have been found to have clonal abnormalities (Table 6.1). Aneuploidy is common.[14,42–44] We have analyzed 166 cases with detailed karyotypes published (Table 6.2),

and the frequency of certain abnormalities was established (Figs 6.1–6.3). As regards numerical abnormalities, some chromosomes tend to be gained (chromosomes 3, 5, 7, 9, 11, 19, and 21), whereas others are commonly lost (chromosomes 13, 8, and X).

Aberrations occurring as a single abnormality are more likely to be of pathophysiologic importance in initiation of the disease, such as the Philadelphia chromosome in chronic myeloid leukemia or the t(14;18) in follicular lymphoma. Single abnormalities in published myeloma cases are listed in Table 6.3. Only the t(11;14) abnormality (*see below*) seems to be a consistent event. Many of the others listed are likely to be part of complex abnormalities that remained unsettled due to a limited number of evaluable metaphases.

Several infrequent structural abnormalities have also been specifically reported. Some of them, such as t(8;14),[22,45,46] t(14;18),[22,47,48] and Philadelphia chromosomes,[49] are well-known to be involved in other hematopoietic malignancies and associated to specific oncogenes (Tables 6.1 and 6.3), whereas others, such as t(1;16)(p11;p11),[23] inv(7)(p11p22),[25] 17p+,[50] t(1;20)(q12.3;p13) and del(3)(p12),[26] t(6;14)(p21.1;q32.3),[33,51] are as yet identified in single studies only.

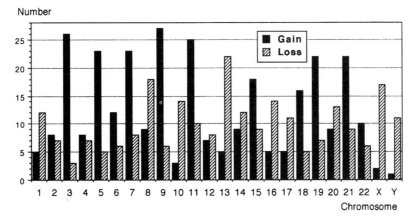

Fig. 6.1 Number of patients with numerical abnormalities of 166 completely reported karyotypes with clonal abnormalities.

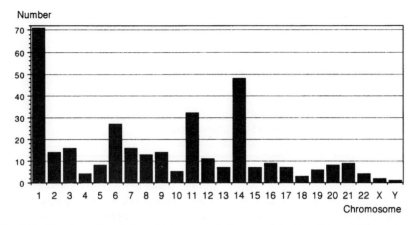

Fig. 6.2 Number of patients with structural abnormalities involving defined chromosomes.

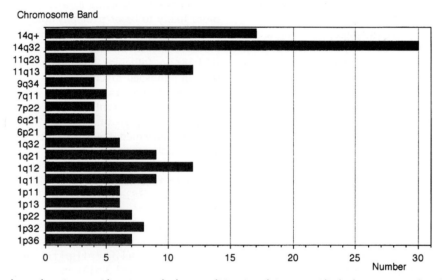

Fig. 6.3 Number of patients with structural abnormalities involving specified chromosome bands. Abnormalities designed 14q+ are likely to be similar to 14q32.

Chromosome 1

Chromosome 1 is by far the most common site for structural abnormalities (Figs 6.2 and 6.3). A large proportion of the defined abnormalities involving chromosome 1 are deletions (Fig. 6.4), mostly terminal, however, with variable breakpoints. It has been argued[17] that these abnormalities are unsignificant findings due to the lack of a specifically involved region, and that chromosome 1 abnormalities are common in other tumor types. However, it seems to us that structural abnormalities of chromosome 1 are non-randomly occuring in myeloma. Common genomic regions may well be deleted even though the breakpoints are different. Analyses of similar widespread breakpoints of chromosome 6 in lymphomas

have recently enabled the identification of candidate gene regions.[52] Further analyses of deletions involving chromosome 1 in myeloma might reveal analogous findings.

The t(11;14)(q13;q32) translocation

The translocation t(11;14)(q13;q32)[13,34,53,54] (Fig. 6.5) is the single most common chromosomal abnormality in myeloma (Table 6.1), and the only one that is convincingly recurrent as the sole chromosomal abnormality[20] (Table 6.3). It also occurs in limited stage disease. Therefore, it seems likely that

Table 6.3 Single clonal abnormalities in myeloma

Numerical

−7	n = 3
−9	n = 2
+12	n = 2
−22	n = 2
−Y	n = 2
+15	
+17	
−18	
+19	
+21	
−21	
−X	

Structural

t(11;14)(q13;q32)	n = 15
inv(7)(p11p22)	n = 2
t(9;22)(q+;q−)	n = 2
del(1)(q42)	
+1q	
t(4;?)(pter–q35;?)	
del(5)(q13q31)	
del(7)(q36)	
t(14;?)(q32;?)	
del(17)(p11)	

Table 6.4 Donor chromosomes for 14q+ markers

Total 14q+	130	
Undefined		60
11q13		51[a]
8q24		9[b]
18q21		3
6p21		3
3p11		1
13q22		1
15		1

[a] 27 of them published by Travis et al.[20]
[b] 5 of them reported by Gould et al.[22]

disorder resembling chronic lymphocytic leukemia. It is now recognized as the specific chromosomal abnormality of intermediate/mantel zone lymphoma,[56] an entity with a poor prognosis identified in European literature by the name of diffuse centrocytic lymphoma.[57] The t(11;14) is also the commonest chromosome abnormality in B-cell prolymphocytic leukemia.[58]

Clonal evolution

Clonal evolution is probably a common phenomenon in myeloma, although few sequential chromosome studies have been performed. This is indicated by a higher incidence of clonal abnormalities in patients with advanced disease and previously treated patients.[24,26,27,44] However, this finding might also be explained by an increasing proliferation rate and a greater sensitivity to mitogenic stimulation of the

the t(11;14) translocation is participating in one of the pathways by which malignant transformation may occur in myeloma, and a role in an early event rather than in disease progression is indicated.

The t(11;14) involves the *bcl-1* oncogene,[55] and is found in other lymphoproliferative disorders. It was originally identified in a case of lymphoproliferative

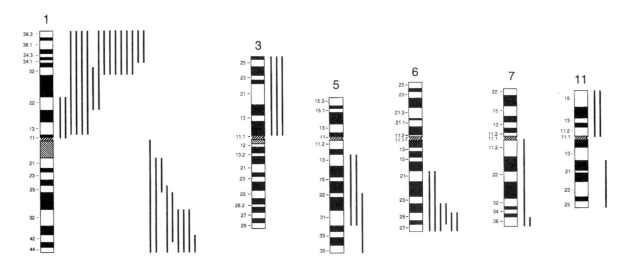

Fig. 6.4 Chromosome deletions in myeloma.

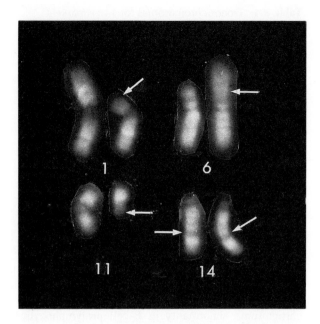

Fig. 6.5 Partial karyotype from a patient with plasma cell leukemia, showing two translocations, i.e. t(11;14)(q13;q32) and t(1;6)(1pter→ 1p34::1p22→ 1qter;1p34→ 1p22::6pter→ 6qter). There is one normal chromosome 11, no normal chromosome 14 but two identical 14q+ chromosomes, probably due to non-disjunction. The patient is previously reported by Gahrton et al.[53] ·

myeloma cells at later stages of the disease. It is likely that the high incidence of normal karyotypes in early-stage myeloma is just a reflection of a higher mitotic rate of the normal cells than of the myeloma cells in the cultures. Complex karyotypes are associated with advanced disease and poor prognosis also in chronic lymphocytic leukemia.[6] In this disease longitudinal chromosome studies have consistently shown a low incidence of clonal evolution.[6,59] The increased incidence of complex karyotypes in advanced chronic lymphocytic leukemia is thus due to the more rapid clinical progression rate of the disease in patients with such karyotypes. Still, the diverse pattern of chromosome changes in early disease, with multiple clones and lack of consistent abnormalities speaks in favor of early clonal evolution in myeloma.

Clinical association to karyotypes

Treated myeloma patients have a higher incidence of clonal abnormalities than untreated cases,[24,26,27,60,61] and patients with clonal abnormalities have a worse prognosis than patients with 'normal' karyotypes,[26,60] i.e. with no clonal abnormalities detected with current cytogenetic techniques, including the use of interleukin-activated cell cultures.[23,26,28] However, patients with advanced disease have more T-cell abnormalities than early-stage patients,[62] which might influence the ability to induce mitosis of myeloma cells *in vitro* in the opposite direction. Hyperdiploidy is also recently suggested to indicate good prognosis in myeloma,[63] similar to the prognostic importance of hyperdiploidy in acute lymphoblastic leukemia.[64]

There have been a few suggested correlations between chromosome abnormalities and myeloma isotype.[25,65] These have been identified in single studies only, and might thus be of limited clinical relevance.

Chromosomes and oncogenes

Myeloma preferentially arise in early B cells with a restricted use of immunoglobulin variable genes.[66,67] This phenomenon has previously been found to be a characteristic feature of other B-cell tumors, such as chronic lymphocytic leukemia.[68]

Oncogene involvement is discussed in Chapter 7, and is thus only briefly referred to here. The *bcl-1* oncogene is also shown to be involved in myeloma by molecular studies,[20,69,70] but the mechanisms of its involvement have not been evaluated. Although the chromosome translocation t(14;18) involving the *bcl-2* oncogene is rare in myeloma, the bcl-2 protein is frequently overexpressed in myeloma.[71] The bcl-2 protein has a definite role in the prevention of apoptosis, and is therefore a potentially relevant pathophysiologic mechanism of chronic lymphoproliferative disorders in addition to follicular lymphoma. Whether the overexpression of *bcl-2* in myeloma is due to *bcl-2* gene rearrangements localized far from the well-known breakpoint regions or if regulatory mechanisms are disturbed is unknown.

There is also a recent finding of retinoblastoma gene (*Rb1*) deletions in myeloma, which was not predicted by the pattern of cytogenetic abnormalities. A relationship between retinoblastoma gene expression and IL-6 is suggested.[72] Retinoblastoma gene deletions are recurrent findings in chronic lymphocytic leukemia, but the cytogenetic correlate, deletions involving 13q14, was not identified until 1987,[73] although it is now recognized as the most common structural abnormality.[6] Small interstitial deletions require optimal quality preparation for their detection, and it might well be that such deletions are also present in myeloma, although not yet identified. Monosomy 13, however, is not uncommon in myeloma (Fig. 6.1). In CLL, recent data have suggested that an adjacent locus, identified by the D13S25 probe, might harbor a more relevant candidate suppressor oncogene than the retinoblastoma gene.[74,75]

Another well-known tumor suppressor gene is *p53*, localized on 17p, which is frequently involved in progression of various tumors. Monosomy of chromosome 17 is less common than monosomy 13 in myeloma, and structural abnormalities of chromosome 17 are also uncommon. However, *p53* has recently been documented in tumor progression of follicular lymphoma,[76] as well as in myeloma.[77–79]

Summary

Chromosome studies since 1959 involving over 1400 patients are summarized. One-third of them had clonal chromosome abnormalities. Five hundred cases with clonal abnormalities are reported, including 166 with extensively presented karyotypes. Studies are hampered by the difficulty in inducing mitosis in myeloma cells or their precursors. This obstacle is still present despite the use of cultures with B-cell mitogens and multiple cytokines. Another technical problem is the complexity of the karyotypes. By the use of molecular and *in situ* hybridization techniques a greater overall success rate might be achieved in the near future.

Aneuploidy is a common finding, with both hypodiploidy and hyperdiploidy. Hyperdiploidy mostly involves chromosomes 3, 5, 7, 9, 11, 19, and 21, whereas hypodiploidy commonly affects chromosomes 13, 8, and X. Structural abnormalities frequently affect chromosome 1, occurring in 42 per cent of the published cases. However, the breakpoints are well distributed over both arms. The only common specific abnormality is the t(11;14)(q13;q32) translocation involving the *bcl-1* oncogene. Deletions of the retinoblastoma and *p53* tumor suppressor genes are also reported. Abnormal karyotypes indicate poor survival, but hyperdiploidy is suggested to indicate good prognosis.

References

1. Boveri T. *Zur Frage der Entstehung Maligner Tumoren*. Fischer Jena, 1914.
2. Nowell PC, Hungerford DA. A minute chromosome in human granulocytic leukemia. *Science* 1960; **132**: 1497 (letter).
3. Baikie AG, Court Brown WM, Jacobs PA et al. Chromosome studies in human leukaemia. *Lancet* 1959, **ii**: 425–8.
4. Castoldi GL, Ricci N, Puntuieri E, Bosi L. Chromosomal imbalance in plasmacytoma. *Lancet* 1963; **i**: 829.
5. Gahrton G, Robèrt K-H, Friberg K et al. Nonrandom chromosomal aberrations in chronic lymphocytic leukemia revealed by polyclonal B-cell mitogen stimulation. *Blood* 1980; **56**: 640–7.
6. Juliusson G, Gahrton G. Cytogenetics in CLL and related disorders. In: Rozman C, Montserrat E eds *Chronic lymphocytic leukemia and related disorders. Baillière's clinical haematology*, vol. 6. London: Baillière Tindall, 1993: 821–48.
7. Pilarski LM, Jensen GS. Monoclonal circulating B cells in multiple myeloma. In: Barlogie B ed. *Hematology/oncology clinics of North America*, vol. 6. Philadelphia, PA: W.B. Saunders, 1992; 297–322.
8. Bottura C. Chromosomal abnormalities in multiple myeloma. *Acta Haematologica* 1963; **30**: 274–9.
9. Lewis FLW, MacTaggart M, Crow RS, Wills MR. Chromosomal abnormalities in multiple myeloma. *Lancet* 1963; **i**: 1183–4.
10. Tassoni EM, Durant JR, Becker S, Kravitz B. Cytogenetic studies in multiple myeloma: a study of fourteen cases. *Cancer Research* 1967; **27**: 806–10.
11. Dartnall JA, Mundy GR, Baikie AG. Cytogenetic studies in myeloma. *Blood* 1973; **42**: 229–39.
12. Jensen MK, Eriksen J, Djernes BW. Cytogenetic studies in myelomatosis. *Scandinavian Journal of Haematology* 1975; **14**: 201–9.
13. Wurster-Hill DH, McIntyre OR, Cornwell GG III. Chromosome studies in myelomatosis. *Virchows Archiv für Cell Pathologische* 1978; **29**: 93.
14. Anday GJ, Fishkin B, Gabor EP. Cytogenetic studies in multiple myeloma. *Journal of the National Cancer Institute* 1974; **52**: 1069–75.
15. Philip P. Chromosomes of monoclonal gammopathies. *Cancer Genetics and Cytogenetics* 1980; **1**: 79–86.
16. Houston EW, Ritzmann SE, Levin WC. Chromosomal aberrations common to three types of monoclonal gammopathies. *Blood* 1967; **29**: 214.
17. Dewald GW, Jenkins RB. Cytogenetic and molecular genetic studies of patients with monoclonal gammopathies. In: Wiernik PH, Canellos GP, Kyle RA, Schiffer CA eds *Neoplastic diseases of the blood*, 2nd edn. Edinburgh: Churchill Livingstone, 1991: 427–38.
18. Spriggs AI, Holt JM, Bedford J. Duplication of part of the long arm of chromosome 1 in marrow cells of a treated case of myelomatosis. *Blood* 1976; **48**: 595.
19. Caspersson T, Lomakka G, Zech L. The 24 fluorescence patterns of the human metaphase chromosomes – distinguishing characters and variability. *Hereditas* 1971; **67**: 89–102.
20. Travis P, Sawyer J, Lary C et al. Translocation 11;14 and *bcl-1* gene abnormalities in multiple myeloma. *Blood* 1993: **82** (suppl.): 261a.
21. Durie BGM, Vela EE, Baum V. Cytogenetic abnormalities in multiple myeloma. In: *Epidemiology and biology of multiple myeloma*. New York: Springer-Verlag, 1991: 137–41.

22. Gould J, Alexanian R, Goodacre A et al. Plasma cell karyotype in multiple myeloma. *Blood* 1988; **71**: 453–6.

23. Facon T, Lai JL, Trillot N et al. Cytogenetic analysis of bone marrow plasma cells after cytokine stimulation in patients with multiple myeloma: a report on 100 patients. *Blood* 1993; **82** (suppl.): 261a.

24. Dermitzaki K, Stamatelou M, Tsaftaridis P et al. Correlation of karyotype with prognosis in multiple myeloma. *Blood* 1990; **76** (suppl.): 346a (abstract).

25. Kokkinou S, Juliusson G, Ohrling M, Gahrton G. Chromosome analysis in B-cell mitogen-stimulated cells from myeloma patients: high incidence of clonal abnormalities in Bence-Jones and non-secretory subtypes. *Blood* 1990; **76** (suppl.): 357a (abstract).

26. Weh HJ, Gutensohn K, Selbach J et al. Karyotype in multiple myeloma and plasma cell leukaemia. *European Journal of Cancer* 1993; **29A**: 1269–73.

27. Dewald GW, Kyle RA, Hicks GA, Greipp PR. The clinical significance of cytogenetic studies in 100 patients with multiple myeloma, plasma cell leukemia, or amyloidosis. *Blood* 1985; **66**: 180–90.

28. Facon T, Lai JL, Nataf E et al. Improved cytogenetic analysis of bone marrow plasma cells after cytokine stimulation in multiple myeloma: a report on 46 patients. *British Journal of Haematology* 1993; **84**: 743–5.

29. Lisse IM, Drivsholm A, Christoffersen P. Occurrence and type of chromosomal abnormalities in consecutive malignant gammapathies: correlation with survival. *Cancer Genetics and Cytogenetics* 1988; **35**: 27–36.

30. Chen K-C, Bevan PC, Matthews JG. Analysis of G banded karyotypes in myeloma cells. *Journal of Clinical Pathology* 1986; **39**: 160–6.

31. Philip P, Drivsholm A, Hansen NE et al. Chromosomes and survival in multiple myeloma. A banding study of 25 cases. *Cancer Genetics and Cytogenetics* 1980; **2**: 243–57.

32. Lewis JP, MacKenzie MR. Non-random chromosomal aberrations associated with multiple myeloma. *Hematology and Oncology* 1984; **2**: 307–17.

33. Nishida K, Yashige H, Maekawa T et al. Chromosome rearrangement, t(6;14)(p21.1;q32.3), in multiple myeloma. *British Journal of Haematology* 1989; **71**: 295–8.

34. Liang W, Hopper JE, Rowley JD. Karyotypic abnormalities and clinical aspects of patients with multiple myeloma and related paraproteinemic disorders. *Cancer* 1979; **44**: 630–44.

35. Ranni NS, Slavutsky I, Wechsler A, Brieux de Salum S. Chromosome findings in multiple myeloma. *Cancer Genetics and Cytogenetics* 1987; **25**: 309–16.

36. Ueshima Y, Fukuhara S, Nagai K et al. Cytogenetic studies and clinical aspects of patients with plasma cell leukemia and leukemic macroglobulinemia. *Cancer Research* 1983; **43**: 905–12.

37. Ferti A, Panani A, Arapakis G, Raptis S. Cytogenetic study in multiple myeloma. *Cancer Genetics and Cytogenetics* 1984; **12**: 247–53.

38. Kowalczyk JR, Dmoszynska A, Chobotow M, Sokolowska B. Cytogenetic studies in patients with multiple myeloma. *Cancer Genetics and Cytogenetics* 1991; **55**: 173–9.

39. Moscinski LC, Papenhausen PR, Ballester. Surface phenotype in plasma cell dyscrasia and the relationship to karyotype and cytomorphology. *Blood* 1991; **78** (suppl.): 116a (abstract).

40. Waldenström J. Incipient myelomatosis or 'essential' hyperglobulinemia with fibrinogenopenia – a new syndrome? *Acta Medica Scandinavica* 1994; **CXVII**: 216.

41. Weh HJ, Fiedler W, Hossfeld DK. Cytogenetics in multiple myeloma. Are we studying the 'right' cells? *European Journal of Haematology* 1990; **45**: 236–7.

42. Lee W, Han K, Drut RM et al. Use of fluorescence *in situ* hybridization for retrospective detection of aneuploidy in multiple myeloma. *Genes Chromosomes Cancer* 1993; **7**: 137–43.

43. Latreille J, Barlogie B, Dosik G et al. Cellular DNA content as a marker of human multiple myeloma. *Blood* 1980; **55**: 403.

44. Bunn Jr PA, Krasnow S, Makuch RW et al. Flow cytometric analysis of DNA content of bone marrow cells in patients with plasma cell myeloma: Clinical implications. *Blood* 1982; **59**: 528–35.

45. Yamada K, Shionoya S, Amano M, Imamura Y. A Burkitt-type 8;14 translocation in a case of plasma cell leukemia. *Cancer Genetics and Cytogenetics* 1983; **9**: 67–70.

46. Selvanayagam P, Blick M, Narni F et al. Alteration and abnormal expression of the c-*myc* oncogene in human multiple myeloma. *Blood* 1988; **71**: 30–5.

47. Nishida K, Taniwaki M, Misawa S et al. Translocation (14;18) found in a patient with multiple myeloma. *Cancer Genetics and Cytogenetics* 1987; **25**: 375–7.

48. Nishida K, Taniwaki M, Misawa S, Abe T. Nonrandom rearrangement of chromosome 14 at band q32.33 in human lymphoid malignancies with mature B-cell phenotype. *Cancer Research* 1989; **49**: 1275–81.

49. van den Berghe H, Louwagie A, Broeckaert-van Orshoven A et al. Philadelphia chromosome in human multiple myeloma. *Journal of the National Cancer Institute* 1979; **63**: 11–16.

50. Manolova Y, Manolov G, Apostolov P, Levan A. The same marker chromosome, mar17p+, in four consecutive cases of multiple myeloma. *Hereditas* 1979; **90**: 307–10.

51. Tominaga N, Katagiri S, Hamaguchi Y et al. Plasma cell leukaemia of the non-producer type with missing light chain gene rearrangement. *British Journal of Haematology* 1988; **69**: 213–18.

52. Offit K, Parsa NZ, Gaidano G et al. 6q deletions define distinct clinico-pathologic subsets of non-Hodgkin's lymphoma. *Blood* 1993; **82**: 2157–62.

53. Gahrton G, Zech L, Nilsson K et al. 2 translocations, t(11;14) and t(1;6), in a patient with plasma cell leukemia and 2 populations of plasma cells. *Scandinavian Journal of Haematology* 1980b; **24**: 42–6.

54. Venti G, Mecucci C, Donti E, Tabilio A. Translocation t(11;14) and trisomy 11q13-qter in multiple myeloma. *Annals of Genetics (Paris)* 1984; **27**: 53–5.

55. Tsujimoto Y, Yunis J, Onorato-Showe L et al. Molecular cloning of the chromosomal breakpoint of B-cell lymphomas and leukemias with the t(11;14) chromosome translocation. *Science* 1984; **224**: 1403–6.

56. Leroux D, Le Marchadour F, Gressin R et al. Non-Hodgkin's lymphomas with t(11;14)(q13;q32): a subset of mantle zone/intermediate lymphocytic lymphoma? *British Journal of Haematology* 1991; **77**: 346–53.

57. Lennert K. Malignant lymphomas other than Hodgkin's disease. In: *Handbuch der speziellen patologischen Anatomie und Histologie.* 1 Band, part 3B. Berlin: Springer, 1978: 111–49.

58. Pittman S, Catovsky D. Chromosome abnormalities in B-cell prolymphocytic leukaemia: a study of nine cases. *Cancer Genetics and Cytogenetics* 1983; **9**: 355–65.

59. Juliusson G, Friberg K, Gahrton G. Consistency of chromosomal aberrations in chronic B-lymphocytic leukemia. A longitudinal cytogenetic study of 41 patients. *Cancer* 1988; **62**: 500–6.

60. Clark RE, Geddes AD, Whittaker JA, Jacobs A. Differences in bone marrow cytogenetic characteristics between treated and untreated myeloma. *European Journal of Cancer and Clinical Oncology* 1989; **25**: 1789–93.

61. Gutensohn K, Weh HJ, Walter TA, Hossfeld DK. Cytogenetics in multiple myeloma and plasma cell leukemia: simultaneous cytogenetic and cytologic studies in 51 patients. *Annals of Hematology* 1992; **65**: 88–90.

62. San Miguel JF, Gonzales M, Gascon A et al. Lymphoid subsets and prognostic factors in multiple myeloma. *British Journal of Haematology* 1992; **80**: 305–9.

63. García-Sanz R, Orfao A, Moro MJ et al. Hyperdiploidy identifies subgroup of multiple myeloma patients with good prognosis. *Blood* 1993; **80** (suppl.): 564a (abstract).

64. Bloomfield CD, Goldman AL, Alimena G et al. Chromosome abnormalities identify high-risk and low-risk patients with acute lymphoblastic leukemia. *Blood* 1986; **67**: 415–20.

65. van den Berghe H. Chromosomes in plasma-cell malignancies. *European Journal of Haematology* 1989; suppl. 7: 1–5.

66. Berenson JR, Cao J, Newman R, Lichtenstein AK. VH gene family usage is nonrandom in multiple myeloma. *Blood* 1991; **78** (suppl.): 122a (abstract).

67. Biggs DD, Anderson KC, Silberstein LE. Molecular characterization of variable region (V) gene usage in multiple myeloma. *Blood* 1991; **78** (suppl.): 119a (abstract).

68. Mayer R, Logtenberg T, Strauchen J et al. CD5 and immunoglobulin V gene expression in B-cell lymphomas and chronic lymphocytic leukemia. *Blood* 1990; **75**: 1518–24.

69. Ladanyi M, Feiner H, Niesvizky R, Michaeli J. Protooncogene analysis in multiple myeloma. *Blood* 1991; **78** (suppl.): 111a (abstract).

70. Fiedler W, Weh HJ, Hossfeld DK. Comparison of chromosome analysis and bcl-1 rearrangement in a series of patients with multiple myeloma. *British Journal of Haematology* 1992; **81**: 58–61.

71. Durie BGM, Mason DY, Giles F et al. Expression of the bcl-2 oncogene protein in multiple myeloma. *Blood* 1990; **76** (suppl.): 347a (abstract).

72. Tricot G, Dao D, Gazitt Y et al. Deletion of the retinoblastoma gene (Rb-1) in multiple myeloma (MM). *Blood* 1993: **82** (suppl.): 261a.

73. Fitchett M, Griffiths MJ, Oscier DG et al. Chromosome abnormalities involving band 13q14 in hematologic malignancies. *Cancer Genetics and Cytogenetics* 1987; **24**: 143–50.

74. Juliusson G, Gahrton G, Einhorn S et al. Chromosome abnormalities and RB1 gene deletions in chronic lymphocytic leukemia. *Blood* 1993; **82**: 138–9.

75. Liu Y, Szekely L, Grandér D et al. Chronic lymphocytic leukemia cells with allelic deletions at 13q14 commonly have one intact RB1 gene: evidence for a role of an adjacent locus. *Proceedings of the National Academy of Sciences* 1993; **90**: 8697–701.

76. Sander CA, Yano T, Clark HM et al. *p53* mutations is associated with progression in follicular lymphomas. *Blood* 1993; **82**: 1994–2004.

77. Preudhomme C, Facon T, Zandecki M et al. Rare occurrence of *p53* gene mutations in multiple myeloma. *British Journal of Haematology* 1992; **81**: 440–3.

78. Willems PMW, Mensink EJBM, Meijerink JPP et al. Sporadic mutations of the *p53* gene in multiple myeloma bone marrow samples, but no evidence for germline mutations of the *p53* gene in familial cases of multiple myeloma. *Blood* 1992; **80** (suppl.): 445a (abstract).

79. Neri A, Baldini L, Trecca D et al. *p53* gene mutations in multiple myeloma are associated with advanced forms of malignancy. *Blood* 1993; **81**: 128–35.

CHAPTER 7

Oncogenes related to multiple myeloma

PAOLO CORRADINI, MARIO BOCCADORO, ALESSANDRO PILERI

Introduction	64	The *bcl–6* locus	67
General mechanisms of oncogenesis	64	The family of *ras* oncogenes	67
The *c-myc* oncogene	65	The *p53* tumor suppressor gene	68
The Moloney leukemia virus integration–4 locus	66	The *Rb1* tumor suppressor gene	69
The *bcl–1* locus	66	Conclusions	69
The *bcl–2* oncogene	67	References	70

Introduction

Multiple myeloma is a B-cell neoplasia characterized by the expansion, primarily in bone marrow, of slowly proliferating malignant plasma cells producing monoclonal immunoglobulins. The disease spans a spectrum of clinical entities, including localized and disseminated, as well as indolent and aggressive forms. Multiple myeloma is sometimes preceded by a premalignant disorder referred to as monoclonal gammopathy of undetermined significance (MGUS) with a typically indolent course, although a prolonged follow-up has shown a significant rate of progression (24.5 per cent) into overt myeloma. A few patients suffer from a related disorder characterized by massive passage of plasma cells into the peripheral blood, and by a rapidly fatal clinical course, defined as plasma cell leukemia.[1]

This chapter summarizes the recent knowledge on the role of oncogene activation and tumor suppressor gene loss/inactivation in multiple myeloma pathogenesis.

General mechanisms of oncogenesis

During the past 5 years, new insights have been gained into the cellular and molecular biology of B-cell neoplasias, and the role of some cellular oncogenes in many hematological neoplasms is now evident. The molecular lesions of lymphoid neoplasia can be divided into broad categories according to their mechanisms, which include activation of proto-oncogenes by chromosomal translocation or point mutation, proto-oncogene amplification and tumor suppressor gene inactivation.

Proto-oncogene activation by chromosomal translocations is the common result of a subset of molecular lesions typical of lymphoid neoplasia, very frequently associated with a specific subtype of lymphoproliferation, and possibly the sole genomic alteration in a given case. Translocation of lymphoid tumors can have one of two effects. First, many translocations involve an antigen receptor locus (the immunoglobulin loci or the T-cell receptor loci in B- and T-cell malignancies, respectively) on one partner chromosome and a proto-oncogene on the other. Once the proto-oncogene is juxtaposed to this locus, its expression becomes regulated by immunoglobulin or T-cell receptor promoters/enhancers and deregulation occurs. Alternatively, translocations may cause the formation of novel transcriptional units deriving from the fusion of the genes at the breakpoint sites (fusion transcripts). Proto-oncogene activation by fusion transcript formation leads to chimeric proteins displaying novel biochemical properties distinct from those of wild type proteins. The proto-oncogenes involved in lymphoid neoplasia translocations belong to different classes of oncogenes. The overwhelming majority are transcription factors directly involved in the control of cell proliferation and differentiation. However, tyrosine kinases and growth factors are also implicated, though at a lower frequency. The most frequent translocation in adult lymphoid tumors involves a novel class of proto-oncogene (*bcl-2*), which does not directly influence cell proliferation, but prevents cell death.

Activation of proto-oncogenes by mutation in lymphoid neoplasia is restricted to the *ras* gene family. These lesions are found in a variety of human cancers. In lymphoid neoplasia, *ras* mutations usually accompany other molecular abnormalities typical of a given tumor type.

Gene amplification as a mechanism of proto-oncogene activation seems of little importance in lymphoid neoplasia, though it is a relatively common mechanism in the molecular pathogenesis of solid tumors.

Finally, disruption of tumor suppressor genes by chromosomal deletion of one allele and inactivation of the other is an additional general mechanism of pathogenesis in lymphoid tumors, best exemplified by the case of *p53*.

Overall, lymphoid malignancies are characterized by a relatively stable genome in comparison to solid tumors. Indeed, many solid tumors carry random alterations involving >20 per cent of the total genome, whereas leukemias and lymphomas mostly carry non-random chromosomal breaks on a low background of random genetic disruption. The biological reason for this difference in genetic instability between solid and lymphoid tumors is not known.[2]

While advances in chromosomal banding and molecular techniques have clarified the role of some cellular oncogenes in many lymphoid neoplasms, e.g. Burkitt's lymphoma and follicular lymphoma, the difficulty of obtaining adequate metaphase chromosomes and the dearth of specific karyotype aberrations have greatly slowed similar progress in plasma cell dyscrasias.[3] Flow cytometry detects aneuploidy in about 80 per cent of multiple myeloma patients, whereas cytogenetic analysis can only detect a minority of karyotype aberrations.

In spite of these difficulties, it is now evident that the wide clinical and pathological heterogeneity of plasma cell dyscrasias is reflected, at the molecular level, by an analogous heterogeneity in the incidence of all the most important genetic lesions considered. This probably corresponds to the progressive genetic deregulation of the neoplastic population. In particular, both activation of dominantly acting oncogenes, and loss/inactivation of tumor suppressor genes are more frequently involved in aggressive and disseminated, rather than indolent plasma cell dyscrasias. Interestingly, the molecular lesions most frequently detected in these dyscrasias are point mutations,[2] and not gene rearrangements, as frequently seen in other differentiated B-cell neoplasms[4] (Table 7.1).

The c-myc oncogene

The c-*myc* oncogene is one of the best characterized proto-oncogenes. It codes for a protein which acts as a transcriptional factor and regulates the activity of other genes by binding to specific DNA sequences. Its stimulation of B-cell proliferation and repression of B-cell differentiation are important steps in the pathogenesis of human B-cell malignancies, and its involvement in several lymphoid neoplasms,

Table 7.1 Genetic alterations described in multiple myeloma

Involved oncogene	Chromosome location	Frequency (%)	Mechanism of activation	References
c-*myc*	8q24	0–5	Rearrangement	3,8,9,13
MLVI-4	8q	16	Rearrangement	13
bcl-1	11q13	0–4	Rearrangement	3,22–24
bcl-2	18q21	0–5	Rearrangement	3,14
N-ras	1p11–13	9–27	Point mutation	9,24,41,47
K-ras	12p11–12	0–7	Point mutation	9,24,41,47
p53	17p13	6–13	Point mutation/deletion	46–48
Rb1	13q14	3	Deletion	48

especially in Burkitt's lymphoma, has been postulated.[2] It is activated through a translocation that juxtaposes it on chromosome 8 and the immunoglobulin genes on chromosome 2 or 14 or 22. In the more frequent t(8;14)(q24;q32), breakpoints located 5' and centromeric to c-*myc* lead to its translocation into the immunoglobulin heavy-chain locus on chromosome 14, thus placing it under the control of the immunoglobulin transcriptional enhancer regions. In the less frequent t(2;8) and t(8;22) translocations, an immunoglobulin light-chain locus is translocated 3' and telomeric to the c-*myc* locus, which remains on chromosome 8. Breakpoint sites on this chromosome are heterogeneous and may occur within the first exon or intron, immediately (< 3 kb) 5' to the c-*myc* promoter or at an undefined distance (> 100 kb) 5' to the c-*myc* locus. In addition, both the J_H and switch region of the heavy-chain locus on chromosome 14 may be involved in the translocation event.[2] Several studies also indicate that all translocated c-*myc* alleles contain structural alterations in putative 5' regulatory sequences of the gene which are either removed or mutated.[5,6] The combination of proximity of the immunoglobulin transcriptional regulatory elements and the presence of such structural lesions is thought to lead to the deregulated expression of c-*myc* translocated alleles with the constitutive production of a normal c-*myc* protein.[2]

Because of its frequent involvement in other human B-cell neoplasms and constant deregulation in murine plasmacytoma,[7] c-*myc* was the first oncogene whose involvement was evaluated in multiple myeloma.

Alterations in the c-*myc* locus were only sporadically reported in multiple myeloma, by Southern blot analysis, with a probe representative of the third exon of the c-*myc* gene.[8,9] The presence of mutations in the 3' region of the first exon (*Pvu*II and *Alu*I endonuclease sites) was reported by Meltzer et al.[10] in 10 of 16 human myeloma and cell lines, but these data have not been confirmed.[3,9] Overexpression of c-*myc* RNA was observed in 25 per cent of patients, without apparent abnormalities of c-*myc* RNA transcript size.[8] It was only rarely ascribable to DNA rearrangements,[8] and no point mutations were found in the first intron/exon putative regulatory sequences.[11] A plausible genetic explanation for this increase has yet to be found.

Involvement of c-*myc* has been carefully investigated in a human myeloma cell line LP-1 derived from a patient in terminal leukemic phase. It displayed increased expression of c-*myc* and abnormal RNA transcripts. No rearrangements were detected with Southern blot analysis but point mutations were detected in exons 1 and 2.[12]

These data cannot rule out the possibility of direct involvement of c-*myc* in some cases of multiple myeloma, for example through alterations in its regulatory sequences. However, no convincing evidence of this kind has been reported for other types of tumors, and the possibility itself is no more than speculative in the absence of clear parameters to define normal levels of c-*myc* expression.

The Moloney leukemia virus integration–4 locus

A particular molecular lesion that could be implicated in the pathogenesis of multiple myeloma, possibly through deregulation of the c-*myc* gene, is rearrangement of the human homologue of the Moloney leukemia virus integration–4 (*MLVI–4*) locus. *MLVI–4* maps 20 kb 3' from the c-*myc* locus. The oncogenetic role of genetic alterations of this locus has been carefully analyzed in rodent tumors. *MLVI–4* is targeted by provirus integration in rodent retrovirus-induced T-cell lymphomas, which activates c-*myc*, and two additional genes, *MLVI–4* and *MLVI–1* through a *cis*-acting mechanism. Palumbo et al. described rearrangements in the human homologue of *MLVI–4* in 16 per cent of multiple myeloma patients[13] (7/42): two were detected at initial diagnosis, four during treatment, one at relapse. Their presence correlated with unresponsiveness to therapy, suggesting that they may be involved in the late stages of oncogenesis, and therefore associated with tumor progression. They also seemed to identify a subset of patients with a poor prognosis. A precise demonstration of the regulatory role of *MLVI–4* on c-*myc* gene in humans, however, is still awaited.

Another regulatory sequence of the c-*myc* gene, the *MLVI–1/PVT–1* locus, has been analyzed in multiple myeloma. This is located 270 kb 3' from the c-*myc* gene. It, too, is targeted by provirus integration in retrovirus-induced lymphomas and activates c-*myc*, but no rearrangements have been detected in human myelomas.[14]

The bcl-1 *locus*

The *bcl-1* locus was originally identified as the breakpoint site on chromosome 11 of the t(11;14)(q13;q32) translocation, a cytogenetic lesion detectable in approximately 50 per cent of intermediate lymphocytic (centrocytic) lymphomas.[15] This translocation results in juxtaposition of the immunoglobulin heavy-chain locus on chromosome 14 to sequences from chromosome 11. Recently, it has been shown that a gene located approximately 200 kb from the *bcl-1* locus is overexpressed in lymphoproliferative disorders displaying *bcl-1* rearrangements. This putative oncogene, called *PRAD–1*, displays homology to cyclins, a family of genes involved in the regulation of cell cycle progression.[16] Its contribution to lymphomagenesis remains to be established by *in vitro* and *in vivo* investigation of its pathogenetic effects.

There have been sporadic reports of multiple myeloma possessing the t(11;14) translocation. In the three largest cytogenetic series, 4 out of 136 cases (3 per cent) were positive.[17–19] It has also been observed in two cell lines derived from patients with multiple myeloma or plasma cell leukemia.[20,21] Rearrangements of *bcl-1* have been described in 5 of 120 cases of multiple myeloma (4 per cent),[22] though none were found in a total of 37 patients in two later studies.[23,24] Their incidence in multiple myeloma thus appears to be very low.

The bcl-2 *oncogene*

The *bcl-2* oncogene belongs to a new class of proto-oncogenes that apparently control cell survival by blocking programmed cell death. The bcl-2 protein is also distinct from other proto-oncogene products by being an integral inner mitochondrial membrane protein. Its physiological function seems to be that of ensuring a longer cell lifespan without division in a number of tissues.[2,25] In the lymphoid system, for instance, it appears to maintain memory B cells, and antibody-secreting plasma cells.[26] The *bcl-2* gene was identified by molecular cloning of the t(14;18) (q32;q21) translocation,[27] which is the most frequent chromosomal abnormality in B-cell non-Hodgkin's lymphoma (present in 85 per cent of follicular lymphomas, and in 25 per cent of diffuse large cell lymphomas).[28] The translocation joins the *bcl-2* gene at its 3′ untranslated region to immunoglobulin heavy-chain sequences,[26] resulting in *bcl-2* deregulation because of the nearby presence of immunoglobulin transcriptional enhancer elements. The presence within the cells of constitutively high levels of the normal *bcl-2* protein prevents the programmed cell death.[29] Evidence that these alterations may contribute to the pathogenesis of follicular lymphoma comes from studies of transgenic mice bearing an activated *bcl-2* gene. These showed development of a disorder characterized by the accumulation of long-lived B cells analogous to that seen in human follicular lymphoma. As in the human disease, some of these mice progress, and develop a more aggressive clonal B-cell lymphoma related to the occurrence of additional genetic alterations.[30]

The t(14;18) translocation has rarely been detected in myeloma patients. Increased expression of *bcl-2* protein has been described by Durie et al. in about 75 per cent.[31,32] Such findings are of particular interest because of the potential effect of *bcl-2* activation in multiple myeloma, since it is characterized by a slow proliferative activity, like follicular lymphoma. None of 60 patients with myeloma showed rearrangement of *bcl-2* gene (within major breakpoint and minor cluster regions).[3] This finding is consistent with the rarity of t(14;18) in multiple myeloma.[32] To further elucidate the molecular mechanism of *bcl-2* overexpression, Palumbo et al. investigated the 5′ region of the *bcl-2* gene, which is rearranged in a significant fraction (10–30 per cent) of chronic lymphocytic leukemias.[33] Rearrangements were detected in 2 out of 35 patients.[14]

The bcl-6 *locus*

The *bcl-6* locus is a recently identified genomic region cloned from the t(3;14), a translocation present in diffuse large-cell non-Hodgkin's lymphomas.[34] It is expressed in mature B cells and codes for a 79 kDa protein containing regions of homology with zinc-finger transcription factors. Its pathogenetic role is uncertain, even though it is regarded as a candidate proto-oncogene whose activation takes place through heterogeneous translocation mechanisms involving band 3q27. Since 14q+ is a frequent cytogenetic abnormality in multiple myeloma (20–25 per cent),[32] the possibility that it contains part of 3q prompted us to study the *bcl-6* locus in primary myeloma. Rearrangements of *bcl-6* were evaluated in 26 patients through hybridization analysis with the Sac4 probe. No alterations were detected.[35]

The family of ras *oncogenes*

The *ras* family consists of three related oncogenes: H-, K- and N-*ras* that encode proteins of 21 kDa (p21) with GTPase activity located at the inner surfaces of the cell membrane and involved in growth signal transduction mechanisms. Their transforming potential is acquired when point mutations occur in codons 12, 13, and 61, resulting in single amino acid substitutions.[36] Screening for somatic point mutations has became much easier in recent years with new simple, sensitive strategies allowing the analysis of large numbers of samples. The most widely used is single-strand conformation polymorphism (SSCP) analysis, which detects mutated

Table 7.2 Incidence of *ras* and *p53* mutations in MGUS, multiple myeloma (MM) and plasma cell leukemia (PCL)

	N-*ras*	K-*ras*	H-*ras*	*p53*
MGUS	0/30(0%)	0/30(0%)	ND	0/15(0%)
MM	26/147(17.6%)	5/147(3.4%)	0/70(0%)	10/120(8.3%)
PCL	8/31(25.8%)	6/31(19.3%)	0/3(0%)	10/40(25%)

Data were derived from five of the most recent works on the subject: refs 9, 41, 46–48.
Three myeloma and one plasma cell leukemia patients were harboring both N- and K-*ras* mutations.
ND, not determined.

DNA strands because of their different electrophoretic mobility under non-denaturing running conditions.[37]

Activated *ras* genes have been identified in several human cancers,[38] including both myeloid and lymphoid malignancies.[39] Mutations, however, have been described in only two B-cell disorders, namely acute lymphoblastic leukemia and multiple myeloma, and are constantly absent in all other differentiated lymphoid tumors, such as chronic lymphocytic leukemia, hairy cell leukemia, and non-Hodgkin's lymphoma.[9]

High levels of the H-*ras* gene product (p21) described in 17 of 23 patients with active myeloma were associated with poor survival.[40] However, since no H-*ras* gene amplification, rearrangement or mutation has been detected in multiple myeloma,[3,9] the true pathogenetic weight of such overexpression still remains to be clarified.

The presence of activating point mutations of N- and K-*ras* genes is a well-documented molecular lesion in multiple myeloma[9,24,41] (Table 7.2). Their overall incidence ranges from 9 per cent[24,41] to 31 per cent,[9] probably due to differences in the clinical status of the patients examined. Mutations of *ras* genes are not randomly distributed. They are not present in MGUS, solitary plasmacytomas and Stage I multiple myelomas, whereas they are detected in a sizable fraction of patients with advanced stage disease and adverse prognostic factors (30 per cent). A similar incidence (30 per cent) has been found in patients with plasma cell leukemias. The absence of a clear difference in incidence between diagnosis and relapse rules out any mutagenic effect on the part of alkylating agents. Mutations affecting the N-*ras* gene at codon 61 are the most frequent finding. A heterogeneous pattern of mutation is observed, including C to A, G to C, G to T and A to T transversions, and G to A transitions. Purine–pyrimidine transversions account for 80 per cent.[41]

In vitro transfection studies also suggest that *ras* oncogenes are involved in myeloma pathogenesis, since their transfection confers both malignancy and terminal differentiated morphology to Epstein–Barr virus infected B cells.[42] This biological effect seems to be peculiar to the *ras* oncogene family, since c-*myc* transfection induces malignant transformation without substantial changes in the differentiation phenotype of target cells, and is particularly intriguing, since plasma cell dyscrasias are typically characterized by the coexistence of a differentiated morphology and full malignancy. However, *ras* mutations are not constantly detected, and thus cannot be considered as a general pathogenetic mechanism in multiple myeloma.

The p53 *tumor suppressor gene*

The *p53* gene encodes a 53 kDa nuclear phosphoprotein that controls the normal cell cycle by regulating transcription and possibly DNA replication.[43] Loss of this growth inhibitor activity is usually the result of point mutations of one allele associated with the loss of the other. The majority of mutations occur between codons 110 and 307 encompassing exon 5–9, and are clustered in four regions that are highly conserved among several different species.[44] They have been detected in many human malignancies, and are now considered the most frequent molecular lesions in human cancer. An appreciable frequency has been observed in some lymphoid malignancies: chronic lymphocytic leukemia, L3-type acute lymphoblastic leukemia, Burkitt's lymphoma[45] and multiple myeloma[46–48] (Table 7.2).

Preliminary observations showed increased levels of p53 mRNA in myeloma cells.[49] In addition, point mutations in exons 5 and 8 were detected in 8 of 10 human myeloma cell lines. All were single base substitutions with a predominance of G to A transitions. Interestingly, in the XG–2 and XG–4 lines, both the wild-type and the mutated allele were concomitantly expressed at the mRNA level, suggesting a dominant negative acting fashion.[50]

The incidence of *p53* mutations in bone marrow samples of patients with plasma cell dyscrasias ranges from 10[48] to 20 per cent.[47] As in chronic lymphocytic leukemia, where they are especially described in the aggressive phase defined as Richter's transformation,[45] they are typically associated with advanced and clinically aggressive forms of multiple myeloma.[47]

These mutations were not found during the chronic phase in three patients positive for *p53* mutations during the terminal phases,[46] and were more frequent in plasma cell leukemia (22 per cent) than in aggressive multiple myeloma.[48] The presence of concomitant *ras* and *p53* mutations has been sporadically described.[48] Mutations in the *p53* gene were similar to those described in other hematological tumors (with prevalence of C to T, G to A and T to C transitions), giving rise both to amino acid substitutions and stop codons.[46]

The Rb1 *tumor suppressor gene*

The *Rb1* tumor suppressor gene encodes a 110 kDa phosphoprotein which accumulates in the nucleus and is associated with DNA binding activity. Preliminary findings suggest that Rb1 protein is involved in the control of transition of normal cells from G^0/G^1 into S-phase of the cell cycle.[51] Rb1 gene product is also expressed in human hematopoietic cells, where it is associated with the control of proliferation and terminal differentiation. Expression of normal protein is lacking in retinoblastomas, and somatic mutations of *Rb1* disrupting normal gene expression have been found in these tumors, as well as in carcinoma of the breast, prostate, lung and, rarely, in hematological malignancies.[52,53]

Various mechanisms of *Rb1* inactivation have been described in human tumors: point mutations, gross rearrangements, large intragenic deletions, and complete deletions. All these lesions cause a lack of protein expression or, in rare cases, the expression of an abnormal protein. Immunohistochemical analysis has shown complete absence of the protein in 17 per cent of multiple myeloma patients, and 18 per cent of plasma cell leukemia patients.[48] Plasma cells normally express Rb1 protein and its absence thus leads support to the idea of a primary defect in *Rb1* gene structure or expression. The molecular basis of such loss of protein expression is not clearly defined. Southern blot analysis of patients lacking protein expression with two probes spanning from exon 1 to 9 and from exon 10 to 27 showed a deletion in only one case. The absence of rearrangements or deletions in the remaining cases suggests that the lesions may be point mutations, as observed in other tumor types.[54] It is noteworthy that patients with a lack of Rb1 protein presented extramedullary masses, indicating an aggressive clinical phase.[48] It has been

proposed, in fact, that *Rb1* gene inactivation may contribute to progression rather than initiation of solid tumors.[52] Such findings are also consistent with its role in progression of malignancy in plasma cell dyscrasias. No clear association between *Rb1* loss or inactivation and *ras* and *p53* mutations was detected.[48]

Immunohistochemical analysis has also been performed on three cell lines. U266 cell line showed lack of protein expression. Sequencing analysis of the *Rb1* cDNA revealed two abnormal mRNA species carrying the deletion of both exon 13 and 14, or only exon 14, respectively.[48] Since a normally sized mRNA species was detected, the presence of these two mRNAs was explained as the result of an alternative splicing mechanism, while additional alterations preventing normal protein expression may be present on the second allele.[48] Deletions of these exons are frequently observed in various types of tumors, including retinoblastoma, breast cancer and osteosarcoma.

Conclusions

Many data have been accumulated on the genetic lesions involved in myeloma pathogenesis.[55] Even so the precise molecular pathway leading to malignant proliferation cannot be clearly defined. However, some general considerations can be drawn.

First, none of the oncogenes and tumor suppressor genes examined were altered in the vast majority of patients, by contrast with several neoplasms where a specific molecular lesion is present in virtually all cases (*bcl-2* in follicular lymphoma, or c-*myc* in Burkitt's lymphoma) (Table 7.1). This can be explained by the presence of some still unknown molecular lesions characteristic of plasma cell dyscrasias or, alternatively, by the presence of several pathways of genetic deregulation leading to the same malignant transformation.

Second, the molecular lesions in multiple myeloma seem to be quite different to that present in other differentiated B-cell neoplasms, such as non-Hodgkin's lymphomas. Point mutations are more frequent than chromosomal translocations. In addition, *bcl-1*, *bcl-2*, *bcl-6* and c-*myc* are frequently involved in non Hodgkin's lymphoma, whereas *ras* and *p53* genes are more frequent in multiple myeloma (Table 7.1).

Third, none of the best-defined molecular lesions, such as *ras* or *p53* mutations, have been detected in MGUS or in multiple myelomas with indolent behavior,[41,48] whereas they become frequent in clinically aggressive forms. Furthermore, mutations are more frequent in plasma cell leukemias (Table 7.2). These

Table 7.3 Correlation between *ras, p53* mutations, and prognostic parameters in multiple myeloma patients

Patient	Involved gene	Stage	Albumin (g/l)	LI (%)	CRP (mg/l)	β_2-Microglobulin (mg/l)
M22	*ras*	IIIA	29	0.5	117	5.5
M127	*ras*	IIIA	31	4	13	7.9
M91	*ras*	IIIB	34	0.8	50	41.9
M19	*ras*	IIIA	25	ND	121.9	13
M89	*ras*	IIIB	50	2.7	6	11.5
M90	*ras*	IIIA	34	1.2	10.7	3.8
M25	*p53*	IIIA	38	2	43	59
M100	*p53*	IIIA	44	4.5	0	4.1
M114	*p53*	IIIB	34	3	4.5	40.5

Underlining indicates abnormal values.
Adverse prognostic factors: Albumin < 30 g/l; LI > 1%; CRP > 6 mg/l; β_2-microglobulin > 6 mg/l.
Most of patients with *ras* or *p53* mutations showed more than one concordant adverse prognostic parameter. From ref. 48.

findings have suggested that *ras* and *p53* mutations contribute to tumor progression rather than initiation. This is supported by a longitudinal analysis of the *p53* gene. Its mutations were absent during the indolent phase, then appeared during the fulminating phase.[46]

Finally, correlations between *ras, p53* mutations, and several prognostic parameters were analyzed, including both prognostic parameters reflecting disease aggressiveness (labeling index and C-reactive protein) and tumor burden (Stage III and β_2-microglobulin)[56–58] (Table 7.3). These mutations were significantly more frequent in patients with high tumor burden, and seem to identify a small subset of multiple myeloma patients with multiple concordant, adverse prognostic factors. This suggests that *ras* and *p53* mutations may confer a proliferative advantage on tumor cells.[41,48]

References

1. Kyle RA. Diagnostic criteria of multiple myeloma. *Hematology/Oncology Clinics of North America* 1992; **6**: 347–58.
2. Gaidano G, Dalla Favera R. Protooncogenes and tumor suppressor genes. In: Knowles DM ed. *Neoplastic hemopathology*. Baltimore: Williams & Wilkins, 1992: 245–61.
3. Barlogie B, Epstein J, Selvanaygam P, Alexanian R. Plasma cell myeloma – New biological insights and advances in therapy. *Blood* 1989; **73**: 865–79.
4. Gaidano G, Dalla Favera R. Biologic and molecular characterization of non Hodgkin's lymphoma. *Current Opinion in Oncology*; 1993; **5**(5): 776–84.
5. Cesarman E, Dalla Favera R, Bentley D, Groudine M. Mutations in the first exon are associated with altered transcription of c-*myc* in Burkitt lymphoma. *Science* 1987; **238**: 1272–5.
6. Murphy JP, Neri A, Richter H et al. C-*myc* regula-

tory sequences are consistently mutated in Burkitt's lymphoma (in preparation).
7. Fahrlender PD, Sumegi J, Yang J-Q et al. Activation of the c-*myc* oncogene by the immunoglobulin gene enhancer after multiple switch region-mediated chromosomal rearrangements in a murine plasmacytoma. *Proceedings of the National Academy of Sciences* 1985; **82**: 3746–50.
8. Selvanaygam P, Blick M, Narni F et al. Alteration and abnormal expression of the c-*myc* oncogene in human multiple myeloma. *Blood* 1988; **71**: 30–5.
9. Neri A, Murphy J, Cro L et al. *Ras* oncogene mutation in multiple myeloma. *Journal of Experimental Medicine* 1989; **170**: 1715–25.
10. Meltzer P, Shadle K, Durie B. Somatic mutation alters a critical region of the c-*myc* gene in multiple myeloma. *Blood* 1987; **70**: 985A.
11. Corradini P, Ladetto M, Voena C et al. Multiple genetic lesions in multiple myeloma and plasma cell leukemia (in preparation).
12. Pegoraro L, Malavasi F, Bellone G et al. The human myeloma cell line LP–1: a versatile model in which to study early plasma-cell differentiation and c-*myc* activation. *Blood* 1989; **73**: 1020–7.
13. Palumbo AP, Boccadoro M, Battaglio S et al. Human homologous of Moloney leukemia virus integration–4 locus (*MLVI–4*), located 20 kilobases 3′ of the c-*myc* gene, is rearranged in multiple myelomas. *Cancer Research* 1990; **50**: 6478–82.
14. Palumbo AP, Lasota J, Battaglio S et al. Molecular analysis of the c-*myc* locus, its flanking chromosomal region, and the *bcl-2* oncogene in multiple myeloma. In: Boccadoro M, Pileri A eds *Multiple myeloma from biology to therapy. Abstract book of III International Workshop*. Torino, 1991: 29–30.
15. Raffeld M, Jaffe ES. *Bcl-1* t(11;14), and mantle cell-derived lymphomas. *Blood* 1991; **78**: 259–63.
16. Motokura T, Bloom T, Goo KH et al. A novel cyclin encoded by a *bcl-1* linked candidate oncogene. *Nature* 1991; **350**: 512–15.
17. Nishida K, Taniwaki M, Misawa S, Abe T. Nonrandom rearrangement of chromosome 14 at band

q32.33 in human lymphoid malignancies with mature B-cell phenotype. *Cancer Research* 1989; **49**: 1275–81.

18. Van den Berghe H, Vermaelen K, Louwagie A et al. High incidence of chromosome abnormalities in IgG3 myeloma. *Cancer Genetic Cytogenetic* 1984; **11**: 381–6.

19. Dewald GW, Kyle RA, Hicks GA, Greipp PR. The clinical significance of cytogenetic studies in 100 patients with multiple myeloma, plasma cell leukemia or amyloidosis. *Blood* 1985; **66**: 380–90.

20. Nacheva E, Fischer PE, Sherrington PD et al. A new human plasma cell line, Karpas 620, with translocations involving chromosomes 1,11 and 14. *British Journal of Haematology* 1990; **74**: 70–6.

21. Ohtsuki T, Yawata Y, Wada H et al. Two human myeloma cell lines, amylase producing KMS–12-PE and amylase-non-producing KMS–12-BM, were established from a patient, having the same chromosome marker t(11;14)(q13;q32). *British Journal of Haematology* 1989; **73**: 199–204.

22. Selvanaygam P, Goodacre A, Strong L et al. Alterations of *bcl-1* oncogene in human multiple myeloma. *American Association for Cancer Research* 1987; Abstract no. 76.

23. Williams ME, Meecker TC Swerdlow SH. Rearrangement of the chromosome 11 *bcl-1* locus in centrocytic lymphoma: Analysis with multiple breakpoint probes. *Blood* 1991; **78**: 493–8.

24. Paquette, RL, Berenson J, Lichtenstein A et al. Oncogenes in multiple myeloma: point mutations of N-*ras*. *Oncogene* 1990; **5**: 1659–63.

25. McDonnel TJ, Deane N, Platt FM et al. *Bcl-2*-immunoglobulin transgenic mice demonstrate extended B cell survival and follicular lymphoproliferation. *Cell* 1989; **57**: 79–88.

26. Nunez G, Hockenberry D, McDonnel TJ et al. *Bcl-2* maintains B cell memory. *Nature* 1991: **353**: 71–3.

27. Bakshi A, Jensen JP, Goldman P et al. Cloning the chromosomal breakpoint of t(14;18) human lymphomas: clustering around J_H on chromosome 14 and near a transcriptional unit on 18. *Cell* 1985; **41**: 889–906.

28. Lee M-S, Blick MB, Pathak S et al. The gene located at chromosome 18 band q21 is rearranged in uncultured diffuse lymphomas as well as follicular lymphomas. *Blood* 1987; **70**: 90–5.

29. Hockenberry D, Nunez G, Milliman C et al. Bcl-2 is an inner mithocondral membrane protein that blocks programmed cell death. *Nature* 1990; **348**: 334–6.

30. McDonnel TJ, Korsmeyer SJ. Progression from lymphoid hyperplasia to high-grade malignant lymphoma in mice transgenic for the t(14;18). *Nature* 1991; **349**: 254–6.

31. Durie BGM, Mason DY, Giles F et al. Expression of the bcl-2 oncogene protein in multiple myeloma. *Blood* 1990; **76**: 347A.

32. Durie BGM. Cellular and molecular genetic features of myeloma and related disorders. *Hematology/Oncology Clinics of North America* 1992; **6**: 463–77.

33. Adachi M, Tefferi A, Greipp PR et al. Preferential linkage of *bcl-2* to immunoglobulin light chain gene in chronic lymphocytic leukemia. *Journal of Experimental Medicine* 1990; **171**: 559–64.

34. Bihui HY, Rao PH, Chaganti RSK, Dalla Favera R. Cloning of bcl-6, the locus involved in chromosome translocations affecting band 3q27 in B-cell lymphoma. *Cancer Research* 1993; **53**: 2732–5.

35. Lo Coco F, Bihui HY, Lista F et al. Rearrangements of the *bcl-6* gene in diffuse large-cell non Hodgkin's lymphoma. *Blood* 1994; **83**: 1757–9

36. Barbacid M. *Ras* genes. *Annual Reviews in Biochemistry* 1987; **56**: 779–827.

37. Orita M, Suzuki Y, Sekiya T, Hayashi K. Rapid and sensitive detection of point mutations and DNA polymorphism using the polymerase chain reaction. *Genomics* 1989; **5**: 874–82.

38. Bos JL. Ras oncogenes in human cancer: A review. *Cancer Research* 1989; **49**: 4682–9.

39. Ahuja HG, Foti A, Bar-Eli M, Cline MJ. The pattern of mutational involvement of ras genes in human hematologic malignancies determined by DNA amplification and direct sequencing. *Blood* 1990; **75**: 1684–90.

40. Tsuchya H, Epstein J, Selvanaygam P et al. Correlated flow cytometric analysis of H-ras p21 and nuclear DNA in multiple myeloma. *Blood* 1988; **72**: 796–800.

41. Corradini P, Ladetto M, Voena C et al. Mutational activation of N- and K-*ras* oncogenes in plasma cell dyscrasias. *Blood* 1993; **81**: 2708–13.

42. Seremetis S, Inghirami G, Ferrero D et al. Transformation and plasmacytoid differentiation of EBV-infected human B lymphoblasts by *ras* oncogenes. *Science* 1989; **243**: 660–3.

43. Soussi T, Caron de Fromentel C, May P. Structural aspects of the p53 protein in relation to gene evolution. *Oncogene* 1990; **5**: 945–52.

44. Hollstein M, Sidransky D, Vogelstein B, Harris CC. *p53* mutations in human cancers. *Science* 1991; **253**: 49–53.

45. Gaidano G, Ballerini P, Gong JZ et al. *p53* mutations in human lymphoid malignancies: association with Burkitt lymphoma and chronic lymphocitic leukemia. *Proceedings of the National Academy of Sciences* 1991; **88**: 5413–17.

46. Neri A, Baldini L, Trecca D et al. *p53* mutations in multiple myeloma are associated with advanced forms of malignancy. *Blood* 1993; **81**: 128–35.

47. Portier M, Molès JP, Mazars GR et al. *p53* and *ras* gene mutations in multiple myeloma. *Oncogene* 1992; **7**: 2539–43.

48. Corradini P, Inghirami G, Astolfi M et al. Inactivation of tumor suppressor genes, *p53* and *Rb1* in plasma cell dyscrasias. *Leukemia* 1994; **8**: 758–67.

49. Palumbo AP, Pileri A, Dianzani U et al. Altered expression of growth-regulated proto-oncogenes in human malignant plasma cells. *Cancer Research* 1989; **49**: 4701–4.

50. Mazars GR, Portier M, Zhang XG et al. Mutations of the *p53* gene in human myeloma cell lines. *Oncogene* 1992; **7**: 1015–18.

51. Goodrich DW, Wang NP, Qian YW et al. The

retinoblastoma gene product regulates progression through the G1 phase of the cell cycle. *Cell* 1991; **66**: 293–302.

52. Benedict WF, Xu HJ, Hu SX, Takahashi R. Role of the retinoblastoma gene in the initiation and progression of human cancer. *Journal of Clinical Investigation* 1990; **85**: 988–93.

53. Tang JL, Yeh SH, Chen PJ. Inactivation of the retinoblastoma gene in acute myelogenous leukaemia. *British Journal of Haematology* 1992; **82**: 502–7.

54. Stanbridge EJ. Human tumor suppressor genes. *Annual Reviews in Genetics* 1990; **24**: 615–57.

55. Corradini P, Boccadoro M, Voena C, Pileri A. Evidence for a bone marrow B cell transcribing malignant plasma cell VDJ joined to Cl sequence in immunoglobulin (IgG)- and IgA-secreting multiple myelomas. *Journal of Experimental Medicine* 1993; **178**: 1091–6.

56. Boccadoro M, Marmont F, Tribalto M et al. Early responder myeloma: kinetic studies identify a patient subgroup characterized by very poor prognosis. *Journal of Clinical Oncology* 1989; **7**: 119–25.

57. Durie BGM, Stock-Novack D, Salmon SE et al. Prognostic value of pretreatment serum beta–2 microglobulin in myeloma: a Southwest Oncology Group study. *Blood* 1990; **75**: 823–30.

58. Bataille R, Boccadoro M, Klein B et al. C-reactive protein and β2 microglobulin produce a simple and powerful myeloma staging system. *Blood* 1992; **80**: 733–7.

CHAPTER 8

Growth factors in the pathogenesis of multiple myeloma

BERNARD KLEIN

Introduction	73	Granulocyte colony-stimulating factor (G-CSF)	77
Interleukin-6	74	Interleukin-10 (IL-10)	77
Other cytokines involved with myeloma growth	76	Therapeutic implications of cytokine and growth	
IFN-α and TNF induce autocrine production of		factor data	78
IL-6 in myeloma cell lines	77	Conclusions	80
IFN-γ acts as a potent inhibitor of myeloma cell		References	80
proliferation by blocking IL-6-receptor expression	77		

Introduction

Multiple myeloma (MM) is a B-cell neoplasia affecting terminal cells of B-cell differentiation (i.e. plasma cells). It develops in the bone marrow, in close contact with stromal cells (myeloid cells, monocytes, fibroblasts, bone cells, etc.). The particular microenvironment of the bone marrow is especially relevant to the evolution of myeloma. Several growth factors such as IL-6, IL-1β G/GM-CSF, IL-3 and others to be discussed, clearly influence myeloma growth. IL-6 is a crucial cytokine for plasma cells, the first shown to selectively promote myeloma cell growth both *in vitro* and *in vivo*.

None the less, two aspects of IL-6 pathophysiology must be emphasized:

1. High levels of endogenous IL-6 do not (in themselves) result in multiple myeloma. For example in systemic lupus erythrematosus (SLE) IL-6 levels are very high, but myeloma is very rare.
2. The disease which is characteristically associated with high serum IL-6 is Castleman's disease (a lymphoma), not myeloma. It seems, therefore, that some additional special aspects are necessary to explain the evolution of myeloma.

Although the discovery of cytokines, such as IL-6, has revolutionized the understanding of myeloma, it is important to concede that other factors are involved in the pathogenesis of both benign (MGUS) and malignant (MM) plasma cell proliferation. For example the monoclonal expansion in MGUS does not appear to involve any aberrancy of IL-6 pathophysiology. Likewise the transition from MGUS to early myeloma involves cell kinetics, cell adhesion and possibly triggering an activation of IL-1β production. The sequence of molecular events involving several oncogenes (e.g. *Myc*, N-*ras*, *bcl-1/2*, etc. and

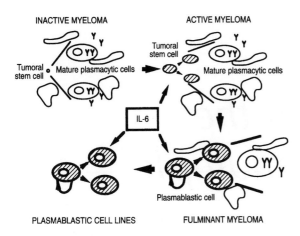

Fig. 8.1 Multiple myeloma is a multistage disease.

anti-oncogenes (e.g. *Rb*) associated with receptor, extracellular matrix (e.g. fibrinectin) and cytokine changes remains to be fully explained. Ultimately, in approximately 30–40 per cent of myeloma patients at the time of presentation IL-6 is clearly involved. In plasma cell leukemia IL-6 is a crucial growth factor and this correlates well with the presence of mutations in the *Rb* oncogene. Based on this example it seems likely that different subsets of patients (e.g. 'smoldering', limited bone disease, aggressive bone

disease, etc.) will have different molecular and cytokine patterns. The model of RG Hawley et al. is especially interesting in this respect. According to this, no expression of IL-1β and IL-6 in plasma cells is associated with an aggressive phenotype and active bone disease. The current understanding with respect to each cytokine will now be discussed separately.

Interleukin-6

Interleukin-6 is a major growth factor for malignant plasmablastic cells.[1–4] Spontaneous myeloma cell proliferation is observed for several days in cultures from myeloma tumor tissue in about half of patients with myeloma, i.e. in those with active disease and proliferating myeloma cells *in vivo*.[1,4,5] Numerous cytokines are produced in these short-term cultures, including IL-6, GM-CSF, G-CSF, IL-11, IL-1, TNF, and probably others as well. All these cytokines may contribute to spontaneous myeloma cell proliferation[1,6–10] (Klein, unpublished results). However, the almost complete inhibition of proliferation by the anti-IL-6 monoclonal antibody (mAb) emphasizes that IL-6 may be the most important myeloma growth factor *in vitro*. We studied 30 patients with active disease, including 14 with plasma cell leukemia.

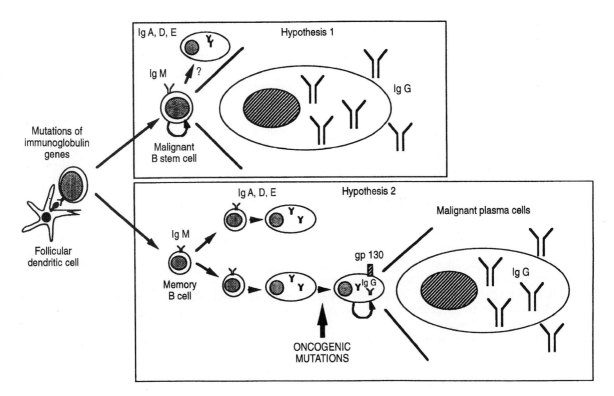

Fig. 8.2 Tumor stem cell in multiple myeloma.

Anti-IL-6 mAb inhibited *in vitro* myeloma cell proliferation in all patients.[9] Most studies have confirmed that IL-6 is a major malignant plasmablastic growth factor[1-4] in these types of patients.

GM-CSF, IL-3 AND IL-5 – RELATION TO IL-6

GM-CSF stimulates the proliferation of myeloma cells,[6] but is not a direct myeloma cell growth factor. It acts on myeloma cells by amplifying their response to IL-6. This property of GM-CSF can be used to increase the proportion of abnormal metaphases in short-term cultures of bone marrow cells from MM patients.[10] IL-3 (Klein, unpublished results) and IL-5,[3] which share the same KH97 transducer chain with GM-CSF, also stimulate the response of myeloma cells to IL-6. The explanation why these cytokines using the same KH97 transducer chain[11] have a synergistic action with IL-6 seems to be that IL-3 increases the expression of IL-6 high-affinity receptors several-fold as demonstrated recently in myeloma cell lines.[12] One important consequence of this observation is that GM-CSF can be used to establish IL-6-dependent plasmablastic cell lines from patients with extramedullary MM.[13] However, IL-6 alone is not sufficient to obtain such cell lines. Instead, activation of the KH97 transducer chain through GM-CSF, IL-3, or IL-5 in the early phases of culture seems critical to prevent downregulation of IL-6 receptors by IL-6.[13,14] These human myeloma cell lines are plasmablastic cell lines that secrete immunoglobulins, express the plasma cell antigens CD38 and B-B4, and lack the usual B-cell antigens (CD19 and CD20). For several years their growth has remained dependent upon addition of exogenous IL-6.[13]

Thus, several hematopoietic cytokines increase IL-6-dependent myeloma cell proliferation and permit the establishment of IL-6-dependent human myeloma cell lines.

The pattern of cytokine effect for medullar ('classic') myeloma versus extramedullary meyloma are quite different, with IL-6 being crucially involved with extramedullary plasma cells, but much less so with classic medullary plasma cells. As already noted, this may relate to effects of *Rb* mutations, but other factors, especially stromal cells, must also be involved. Whether or not the pattern of somatic mutations in the rearranged immunoglobulin genes, increased expression of *bcl-2* or aberrant signalling proteins stimulated by the IL-6 receptor are pertinent to these differences remains to be clarified.

IL-6 AND PROGNOSIS

The serum IL-6 level in patients with active MM is increased as compared to the level in those with stable MM, benign monoclonal gammopathies, or healthy donors. The highest levels are found in patients with terminal disease.[15] Also, the serum IL-6 level appears to be a strong prognostic factor.[16] Overproduction of IL-6 in MM patients is directly related to increased C-reactive protein (CRP) production in these patients.[17] CRP production by human hepatocytes in primary culture is mainly controlled by IL-6[18] and, throughout the treatment of MM patients with anti-IL-6 mAb, CRP production is blocked, but is reversible at the end of the therapy period.[19] This is evidence that CRP production is controlled by IL-6 *in vivo* and is an easy indicator of the total production of IL-6.

Based on CRP measurements, overproduction of IL-6 was found in 37 per cent of MM patients at diagnosis, particularly in patients with a high proportion of proliferating myeloma cells.[17] Moreover, these patients had a 2.3-fold shorter median survival time compared to patients with normal plasma CRP levels. Thus, overproduction of IL-6 is associated with progressive disease.

IL-6 – SITE OF PRODUCTION

IL-6 is produced predominantly or exclusively by bone marrow stromal cells (paracrine production) and not as previously suggested by the myeloma cells (autocrine production).[2] Although the view of autocrine production found some support from studies of IL-6 production in autonomously growing myeloma cell lines, i.e. the U266 cell line,[20,21] it was later shown that secondary *in vitro* mutations had occurred.[22] Later studies showed that enriched myeloma cell populations produced only small amounts of IL-6, probably due to contamination with stromal cells,[1] and subsequently studies of separated myeloma and bone marrow stromal cells showed that IL-6 mRNA was present only in the stromal cells,[23] i.e. the best evidence of a paracrine production.

CYTOKINES USING THE gp130 IL-6 TRANSDUCER CHAIN

Four cytokines were recently found to use the same transducer chain as IL-6, i.e. ciliary neurotrophic factor (CNTF), interleukin-11 (IL-11), leukemia inhibitory factor (LIF), and oncostatin M (OM).[24] These cytokines were also shown to share the major functions of IL-6, and were potent growth factors for some IL-6-dependent myeloma cell lines.[25] Their growth-inducing activity was completely abrogated by antibodies to the gp130 IL-6 transducer.[25] Thus, these growth factors may have an important role in myeloma cell growth regulation.

Other cytokines involved with myeloma growth

IL-1, IL-1β, CSF-1, TFN AND IL-4

Several investigators have indicated that populations enriched in myeloma cells produced IL-1β.[7,8] However, as stromal cells stick to myeloma cells, it is difficult to ascertain from these studies whether myeloma cells definitely produced IL-1β. Moreover, purification of myeloma cells requires *in vitro* manipulations that may lead to activation of IL-1β gene transcription *in vitro*. To assess whether myeloma cells produce IL-1β *in vivo* or not, we performed *in situ* hybridization on freshly explanted myeloma cells. A weak IL-1β gene expression was found in the bone marrow plasma cells of all eight patients studied (Klein, unpublished results) (Fig. 8.3). Interestingly, no IL-1β gene expression was found in circulating myeloma cells of patients with plasma cell leukemia or in ten human myeloma cell lines,[26] which suggests that IL-1β production by myeloma cells is not due to constitutive activation of the IL-1β gene but to activation of the myeloma cells trapped in the bone marrow environment. A strong IL-1β expression was also found in myeloid cells and polymorphonuclear cells of the bone marrow environment (Klein, unpublished observations).

By using antibodies to IL-1 or the IL-1 receptor antagonist, several groups have shown that IL-1β is the main cytokine controlling endogenous IL-6 production in short-term culture of bone marrow cells from patients with MM[7] (Klein, unpublished results). In addition, we showed that IL-1 induces production of prostaglandin E_2 (PGE_2) that controls IL-6 production (Klein, unpublished results) (Fig. 8.3).

However, IL-1 antagonists cannot completely inhibit production of endogenous IL-6, which suggests that other cytokines control IL-6 production. CSF-1, which is produced by myeloma cells,[27] could be one of these cytokines. TNF could also be involved, but its role remains to be specified.

Inhibition of IL-6 production is of potential therapeutical interest in MM.[28] Among potential

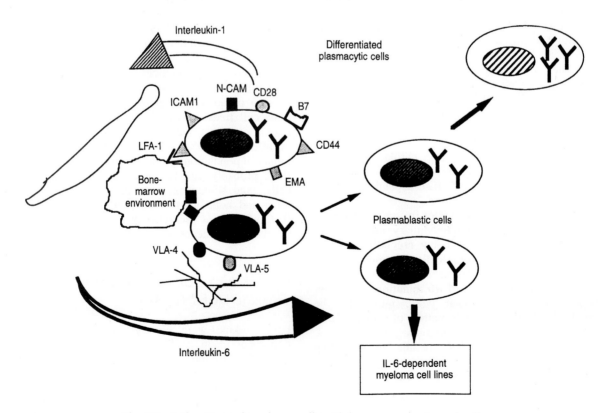

Fig. 8.3 Interactions of myeloma cells with bone marrow stromal cells.

inhibitors, interleukin-4 (IL-4) was shown to block IL-6 production and myeloma cell proliferation.[4]

IFN-α and TNF induce autocrine production of IL-6 in myeloma cell lines

Myeloma cell lines which are completely dependent on addition of exogenous IL-6 for their proliferation can be obtained reproducibly. We recently demonstrated that several cytokines (such as IFN-α and tumor necrosis factor (TNF)) can stimulate the growth of these cell lines by inducing autocrine production of IL-6 in myeloma cells and that IFN-α-dependent myeloma cell lines can thus be obtained.[29]

Interestingly, autonomously growing cell lines producing their own IL-6 as an autocrine growth factor can be established very quickly (within several weeks) from these IFN-α- or TNF-dependent cell lines. This may explain why most myeloma cell lines reported in the literature are autonomously growing ones and why some, such as the U266 cell lines, produce IL-6 as an autocrine growth factor.[20,21] The original U266 line depended on addition of exogenous IL-6 for its proliferation.[22]

It has recently been shown that IFN-α increases plateau-phase duration in patients responsive to chemotherapy.[30] These *in vivo* results do not necessarily contradict other current findings. We have shown that myeloma cells from patients with inactive disease and no proliferating myeloma cells *in vivo* did not respond to IL-6 *in vitro*.[5] However, continuous administration of INF-α in patients in early relapse might allow the rapid emergence of autocrine IL-6-producing myeloma subclones and thus worsen the disease. This issue should be addressed in future studies.

IFN-γ acts as a potent inhibitor of myeloma cell proliferation by blocking IL-6-receptor expression

We found that IFN-γ was able to inhibit IL-6-dependent myeloma cell proliferation completely in all twenty-five MM patients studied.[31] IFN-γ did not affect endogenous IL-6 production but acted directly on myeloma cells. In addition, it completely inhibited the proliferation of IL-6-dependent myeloma cell lines.[31] IFN-γ is able to downregulate expression of

the α chain (IL-6R) of the IL-6 receptor complex.[31] These results indicate the potential therapeutic use of IFN-γ in MM since this cytokine also inhibits cytokine-mediated bone resorption that is a major clinical problem in MM.

Granulocyte colony-stimulating factor (G-CSF)

The gene of the hematopoietic cytokine G-CSF is structurally homologous with the IL-6 gene.[32] Moreover, the G-CSF receptor is a dimer of two gp130 chains, which shares strong homologies with the gp130 IL-6 transducer.[32] Also, the signalling events induced by the G-CSF receptor are related to those induced by the gp130 IL-6 transducer,[33] and both G-CSF and IL-6 induce activation of the NF-IL-6 nuclear factor. These properties of G-CSF might explain that it is a potent growth factor for freshly explanted myeloma cells[34] and that G-CSF-sensitive myeloma cell lines can be established. The myeloma cell growth factor activity of G-CSF is abrogated by anti-IL-6 and anti-IL-6R antibodies, which indicates that it is mediated by IL-6. However, G-CSF does not control endogenous IL-6 production and does not increase IL-6 receptor expression. Thus, G-CSF might cooperate with IL-6, perhaps at the level of signalling proteins.

The role of G-CSF in the physiopathology of MM *in vivo* is unclear. However, the G-CSF gene is expressed in freshly obtained myeloma cells, and G-CSF is found in the serum of most patients. Also, treatment of one patient with advanced myeloma with recombinant G-CSF (1 µg/kg/day) for 7 days without any associated treatment resulted in a dramatic increase in tumor cell proliferation *in vivo*.[34] Thus, caution in using G-CSF to accelerate hematopoietic recovery after high-dose chemotherapy treatment is warranted, as recently emphasized by three studies.[35–37]

Interleukin-10 (IL-10)

Interleukin-10 (IL-10) is a potent differentiation factor of normal B cells into immunoglobulin-secreting cells.[38] Recently, it has also been shown that recombinant (r)IL-10 increases the spontaneous proliferation of freshly explanted bone marrow myeloma cells in short-term cultures.[39] This stimulatory effect is probably not mediated by IL-6 since rIL-10 inhibits endogenous IL-6 production (50 per cent) and anti-IL-6 mAb only slightly inhibits (30 per cent) IL-10-induced myeloma cell proliferation. Also, IL-10 does not suppress endogenous IFN-γ, a potent inhibitor of myeloma cell proliferation. IL-10 stimulates the proliferation of long-term growth of some IL-6-dependent human myeloma cell lines. This activity of IL-6 is unaffected by anti-IL-6 and anti-IL-6 antibodies (*see below*). Differentiation of these plasmablastic cell lines is not induced by IL-10 and the immunoglobulin production is not increased. IL-10 (< 1 pg/ml) is rarely detected in the serum of patients with MM (3 out of 100), except for patients with plasma cell leukemia[39] or solitary plasmacytoma. Thus, IL-10 is a potent IL-6-unrelated growth factor for multiple myeloma cells *in vitro* and might be involved in the pathogenesis of MM *in vivo*.

Therapeutic implications of cytokine and growth factor data

TREATMENT WITH ANTI-IL-6 ANTIBODIES

The experimental evidence that IL-6 might promote myeloma cell growth encouraged us to treat ten patients with terminal disease with murine anti-IL-6 monoclonal antibodies.[19] Anti-IL-6 mAb was given intravenously to ten patients on a daily basis for 4–60 days. A patient with malignant pleural effusion received intrapleural therapy. The dosage of anti-IL-6 mAb was chosen to give a concentration that was equivalent to the concentration shown to produce inhibition in cell cultures.

Out of 3 patients who succumbed to progressive disease within one week of treatment, 2 were evalu-

ated. The inhibition of plasmablastic proliferation was 90 per cent and 80 per cent, respectively, in the 2 patients. Among 7 patients who received anti-IL-6 intravenously for 7–60 days, complete inhibition of CRP was observed in 3 patients. A 30–50 per cent reduction of myeloma cell mass was obtained in these patients, while in the 4 patients not achieving normalization of serum CRP levels the disease progressed rapidly. Anti-IL-6 was associated with resolution of low-grade fever in all the patients; aggravation of thrombocytopenia occurred in 7 patients and mild neutropenia in 5 patients.

Immunization to the murine anti-IL-6 mAb was detected in 3 out of 9 patients.[40] The presence of human antibodies to the Fc fragment of the murine mAb in one patient was associated with dramatic relapse. In the other 2 patients, immunization was restricted to F(ab′)$_2$ fragments of the murine mAb and was not associated with treatment resistance.[40]

There may be many explanations for the lack of response to anti-IL-6 mAb in four patients. Table 8.1 illustrates the relation between response, CRP, IL-6 production, and serum IL-10 in six patients who received anti-IL-6 for 10 days or more. High IL-10 values indicate that this cytokine, and perhaps those cytokines that stimulate the same transducer chain as IL-6, i.e. CNTF, IL-11, LIF, and OM, may be responsible for the progression of the disease. Another or additional explanation is that the IL-6 level was too high to be neutralized by anti-IL-6. Recent experiments by us favor this explanation. We have shown that anti-IL-6 mAb, by preventing binding of IL-6 to cell-surface receptors, induces huge amounts of IL-6 to circulate in the form of monomeric immuno complexes.[41] These complexes are very stable and have a half-life of 3–4 days, similar to that of the free anti-IL-6 mAb.[42] Complete inhibition of IL-6 binding to its cell-surface receptor forces IL-6 to stay in the circulation. Quantification of circulating IL-6, when

Table 8.1 Daily production of interleukin-6 and serum IL-10 levels in patients treated with anti-IL-6 antibodies

	Patient 1	Patient 2	Patient 3	Patient 4	Patient 5	Patient 6
Sex/ age	F/ 54	M/ 61	M/ 59	F/ 45	F/ 56	M/ 65
Clinical status	PCL*	PCL	PCL	PCL	PCL	PCL
Duration of treatment (days)	45	60	16	16	10	12
Reduction of tumor cell mass (%)	50	30	86 (transitory)	Progression	Progression	Progression
Inhibition of CRP (%)	100	100	88	50	75	60
IL-6 production (μg/day)	1.7	4	> 47	> 46	> 165	> 341
Serum IL-10 (pg/ml)	< 1	< 1	122	480	Not done	110

* PCL: plasma cell leukemia.

inhibition is complete, is therefore a measure of the total production of IL-6. In the two responder patients, with a complete inhibition of CRP production, IL-6 production ranged from 2 to 4 μg/day, and mathematic modeling indicated that the concentration of the anti-IL-6 mAb was sufficient to neutralize this IL-6 production. In contrast, the total production of IL-6 was larger in four non-responding patients, i.e. > 46 to > 341 μg/day. In these patients, mathematical modeling predicted that the concentration of the antibody was not sufficiently high to neutralize this large IL-6 production, which could be the reason for the lack of response.

The role of IL-6 in preventing response may be further elucidated by the simultaneous injection of different anti-IL-6 antibodies. This will induce formation of polymeric immune complexes with a shorter half-life than that of the monomeric complex (1 h versus 96 h).[43] The efficiency of anti-IL-6 treatment can therefore be increased by 100-fold and *in vivo* IL-6 production will be neutralized in most patients. This confirms that IL-6 is the major myeloma cell growth factor *in vivo*.

CLINICAL AND THERAPEUTIC IMPLICATIONS OF OTHER CYTOKINES

Two categories of cytokines are growth factors for malignant plasmablastic cells *in vitro*, i.e. the cytokines stimulating the gp130 IL-6 transducer or related transducer chains and IL-10 (Fig. 8.4).

The almost complete inhibition of the spontaneous proliferation of freshly explanted myeloma cells by anti-IL-6 antibodies indicates that IL-6 is the major myeloma cell growth factor *in vitro*. In addition, this cytokine is produced in large amounts in patients with MM who have a poor prognosis. However, the other four myeloma cell growth factors activating the gp130 IL-6 transducer (CNTF, IL-11, LIF, OM) may be important at least in a subset of patients.

The recent discovery that IL-10 is a potent myeloma cell growth factor *in vitro* is intriguing.[39] Indeed, the IL-10 receptor does not use the gp130 transducer[44] and the effect of IL-10 on myeloma cells is unrelated to IL-6.[39] It could be related to one of the four cytokines activating the gp130 transducer. If this is the case, activation of the gp130 IL-6 transducer will remain the key signal for induction of proliferation of malignant plasmablastic cells by presently identified myeloma cell growth factors.

Mutations in the gp130 transducer chain or in the proteins involved in the signaling cascade stimulated by gp130 might be important oncogenic events leading to the emergence of malignant plasmablastic cell clones. We recently found mutations in the critical

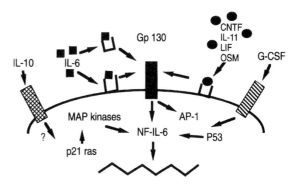

Fig. 8.4 Myeloma cell growth factors and signal transduction.

cytoplasmic domain of gp130 transducer in some tumoral samples from patients with MM.[45] Moreover, *ras* and *p53* genes are frequently mutated in patients with terminal disease[46,47] and mutated *ras* or *p53* proteins have recently been shown to increase the signaling cascade induced by gp130 activation[48,49] (Fig. 8.4).

IL-6 is overproduced in patients with MM characterized by a poor prognosis, suggesting that this cytokine is involved in MM *in vivo*. However, analysis of the treatments with anti-IL-6 mAb does not permit the conclusion that this cytokine is the same major tumor growth factor *in vivo* as it is *in vitro*. The treatment was efficient only in some patients in whom *in vivo* IL-6 production was low enough to be neutralized by the anti-IL-6 antibodies. In non-responders IL-6 production was too large to be inhibited by the anti-IL-6 antibodies, but another myeloma cell growth factor, IL-10, was produced in high amounts in these patients, and other myeloma cell growth factors which were not investigated may as well be active. It is therefore difficult to predict the relative importance of IL-6 and other myeloma cell growth factors in non-responder patients.

It has been established that IL-6 is a paracrine myeloma cell growth factor produced by the bone marrow stromal cells, mainly monocytes, and not by myeloma cells.[23,50,51] Stromal cells from MM patients can be abnormal and produce high amounts of IL-6.[52] However, *in vitro* models indicate that activation of the bone marrow environment to produce IL-6 is mediated by interacting adhesion molecules and cytokines between the tumor and its environment.[50,51] IL-1β is the main cytokine involved in this interaction.[7] CSF-1, which is produced by myeloma cells *in vitro*,[27] and TNF could also be involved. The possibility that bone marrow stromal cells from MM patients produce other myeloma growth factors (IL-10, IL-11, LIF, OM, CNTF, G-CSF) remains to be explored. Indeed, these cytokines are known to be produced by activated monocytes, the major producer of IL-6 in MM patients, and by bone cells. In a recent study, we found LIF gene expression in bone marrow cells of 2 out of 8 MM patients.[26]

The other cytokines that control myeloma cell growth *in vitro* act by inducing autocrine production if IL-6 (IFN-α or TNF) in myeloma cells, or by synergizing with IL-6 (hematopoietic cytokines), or by downregulating the IL-6 receptor (IFN-γ). Again, we have to explore whether these cytokines could interact with the other myeloma cell growth factors. The seemingly positive effects of hematopoietic cytokines and IFN-α for treatment of patients with MM are difficult to explain, considering the stimulatory effect *in vitro*. Treated patients should be included in clinical trials and followed closely. In one patient treated for 5 days with recombinant G-CSF, an 8-fold increase in proliferating plasmablastic cells was observed. Thus, although perhaps hazardous, in some patients certain cytokines (IL-6, G-CSF, IL-3, or GM-CSF) could be tried as inducers of myeloma cell proliferation to enhance subsequent chemotherapy. Other cytokines could be used as inhibitors of myeloma cell proliferation (IFN-γ).

It is essential to determine which of these cytokines are involved in the emergence of myeloma disease *in vivo* and how their production or activity *in vivo* can be antagonized. In this review, we have presented evidence clearly assessing the major role of IL-6 in the development of MM *in vivo*. The involvement of the other four cytokines able to activate the gp130 transducer chain (CNTF, IL-11, LIF, OM) needs to be rapidly clarified. Serum IL-10 is detected in patients with terminal disease. No detectable mRNA levels for GM-CSF and IL-3 were found in the bone marrow of MM patients *in vivo*, but significant levels were found for G-CSF and IL-1.[26] In addition, increased serum CSF-1 levels have been found in MM patients.[53] Thus, these cytokines (C-CSF, IL-1β, IL-10, CSF-1) could also be involved in the development of MM.

It seems essential to keep inflammatory cytokine production *in vivo* as low as possible, particularly by using CRP to monitor IL-6 production. Over the last 2 years, several agents have been shown to inhibit inflammatory cytokine production, including corticoids, estrogens, IL-4 and inhibitors of PGE$_2$, some of which are already useful in treatment of MM.[3,54] In addition, new IL-6 antagonists useful for blocking myeloma cell proliferation (humanized anti-IL-6 or anti-IL-6R antibodies, mutated IL-6, soluble receptor, natural antagonists, etc.) should soon be available. This research needs to identify which motifs of IL-6 or IL-6 receptor molecules are important in the activation of the receptor complex.[55–57] The choice of IL-6 receptor inhibitors should take into account the high level of circulating soluble IL-6 receptors in patients with MM.[58] Research on IL-6 signaling and its abnormalities in MM should also lead to therapeutical applications.

Conclusions

Much has been learned in a very short span of time (≈5 years) about cytokines and growth factors relevant to myeloma growth *in vivo*. Major activation of the so-called 'cytokine loops' with increase in serum levels occurs in patients with active (labeling index per cent > 1 per cent) usually aggressive disease. Strangely, although high levels of cytokines such as IL-6 and IL-1β are present in bone marrow, these cytokines are predominantly crucial growth factors in patients with extramedullary disease and/or plasma cell leukemia. For the majority of patients with classic myeloma of bone, the cytokine loops are more pertinent to the local bone resorptive process in the marrow than myeloma cell proliferation *per se*. The absence of excess IL-6 in the serum does not necessarily reflect the powerful interactions involving IL-6, IL-1β, M-CSF, PGE$_2$ and other factors associated with local bone destruction in the marrow as discussed by Bataille in Chapter 5. Interruption of this complex 'looping' in the marrow is clearly less influenced by systemic anti-IL-6 therapy. More effective IL-6 inhibition, perhaps combined with inhibition of other cytokines such as IL-1β or the use of biphosphonate to block bone marrow resorption may prove more effective in this setting.

Clearly it will take time to assess the intricacies of the pathophysiology of all the recently discovered cytokines plus cell–cell and cell–matrix interactions and other influences on myeloma growth and evolution. But, finally there are many new exciting avenues to explore and it seems reasonable to expect that important new therapeutic strategies will soon emerge.

References

1. Klein B, Zhang XG, Jourdan M et al. Paracrine rather than autocrine regulation of myeloma-cell growth and differentiation by interleukin-6. *Blood* 1989; **73**: 517–26.
2. Kawano M, Hirano T, Matsuda T et al. Autocrine generation and requirement of BSF-2/IL-6 for human Multiple Myelomas. *Nature* 1988; **332**: 83–5.
3. Anderson KC, Jones RM, Morimoto C et al. Response patterns of purified myeloma cells to hematopoietic growth factors. *Blood* 1989; **73**: 1915–24.

4. Herrmann F, Andreeff M, Gruss HJ et al. Interleukin-4 inhibits growth of multiple myelomas by suppressing interleukin-6 expression. *Blood* 1991; **78**: 2070–4.

5. Zhang XG, Klein B, Bataille R. Interleukin-6 is a potent myeloma-cell growth factor in patients with agressive multiple myeloma. *Blood* 1989; **74**: 11–13.

6. Zhang XG, Bataille R, Jourdan M et al. Granulocyte-macrophage colony stimulating factor synergizes with interleukin-6 in supporting the proliferation of human myeloma cells. *Blood* 1990; **76**: 2599–605.

7. Carter A, Merchav S, Silvian-Draxler I, Tatarsky I. The role of interleukin-1 and tumor necrosis factor-α in human multiple myeloma. *British Journal of Haematology* 1990; **74**: 424–31.

8. Cozzolino F, Torcia M, Aldinucci D et al. Production of interleukin 1 by bone marrow myeloma cells. *Blood* 1989; **74**: 380–7.

9. Zhang XG, Bataille R, Wijdenes J, Klein B. Interleukin-6 dependence of advanced malignant plasma cell dyscrasias. *Cancer* 1992; **69**: 1373–6.

10. Facon T, Lai JL, Nataf E et al. Improved cytogenetic analysis of bone marrow plasma cells after cytokine stimulation in multiple myeloma: a report on 46 patients. *British Journal of Haematology* 1993; **84**: 743–8.

11. Miyajima A, Hara T, Kitamura T. Common subunits of cytokine receptors and the functional redundancy of cytokines. *Trends in Biochemical Science* 1992; 378–82.

12. Kobayashi M, Tanaka J, Imamura M et al. Up-regulation of IL-6 receptors by IL-3 on a plasma cell leukemia cell line which proliferates dependently on both IL-3 and IL-6. *British Journal of Haematology* 1993; **83**: 535–8.

13. Zhang XG, Gaillard JP, Robillard N et al. Reproducible obtaining of human myeloma cell lines as a model for tumor stem cell study in human multiple myeloma. *Blood* 1994; **83**: 3654–63.

14. Portier M, Lees D, Caron E et al. Up-regulation of interleukin (IL)-6 receptor gene expression *in vitro* and *in vivo* in IL-6 deprived myeloma cells. *FEBS* 1992; **302**: 35–8.

15. Bataille R, Jourdan M, Zhang XG, Klein B. Serum levels of interleukin-6, a potent myeloma cell growth factor, as a reflect of disease severity in plasma cell dyscrasias. *Journal of Clinical Investigation* 1989; **84**: 2008–11.

16. Reibnegger G, Krainer M, Herold M et al. Predictive value of interleukin-6 and neopterin in patients with multiple myeloma. *Cancer Research* 1991; **51**: 6250–3.

17. Bataille R, Boccadoro M, Klein B et al. C-Reactive Protein and beta-2 microglobulin produce a simple and powerful Myeloma Staging System. *Blood* 1992; **80**: 733–7.

18. Castell JV, Gomez-Lechon MJ, David M et al. Acute phase response of human hepatocytes: regulation of acute phase proteins synthesis by interleukin-6. *Hepatology* 1990; **12**: 1170–85.

19. Klein B, Wijdenes J, Zhang XG et al. Murine anti-interleukin-6 monoclonal antibody therapy for a patient with plasma cell leukemia. *Blood* 1991; **78**: 1198–204.

20. Levy Y, Tsapis A, Brouet JC. Interleukin-6 antisense oligonucleotides inhibit the growth of human myeloma cell lines. *Journal of Clinical Investigation* 1991; **88**: 696–9.

21. Schwab G, Siegall CB, Aarden LA et al. Characterization of an interleukin-6-mediated autocrine loop in the human multiple myeloma cell line, U266. *Blood* 1991; **77**: 587–93.

22. Nilsson K, Jernberg H, Pettersson M. IL-6 as a growth factor for human myeloma cells – a short overview. *Current Topics in Microbiology and Immunology* 1990; **166**: 3–12.

23. Portier M, Rajzbaum G, Zhang XG et al. *In vivo* interleukin-6 gene expression in the tumoral environment in multiple myeloma. *European Journal of Immunology* 1991; **21**: 1759–62.

24. Kishimoto T, Akira S, Taga T. Interleukin-6 and its receptor – A paradigm for cytokines. *Science* 1992; **258**: 5082–4.

25. Zhang XG, Gu ZJ, Lu ZY et al. Ciliary neurotrophic factor, interleukin 11, leukemia inbitory factor, and oncostatin M are growth factors for human myeloma cell lines using the interleukin 6 signal transducer gp130. *Journal of Experimental Medicine* 1994; **179**: 1337–42.

26. Portier M, Zhang XG, Ursule E et al. Cytokine gene expression in human multiple myeloma. *British Journal of Haematology* 1993; **85**: 514–20.

27. Nakamura M, Merchav S, Carter A et al. Expression of a novel 3.5-kb macrophage colony-stimulating factor transcript in human myeloma cells. *Journal of Immunology* 1989; **143**: 3543–7.

28. Klein B, Lu ZY, Gaillard JP et al. Inhibiting IL-6 in Human Multiple Myeloma. *Current Topics in Microbiology and Immunology* 1992; **182**: 237–44.

29. Jourdan M, Zhang XG, Portier M et al. IFN-alpha induces an autocrine production of IL-6 in myeloma cell lines. *Journal of Immunology* 1991; **12**: 4402–7.

30. Mandelli F, Avvisati G, Amadori A et al. Maintenance treatment with recombinant interferon-α 2b in patients with multiple myeloma responding to conventional induction chemotherapy. *New England Journal of Medicine* 1990; **322**: 1430–4.

31. Portier M, Zhang XG, Caron E et al. Gamma interferon in multiple myeloma: inhibition of interleukin-6 (IL-6)-dependent myeloma cell growth and downregulation of IL-6 receptor expression *in vitro*. *Blood* 1993; **81**: 3076–82.

32. Larsen A, Davis T, Curtis BM et al. Expression cloning of a human granulocyte colony-stimulating factor receptor: A structural mosaic of hematopoietin receptor, immunoglobulin, and fibronectin domains. *Journal of Experimental Medicine* 1990; **72**: 1559–65.

33. Ziegler SF, Bird TA, Morella KK et al. Distinct regions of the human granulocyte colony stimulating factor receptor cytoplasmic domain are required

for proliferation and gene induction. *Molecular and Cell Biology* 1993; **43**: 2384–90.

34. Boiron JM, Porteir M, Lu ZY et al. G-CSF stimulates myeloma cell proliferation *in vitro* and *in vivo* through an IL-6 related mechanism. *Blood* 1991; **78** (suppl. 1): 385a.

35. Sawamura M, Sakura T, Miyawaki S. Exacerbation of monoclonal gammopathy in a patient treated with G-CSF. *Annals of Internal Medicine* 1993; **118**: 318.

36. Vora AJ, Hoc TC, Peel J, Greaves M. Use of granulocyte colony-stimulating factor (G-CSF) for mobilizing peripheral blood stem cells: risk of mobilizing clonal myeloma cells in patients with bone marrow infiltration. *British Journal of Haematology* 1994; **86**: 180–2.

37. de la Rubia J, Bonanad S, Palau J et al. Rapid progression of multiple myeloma following G-CSF mobilization. *Bone Marrow Transplantation* 1994; **14**: 475–7.

38. Rousset F, Garcia E, Defrance T et al. Interleukin 10 is a potent growth and differentiation factor for activated human B lymphocytes. *Proceedings of the National Academy of Sciences* 1992; **89**: 1890–5.

39. Lu ZY, Zhang XG, Rodriguez C et al. Interleukin-10 is a proliferation factor but not a differentiation factor for human myeloma cells. *Blood* (in press).

40. Legouffe E, Liautard J, Gaillard JP et al. Human anti-mouse antibody response to the injection of murine monoclonal antibodies against IL-6. *Clinical and Experimental Immunology* 1994; **98**: 323–9.

41. Lu ZY, Brochier J, Wijdenes J et al. High amounts of circulating interleukin (IL)-6 in the form of monomeric immune complexes during anti-IL-6 therapy. Towards a new methodology for measuring overall cytokine production in human *in vivo*. *European Journal of Immunology* 1992; **22**: 2819–24.

42. Lu ZY, Brailly H, Rossi JF et al. Overall interleukin-6 production exceeds 7 mg/day during Gram-negative sepsis. *Cytokine* 1993; **5**: 578–82.

43. Montero-Julian FA, Klein B, Gautherot E, Brailly H. Pharmacokinetic study of anti-interleukin-6 (IL-6) therapy with monoclonal antibodies: enhancement of IL-6 clearance by cocktails of anti-IL-6 antibodies. *Blood* (in press).

44. Liu Y, Wei SHY, Ho ASY et al. Expression cloning and characterization of a human IL-10 receptor. *Journal of Immunology* 1994; **152**: 1821–9.

45. Rodriguez C, Theillet C, Portier M et al. Molecular analysis of the IL-6 receptor in human multiple myeloma, an Il-6-related disease. *FEBS Lett.* 1994; **341**: 156–61.

46. Mazars GR, Portier M, Zhang XG et al. Mutations of the *p53* gene in human myeloma cell lines. *Oncogene* 1992; **7**: 1015–20.

47. Portier M, Molès JP, Mazars GR et al. *p53* and *RAS* gene mutations in multiple myeloma. *Oncogene* 1992; **7**: 2539–43.

48. Nakajima T, Kinoshita S, Sasagawa T et al. Phosphorylation at threonine-235 by a *ras*-dependent mitogen-activated protein kinase cascade is essential for transcription factor NF-IL6. *Proceedings of the National Academy of Sciences* 1993; **90**: 2207–11.

49. Santhanam U, Ray A, Sehgal PB. Repression of the interleukin-6 gene promoter by *p53* and the retinoblastoma susceptibility gene product. *Proceedings of the National Academy of Sciences* 1991; **88**: 7605–9.

50. Caligaris-Cappio F, Bergui L, Gregoretti MG et al. Role of bone marrow stromal cells in the growth of human multiple myeloma. *Blood* 1991; **77**: 2688–93.

51. Uchiyama H, Barut BA, Mohrbacher AF et al. Adhesion of human myeloma-derived cell lines to bone marrow stromal cells stimulates interleukin-6 secretion. *Blood* 1993; **82**: 3712–20.

52. Kaisho T, Oritani K, Ishikawa J et al. Human bone marrow stromal cell lines from myeloma and rheumatoid arthritis that can support murine pre-B cell growth. *Journal of Immunology* 1992; **149**: 4088–95.

53. Janowska-Wieczorek A, Belch AR, Jacobs A et al. Increased circulating colony-stimulating factor-1 in patients with preleukemia, leukemia and lymphoid malignancies. *Blood* 1991; **77**: 1796–803.

54. Klein B, Lu ZY, Bataille R. Clinial application of IL-6 inhibitors. *Research in Immunology* 1992; **143**: 774–7.

55. de Hon FD, Ehlers M, Rose-John S. Development of an interleukin (IL) 6 receptor antagonist that inhibits IL-6-dependent growth of human myeloma cells. *Journal of Experimental Medicine* 1994; **180**: 2395–400.

56. Savino R, Lahm A, Salvati AL et al. Generation of interleukin-6 receptor antagonists by molecular-modeling guided mutagenesis of residues important for gp130 activation. *EMBO Journal* 1994; **13**: 1357–67.

57. Yawata H, Yasukawa K, Natsuka S et al. Structure–function analysis of human IL-6 receptor: dissociation of amino acid residues required for IL-6-binding and for IL-6 signal transduction through gp130. *EMBO Journal* 1993; **12**: 1705–12.

58. Gaillard JP, Bataille R, Brailly H et al. Increased and highly stable levels of functional soluble interleukin-6 receptor in sera of patients with monoclonal gammopathy. *European Journal of Immunology* 1993; **23**: 820–4.

CHAPTER 9

Monoclonal gammopathies of undetermined significance: Their relation to multiple myeloma

PHILIP R GREIPP, JOHN A LUST

Introduction	83	Critique	94
Review of differences between MGUS and MM	84	Conclusion	94
Summary	93	References	94

Introduction

Monoclonal gammopathy of undetermined significance (MGUS) is a disorder in which a monoclonal protein (M-protein) is detected in the serum or urine by electrophoresis and immunoelectrophoresis or immunofixation. The frequency of monoclonal gammopathy increases with age; among patients 70 years of age or older, 3 per cent have an M-protein in the serum. The M-protein is produced by monoclonal plasma cells in the marrow. These plasma cells proliferate slowly but do not accumulate or exhibit malignant behavior. Patients are asymptomatic and have stable M-protein measurements. The term 'MGUS' has been adopted because some patients develop serious disease, most often multiple myeloma (MM).

MM is the malignant expression of IgG, IgA, IgD, IgE, or free immunoglobulin light-chain monoclonal gammopathy. Marrow plasma cells accumulate, and patients die of progressive disease related to bone destruction, renal failure, anemia, and infection or bleeding. Accurate differentiation of MM from MGUS is important because patients with MGUS do not require treatment; unnecessary treatment leads to unacceptable expense and complications. Some patients develop myelodysplastic syndrome or acute leukemia. Conversely, failure to recognize and treat overt MM properly can allow unchecked progression of the disease.

A clear understanding of the differences between MGUS and MM is needed (1) to improve diagnosis, (2) for better patient counseling, (3) for accurate classification of patients entering clinical trials, and (4) to help develop new therapies targeted at biological differences between benign and malignant disease. The purposes of this chapter are to review clinical, biological, and molecular differences between MGUS and MM and to focus on new approaches to diagnose these diseases more

Table 9.1 Criteria for monoclonal gammopathy of undetermined significance (MGUS) and smoldering multiple myeloma (SMM)

MGUS[a]	SMM[a]
Serum M-protein (usually <3 g/dl)	Serum M-protein (usually >3 g/dl)
Fewer than 10% plasma cells, and no aggregates on biopsy	10% or more marrow plasma cells or aggregates on biopsy
No anemia, renal failure, or hypercalcemia	No anemia, renal failure, or hypercalcemia attributable to myeloma
Ancillary tests negative[b]	Ancillary tests negative[b]
Bone lesions absent on radiographic bone survey[c]	Bone lesions absent on radiographic bone survey[c]
Bone marrow contains <10% plasma cells without aggregates on biopsy	Bone marrow contains <10% plasma cells without aggregates on biopsy
Bone marrow plasma cell LI <1.0%	Bone marrow plasma cell LI <1.0%
Plasmablasts absent	Plasmablasts absent

[a] Patients with MGUS and SMM must not have solitary plasmacytoma, amyloidosis, or light-chain deposition disease.
[b] Normal β_2-M, absence of circulating isotype-specific plasma cells, peripheral blood B-cell LI of less than 0.5%, light-chain isotype suppression, urinary light chain less than 500 mg/24 h, and stable M-protein level in the serum or urine during follow-up.
[c] Computed tomography or magnetic resonance imaging may be needed to rule out skeletal lesions.

accurately. Emphasis is placed on recent studies. Schemata for the differential diagnosis of MGUS, smoldering multiple myeloma (SMM), and MM are shown in Tables 9.1 and 9.2.

In addition to MGUS, several syndromes are associated with monoclonal gammopathy (Table 9.3). The clinician involved in the care of patients with monoclonal gammopathy should keep these in mind, but it is beyond the scope of this chapter to review these syndromes in detail.

Review of differences between MGUS and MM

MGUS: RELATION TO MM

Clinical observation suggests that MGUS and MM are two discrete phases in the same disease process. Change from the benign-behaving MGUS to overtly malignant MM is not uncommon. Seventeen per cent of patients with MGUS develop overt MM.[1] The stable MGUS phase may be long (10 years or more), but the transition to MM often occurs in less than 1 year. The transition from MGUS to MM is discrete, so discrete biological events probably cause it.

Most patients with MM probably have undetected prior MGUS. In a retrospective study of patients with MM in Olmsted County, Minnesota, 58 per cent had prior MGUS or plasmacytoma.[2] Because of the retrospective nature of the study, many of the patients with the diagnosis of MM did not have prior studies to detect MGUS.

Table 9.2 Diagnosis of multiple myeloma[a]

M-protein present in serum or urine
10% or more marrow plasma cells or aggregates on biopsy
Ancillary findings (one or more)[b]; must not be attributable to another cause
Anemia
Lytic lesions[c] (osteoporosis satisfies if there are 30% or more plasma cells in marrow)
Bone marrow LI >1%
Renal insufficiency (not due to adult-acquired Fanconi syndrome or light-chain deposition disease)
Hypercalcemia

[a] Patients with MM may have associated amyloidosis or light-chain deposition disease.
[b] Increased β_2-M level, peripheral blood B-cell LI>0.5, circulating monoclonal isotype-specific plasma cells, and loss of light-chain isotype suppression (LCIS).
[c] Computed tomography or magnetic resonance imaging for suspected skeletal lesions.

CLINICAL DIFFERENCES

Differences between MGUS and MM provide clues to the underlying biological changes. Obvious differences include anemia, bone lesions, and renal failure; patients with MM also have a higher percentage of plasma cells, greater amounts of Bence Jones proteinuria, and a higher plasma cell labeling index (PCLI), indicating a higher proliferative rate.

A focus for study of the difference between MGUS and MM is a transitional state called smoldering

Table 9.3 Miscellaneous monoclonal gammopathies[a]

Solitary plasmacytoma of bone (SPB) and extramedullary plasmacytoma (EMP)
    Biopsy-proven plasmacytoma (medullary or extramedullary)
    Exclusion of MM
Idiopathic Bence Jones proteinuria
    Free light chain in the urine >1g/24 h
    Exclusion of MM or amyloidosis
Systemic amyloidosis[b][c]
    M-protein in the serum or urine and tissue confirming amyloidosis
Light-chain deposition disease[c]
    Renal glomerular deposition and excretion of monoclonal light chain (often λ)
Adult-acquired Fanconi's syndrome[c]
    Renal tubular deposition and excretion of monoclonal light chain (usually κ)
Scleromyxedema[c]
    Mucinous deposition in the skin and M-protein in the serum (usually IgGλ)
Monoclonal gammopathy with peripheral neuropathy[c]
    IgM M-protein; may occur with IgG and IgA also
Osteosclerotic myeloma
    Single or multiple osteosclerotic bone lesions
    Usually a λ-isotype M-protein
    May have associated plasma cell dyscrasia with polyneuropathy, organomegaly, endocrinopathy, M-protein and skin
        changes (POEMS) or angiocentric follicular lymph node hyperplasia or Castleman's disease
Heavy-chain disease (γ, α, μ)
    Free heavy chain in the urine in γ; heavy and light chain in μ diverse clinical and pathologic presentations
Monoclonal IgM-associated lymphoproliferative disease or lymphoma
    Includes Waldenström's macroglobulinemia
Cold agglutinin disease
    IgM-κ M-protein on erythrocytes
Type I and II monoclonal cryoglobulinemia
    Type I IgG, IgA, or IgM M-protein
    Type II IgM, or IgG M-protein immune complex with polyclonal immunoglobulin
Systemic capillary leak syndrome
    Usually an IgGκ M-protein with intermittent edema and effusions
Acquired C_1 esterase-inhibitor deficiency
    Usually IgM M-protein with angioedema

[a] MM must be excluded initially but may develop during follow-up, especially in patients with SPB or idiopathic Bence Jones proteinuria.
[b] Serum amyloid A and familial (transthyretin) amyloidosis and senile or localized amyloidosis must be excluded.
[c] Typical MM infrequently may be associated with these conditions.

multiple myeloma (SMM).[3] Patients with SMM do not require treatment because they have not developed anemia, bone lesions, renal failure, or other evidence of progressive disease. Indeed, in these patients the disorder more closely resembles MGUS than MM because they can remain in this transitional state for many years without treatment. Patients with SMM may be an especially fruitful group in which to study the biological changes that take place as patients develop overt MM because (1) a higher percentage of patients with SMM eventually develop overt myeloma than do patients with MGUS and (2) most patients with SMM who develop MM do so in a shorter time than patients with MGUS.

No single test distinguishes MM from MGUS and SMM. We must rely on a combination of clinical and laboratory observations; in the final analysis, all patients with MGUS and SMM must be followed indefinitely with serial M-protein measurements to watch for the development of myeloma or other serious disease. Despite our knowledge of clinical differences between MGUS, SMM, and MM, we do not understand the fundamental biological mechanisms causing the transition. Through better understanding of myeloma biology, we may be able to discriminate better clinically between MGUS, SMM, and MM.

BIOLOGICAL DIFFERENCES: LITERATURE REVIEW

Biological differences between MGUS and MM are just beginning to be elucidated. Investigators have demonstrated kinetic differences between the plasma cells of MGUS and those of MM. Cytokines produced by marrow stromal cells and by the myeloma cells themselves, including interleukin–6 (IL–6), mediate this proliferation. Changes in the phenotype of plasma cells, including cell adhesion markers, have been reported. Plasma cell morphological differences have been observed, and oncogenes have been implicated in the increased proliferation of plasma cells. Immunofluorescence and molecular techniques have shown peripheral blood involvement by malignant plasma cells and precursor B cells in myeloma. Failure of B-cell regulation by T cells and natural killer (NK) cells may be a factor. Loss of apoptosis, or programmed cell death, may allow accumulation of plasma cells. Finally, serum markers that increase during development of MM may be secondary manifestations of underlying changes.

CELL KINETIC DIFFERENCES

In a comprehensive assessment of plasma cell morphological and cell kinetic changes in MM compared with those in MGUS,[4] MM plasma cells showed significantly higher nucleolar size and a higher PCLI (Fig. 9.1). The weight of opinion is that plasma

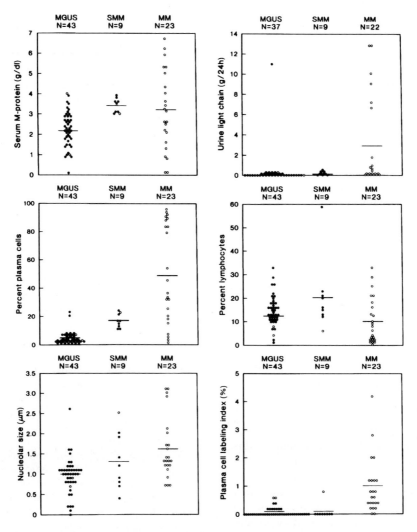

Fig. 9.1 Distribution and mean values (horizontal bars) for clinical, morphological, and cell kinetic parameters in MGUS, SMM, and MM. Open circles indicate patients with MM or patients initially classified as having MGUS or SMM who developed overt MM within 6 months. (Reproduced from ref. 4, *Blood* 1983, 62:166–71, with permission.)

Table 9.4 Mean values for clinical, morphological, and cell kinetic parameters in multiple myeloma (MM), monoclonal gammopathy of undetermined significance (MGUS), and smoldering multiple myeloma (SMM)

Parameter	MM (23 patients)	MGUS (43 patients)	SMM (9 patients)
Serum M-protein (g/dl)	3.2	2.2****	3.4
Urine M-protein (g/24 h)	2.9	0.3**	0.2
Marrow plasma cells (%)	49	5*	17****
Marrow lymphocytes (%)	10	15***	21****
Nucleolar size (μm)	1.6	1.0*	1.3
Grade	1.6	1.1*	1.3
Asynchrony	0.5	0.3***	0.5
Plasma cell labeling index (%)	1.0	0.1*	0.1**

Significant difference from MM value (Kruskal–Wallis test; chi-square approximation): $*p < 0.0001$; $**p < 0.001$; $***p < 0.01$; $****p < 0.05$.

cells in MM are less differentiated and have a higher proliferate rate. The PCLI was the most consistent difference and best discriminated a group of patients with MM from groups with MGUS and SMM (Table 9.4). Published studies confirming the diagnostic value of the PCLI in differentiating MM from MGUS are listed in Table 9.5.[4–10]

Both Ki–67 nuclear staining, which is an index of growth fraction, and BrdUrd PCLI, a measure of the number of plasma cells in DNA S-phase, effectively discriminated between early-stage MM and MGUS in a study of 16 patients with MGUS and 49 with MM (43 treated, 6 relapsed).[11] A high PCLI and an abnormal κ:λ ratio manifest by an excess of B cells with the same surface light chain as that expressed by the malignant plasma cells distinguished all patients with MM from those with SMM in another study. Low PCLI and normal B cell κ:λ ratios were characteristic of MGUS.[8] In a later study by the same investigators, discriminant analysis showed statistically significant differences in κ:λ ratio, percentage of bone marrow plasma cells, PCLI, CD3, CD4, and CD8 lymphocyte absolute values between patients with MM and those with MGUS.[9]

The diagnostic cutoff for a high PCLI using the BrdUrd PCLI is only about 1 per cent. A more sensitive measure of cell proliferation is needed. Such a measure could allow more sensitive detection of progression to MM and measurement of the effects of cytokines such as IL–6 on myeloma cells and myeloma cell subsets including precursor cells. Candidates are proliferating cell nuclear antigen (PCNA),[12] cyclin D1, DNA S-phase by flow cytometry, and Ki–67 nuclear immunostaining. Cyclin D1 is positive in late G_1-phase of the cell cycle. PCNA and DNA staining measure the number of cells in S-phase of the cell cycle, and Ki–67 measures the growth fraction and percentage of cells going through the cell cycle at one time. Assays for proliferation like PCNA, cyclin D1, DNA S-phase flow cytometry, and Ki–67 techniques still need to be better optimized and standardized.

PHENOTYPIC DIFFERENCES

CD56 (N-CAM), a cell-adhesion molecule, is strongly expressed in most myeloma plasma cells (43 of 55 patients).[13] In a confirmatory study, CD56 expression in high density was present in 43 of 57 patients with untreated MM but in none of 23 patients with MGUS.[14] The observation that none had strong CD56 reactivity was confirmed in another study of 23 patients with MGUS.[8] In another study, however, a higher percentage of CD19+/CD56− plasma cells in MGUS was attributed to residual phenotypically normal plasma cells.[15]

It is attractive to suggest that the acquisition of adhesion molecules such as CD56 marks the transition of MGUS to MM. However, the data are not in agreement. A study of plasma cells from the bone marrow of 45 patients with MM and 8 patients with MGUS by another group of investigators revealed no differences in the expression of CD56 in MGUS and MM.[10] CD56 (N-CAM) did not differentiate between MGUS and MM in another study.[16] Our studies were similarly negative. We found a similar percentage of MM and MGUS cases, 55 per cent and 67 per cent, respectively, expressing CD56 strong staining. These investigative discrepancies need to be resolved to understand the role of cell-adhesion markers, particularly CD56, in the transition of MGUS to MM.

In a study of other plasma cell-adhesion molecules, four MGUS cases were all negative for LFA–3; of 18 myeloma cases and normals, 12 were positive for the adhesion molecule LFA–3.[16] Adhesion markers, LFA–1 (CD11a), CD54 (I-CAM–1), and CD44

Table 9.5 The diagnostic value of the bone marrow plasma cell labeling index

Test	First author	Year	No. of cases	Mean LI % MGUS/ SMM	MM	Discriminant %	*p*-value
[³H] Tdr LI	Greipp[4]	1983	75	0.1/0.1	1.0	0.4	<0.001[a] <0.001[b]
[³H]Tdr LI	Boccadoro[5]	1984	52	NA	NA	1.0	<0.001
BrdUrd LI	Greipp[7]	1987	52	0.3/–	1.4	0.8	<0.002
BrdUrd LI	Boccadoro[6]	1987	43	0.3/–	1.4	1.0	0.004[c]
BrdUrd LI	Büchi[8]	1990	43	0.6/0.8	2.5	1.0	<0.05
BrdUrd LI	Girino[11]	1991	59	0.3/–	1.9	–	<0.05
BrdUrd LI	Ahsmann[17]	1992	28	0.4/–	1.4	–	–
BrdUrd LI	Leo[10]	1992	53	0.3/–	1.4	–	–
BrdUrd LI	Büchi[9]	1993	47	0.6/0.8	2.4	1.0	<0.05

[a] For differences between MGUS and MM.
[b] For differences between SMM and MM.
[c] For differences between MGUS and Stage I multiple myeloma (continuous variable).

(H-CAM), however, did not distinguish between MM and MGUS. In another study, however, results for LFA–1 were positive; LFA–1 (CD11a) was increased in active MM but not in MGUS, normal individuals, or patients with MM in a non-active phase of their disease.[17] These authors found a strong correlation between LFA–1 expression and PCLI. As with CD56, results are not concordant, and further studies are warranted.

Other myeloma cell surface markers have been tested, including antibodies selective for plasma cell surface antigens and other markers that are aberrantly expressed on plasma cells. In a study of bone marrow biopsies, plasma cell MB–2 antigen expression was higher in MM than MGUS.[18] High-intensity staining MB–2 plasma cells were noted on only 3 of 41 biopsy specimens from patients with MGUS but in 54 of 59 biopsy specimens from patients with MM. Strong multidrug resistance (MDR) gene expression was found in plasma cells of 32 of 38 untreated MGUS cases, in 33 of 105 untreated MM cases, and none of 10 normal samples.[14] Plasma cells from the bone marrow of 45 patients with MM and 8 patients with MGUS showed no differences in the expression of CD38 and CD54. Co-expression of B-cell antigens CD19 and CD20, and of the myeloid antigen CD33, was uncommon and observed only in MM. Co-expression of the pre-B-cell antigen CD10 and of the monocyte antigen CD14 occurred rarely in MM plasma cells but not in MGUS.[10]

Further study of adhesion molecules, surface markers, MDR, and the molecular and proliferative characteristics of cells bearing these changes are needed to elucidate the role of phenotypic changes in the pathogenesis of myeloma. Investigators must use caution in interpreting the phenotype of plasma cells in patients with MGUS because the residual phenotypically normal plasma cells can dilute the results, especially in MGUS, in which the clonal cells are fewer. At this time, we cannot recommend phenotyping to reliably distinguish MM from MGUS.

CYTOKINES AND CYTOKINE RECEPTORS

Elevated levels of IL–6 were detected in 42 per cent of 47 patients with MM and 13 per cent of 24 patients with MGUS.[19] IL–6 levels were higher in patients with active and advanced-stage disease than in patients with Stage I MM, MGUS, or plateau-phase MM. On radioimmunoassay, IL–6 levels correlated with neopterin, tumor necrosis factor α, β_2-microglobulin (β_2-M), and hemoglobin levels.[20] IL–6 levels in MM were also significantly correlated with bone marrow plasmacytosis, LDH, β_2-M, serum calcium, hemoglobin, and features of aggressive disease.[19]

In another study, measurable serum IL–6 was paradoxically higher in 12 patients with MGUS and 22 with Sjögren's syndrome than in controls, 9 patients with chronic lymphocytic leukemia, and 16 patients with MM.[21] These unexpectedly high results in MGUS are not easily explained. Standardization of IL–6 assays is needed before conclusions can be drawn about their usefulness in distinguishing MM from MGUS.

Myeloma marrow supernatant is rich in cytokines; there is increased secretion of IL–6 and interleukin–8 (IL–8) by bone marrow stromal cells but only low-level production by normal or MGUS bone marrow stromal cells.[22] Among marrow stromal cells, the osteoblast is an important producer of IL–6. Although osteoclasts are the important mediators of lytic bone lesions in myeloma, coincident osteoblastic recruitment correlated with measured increase in bone resorption in 16 patients with early MM and 10 patients with MGUS.[23] This stimulation of osteoblasts was observed before the development of bone lesions. Stimulated osteoblasts produce large amounts of IL–6, a potent myeloma cell growth factor. Among marrow stromal cells, osteoblasts could play a pivotal role early in the transition from MGUS to MM.

In a study of 30 healthy individuals, 32 patients with MGUS, 20 patients with early MM, and 54 patients with overt MM, soluble serum IL–6 receptor (sIL–6R) levels were increased similarly in MGUS and MM – 51 per cent and 44 per cent, respectively. Increases of 116 per cent were observed in patients with overt MM.[24] Increases in sIL–6R were independent of PCLI and β_2-M. The elevated sIL–6R level is unchanged, even after elimination of myeloma cells by the conditioning treatment used for stem cell transplant. This finding suggests stable production and tight regulation of sIL–6R from a regulatory cell, not just myeloma cells.

Proportions of interleukin–2 (IL–2) secreting cells are higher in MM than in MGUS and stable MM.[25] Serum levels of IL–2 are increased in MM compared with controls. IL–2 levels are highest in patients with stable MM.[26] The role of IL–2 in possibly preventing progression from MGUS to MM is not well understood.

Interleukin–1 (IL–1) is a potent bone-resorbing cytokine that probably contributes most to the development of lytic bone disease in MM. Myeloma cells secrete IL–1.[27] Amounts of IL–1 produced by marrow plasma cells are increased in MM compared with MGUS.[28] IL–1β gene expression has been identified in freshly separated myeloma cells.[29] IL–1 transfected plasmacytoma cell lines produce metastatic bone lesions in syngeneic mice compared with untransfected B cells, which do not induce bone abnormalities.[30] IL–1 also induces adhesion molecules such as I-CAM and CD44 on the surface of mouse plasmacytoma cells.[31] In human myeloma, a similar mechanism may be operative because adhesion molecules such as VLA–4, CD54, CD56, and other surface molecules may be expressed in increased numbers on monoclonal plasma cells from patients with MM.[13,32–35] The weight of evidence suggests that IL–1 is a critical early step in the transition from MGUS to MM.

Further evidence of the role of IL–1 comes from studies that show that IL–6 produced by fibroblasts, macrophages, T lymphocytes, and other marrow stromal cells can be induced by IL–1.[36] Carter and colleagues[37] showed that myeloma cells are able to induce IL–6 production in marrow stromal cells through endogenous-released IL–1. This IL–6 stimulatory activity is completely abrogated by an antibody to an IL–1β. Myeloma cells may, by producing IL–1, enslave marrow stromal cells to produce for themselves the critical growth factor IL–6. Measuring IL–1β expression by plasma cells may provide a useful way to distinguish myeloma from MGUS.

PLASMA CELL MORPHOLOGICAL DIFFERENCES

In a study of 295 patients with MGUS and 266 with MM in a multicenter trial, MM differed from MGUS by percentage of bone marrow plasma cells, a shift from plasmacytic to plasmablastic cytology, an increase in bone marrow cellularity and fibrosis, and a change in bone marrow infiltration becoming diffuse rather than interstitial. There was a concomitant decrease in residual hematopoiesis and an increase in osteoclasts.[38] Those observations confirm our own studies[4] and suggest that the development of morphological immaturity and dedifferentiation sometimes marks the change from MGUS to MM (Fig. 9.1). Changes to an immature plasma cell morphology will not be measurable in all cases of overt MM, because MM plasma cells are commonly mature in appearance.[39] Cytological criteria are not adequate for distinguishing MM from MGUS, especially 'mature MM'.

ONCOGENES

The p21 protein (coded by the H-*ras* oncogene) is increased on plasma cells of patients with MM compared with those who have MGUS, especially with advanced (Stage III) disease; p21 positivity has been closely correlated with bone marrow plasma cell infiltration.[40] K-*ras* p21 protein fluorescence is found in aneuploid myeloma cells.[41] K-*ras* and N-*ras* oncogene mutations are found in approximately 30 per cent of patients with myeloma.[42] H-*ras* and N-*ras* transfected into human B lymphoblasts lead to malignant transformation and terminal differentiation into plasma cells.[43] Mutations involving codons 12, 13, or 61 of *ras* genes were identified in 3 of 12 patients with MM and none of 3 patients with MGUS.[44] Although *ras* mutations have not been observed in MGUS, their frequency in MM is low, and they are found mostly in aggressive disease. Such mutations are not likely to help differentiate MM from MGUS clinically.

Mutated *p53* is rare in freshly explanted myeloma cells but is common in human myeloma cell lines.[29] Mutated *p53* genes may be more common in patients with a terminal phase of myeloma. *ras* and *p53* mutations probably occur *after* the initial events that cause the development of MM from MGUS; whether they are associated with increased myeloma cell proliferation and more aggressive disease is being investigated. Mutated c-*myc* is rare in myeloma, but it is critical to the development of myeloma in mice.

Deleted *rb–1* gene was present by fluorescence *in situ* hybridization (FISH) in 12 of 23 patients with myeloma; however, a karyotype abnormality of chromosome 13 where *rb–1* resides was found in only 4 of 23 patients.[45] This difference illustrates the potential advantages of FISH over standard karyotyping. Because both *p53* and *rb–1* function as transcriptional repressors of IL–6 gene expression,[46] loss of *p53* and *rb–1* function could result in increased IL–6 production by myeloma cells (autocrine cell growth). Such patients may be expected to have a high level of proliferation and a poor prognosis.

CIRCULATING MYELOMA CELLS AND PRECURSORS

The ability to grow monoclonal cytoplasmic immunoglobulin (cIg)-positive plasma cells from T cell-depleted non-adherent cells in the presence of interleukin–3 (IL–3) and IL–6 is greater in MM (15 of 27 evaluable patients) than MGUS (0 of 3 patients).[47] This suggests a greater number of myeloma cells or myeloma cell precursors in the blood of MM than in MGUS. Southern blot analysis has shown an increased peripheral blood involvement in MM compared with MGUS.[48]

B cells including CD5 B lymphocytes are decreased in patients with active MM compared with MGUS and stable MM.[49] In contrast, late-stage B cells expressing CD19, CD20, PCA–1, and CD45RO have been reported to be increased in MM and MGUS.[50,51] Most other investigators are unable to find a consistent increase in CD19-positive cells in peripheral blood. These authors also found abnormal B cells defined by CD45RO, CALLA, and PCA–1 positivity more frequently in MM than MGUS. The role of abnormal B cells, bearing markers of late-stage B-cell development, needs to be clarified.

Comparing 18 patients with MGUS and 52 patients with MM, other investigators found no significant phenotypic differences in circulating lymphocytes expressing plasma cell antigens (CD38 and PCA–1).[52] Peripheral blood clonal B-cell excess, defined by κ or λ light-chain excess, could not be demonstrated in patients with MGUS (0 of 8) and was rarely observed in early Stage I myeloma (1 of 7) but was common in Stage II (8 of 8) and Stage III MM (4 of 6). Increases in 'clonal' B cells in MM were associated with a high labeling index of marrow plasma cells.[53] These observations suggest, but do not prove, a role for 'clonal' circulating B cells in the development of MM. Clonality is usually assessed by indirect measures such as surface phenotypic changes or light-chain ratios, which could be the result of abnormal T-cell regulation. Although immunoglobulin gene rearrangement techniques can detect circulating B cells and myeloma plasma cells, it is often difficult to distinguish B cells from plasma cells with simple gene rearrangement studies.

In contrast to the controversy about the role of

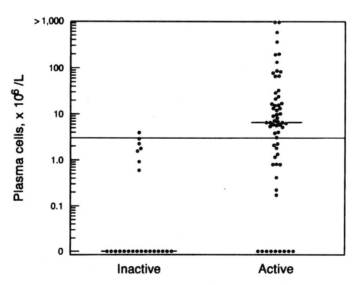

Fig. 9.2 Absolute number of circulating plasma cells × 10⁶/l in relation to disease activity. The horizontal line at 3×10⁶/l is the cutoff value between a low and high number of circulating plasma cells; the short horizontal lines represent median values. (Reproduced from ref. 54 *Cancer* 1993, 72:108–13, with permission.)

small numbers of circulating B cells, circulating monoclonal plasma cells are clearly increased in active myeloma and are an important difference between overt MM and MGUS.[54] In a study of 84 patients, Witzig *et al.* showed that 'clonal' plasma cells (plasma cells expressing cytoplasmic light chain of the same immunoglobulin as plasma cells in the marrow) circulate in increased numbers in patients with myeloma (Fig. 9.2).[54] Fifty-seven per cent of newly diagnosed cases and 81 per cent of relapsed MM cases had $>3\times10^6$ monoclonal plasma cells/l. All patients with MGUS had $<3\times10^6$ plasma cells/l. In our experience, the number of circulating plasma cells was a better discriminator of disease activity than B cell light-chain $\kappa:\lambda$ ratio.

Our assay for detection of circulating myeloma cells is enhanced by removal of T cells from the mononuclear cells prior to analysis. This purification markedly increases the concentration of circulating monoclonal plasma cells on immunofluorescence-stained slides. The presence of clonal cells in the peripheral blood has also been demonstrated using allele-specific oligonucleotide techniques.[55] Correlation with results of immunofluorescence microscopy suggests that many of these cells are clonal plasma cells, but a small proportion may be clonally related B cells, because they are smaller by light scatter characteristics and possess $C\mu$ immunoglobulin rearrangements.[56]

REGULATORY T CELLS AND NK CELLS

Changes in circulating B cell number and their phenotype reviewed in the preceding paragraphs may reflect regulatory T cell changes induced by immunoglobulin idiotype. Proliferation of lymphocytes in response to low concentrations of $F(ab')_2$ immunoglobulin fragments of autologous M-components were found in one patient with MGUS and two patients with MM, Stage I, providing evidence for idiotypic reactive T cells in MGUS and MM.[57] Shared immunoglobulin idiotypes have been observed in both MGUS and MM.[58] Normal B-cell $\kappa:\lambda$ ratio is seen in patients with MGUS but is lost as myeloma develops.[53] Normal immunoglobulin production and primary immune responses are decreased in MM. These data support the suggestion that T cells may act as regulatory cells in MGUS, but this regulatory function may be lost upon progression to myeloma. The percentage of CD4 cells was reduced and CD8 cells expanded during progression of myeloma.[52] Supporting evidence for T-cell abnormalities in MM comes from observations in both MGUS and MM that T cells show selective α/β V-gene segment usage of both CD4+ and CD8+ T cells.[59]

An increased frequency of CD16+/CD3+ peripheral blood mononuclear cells in 7 of 15 patients with MGUS indicated that T lymphocytes with NK-like phenotype are expanded in at least a subset of patients with MGUS,[60] but there was no difference in functional NK activity between patients with MGUS and those with MM. In another study, the ability to induce lymphokine-activated killer (LAK) cells was significantly decreased in MM compared with MGUS despite an increase in CD8+/CD11b+ cells. Purified CD8+/CD11b+ lymphocytes were purified and were intrinsically unable to generate LAK activity following recombinant IL–2 simulation.[61]

Taken together, these data suggest complex dysregulation of T-cell effector functions and NK activities in patients with MM. Despite the data suggesting a role for T-cell dysregulation in MM, it is interesting to note one study that does not. Infection with human immunodeficiency virus (HIV) did not predict a higher rate of conversion from MGUS to MM.[62] Of 11 patients with MGUS followed for a mean of 50 months, 7 showed disappearance of the M-protein, and three developed additional M-proteins. Of the 4 remaining, none developed malignant disease.

LOSS OF APOPTOSIS

The majority of patients with MM appear to have increased *bcl–2* expression.[63] The *bcl–2* gene plays an important role in apoptosis (programmed cell death). Increased *bcl–2* causes failure of apoptosis. The resulting accumulation of tumor cells may be an important factor in increased tumor burden observed in MM compared with MGUS.[64] Other factors that may decrease apoptosis include IL–6 and CD40.

SERUM MARKERS AND MISCELLANEOUS DIFFERENCES

In general, serum markers are not that useful in discriminating MGUS from MM. This is especially true for early-stage MM, in which diagnostic discrimination is most needed. Elevations are most common in overt late-stage MM, in which the diagnosis is already established by the presence of bone lesions, anemia, or renal failure. In a study of 684 patients with newly diagnosed monoclonal gammopathy, including 343 with MGUS and 341 with MM, bone marrow plasma cells (20 per cent), erythrocyte sedimentation rate, percentage reduction in normal immunoglobulin, level of serum M-component, and thymidine kinase (TK) levels each helped discriminate patients with MM from those with MGUS.[65] Osteocalcin is elevated more frequently in MM than in MGUS.[66] In serial studies of 73 patients with MM

(63 at diagnosis, 58 at remission, and 35 at relapse) compared with 56 patients with MGUS, there was a statistically significant difference in the level of serum neopterin.[67] Statistically significant differences in leukocyte alkaline phosphatase (LAP) scores were observed between 20 patients with MM and 18 patients with MGUS. In the myeloma group, 19 of 20 had elevated LAP scores. In the MGUS group, only 6 of 18 had elevated LAP scores, but in 2 of them, a cause other than MGUS was present.[68] Plasma cell cytoplasmic 5'nucleotidase levels were more often positive in patients with MM (46.4 per cent) than those with MGUS (15.3 per cent) and control subjects (1.2 per cent).[69]

Elevated TK levels distinguished some of the patients with Stage I MM in a study of 97 patients with MGUS and 149 with MM.[70] TK levels increased in two patients at the time of progression to overt MM. Serum TK levels did not predict transition to MM in 4 of 5 patients among another series of 35 patients developing MM during a 4-year period.[71] Elevation of C-reactive protein (CRP) >10 mg/l were found in 27 per cent of 51 patients with myeloma and in none of 17 patients with MGUS.[72] Serum albumin and hemoglobin values were considered the best differential clinical diagnostic factor in patients with normal renal function in one series.[73] Differences in haptoglobulin, ceruloplasmin, transferrin, and α_2-macroglobulin did not predict the development of MM.[72] More sensitive and specific markers are needed for discrimination of MGUS and MM.

Computed tomography and magnetic resonance imaging increase the sensitivity of detecting myeloma bone lesions when standard radiographs are negative. Bone scintigraphy may also detect abnormalities in specific sites not fully visualized by radiographs in patients with suspected MM because of its ability to detect early bone lesions.[74] Judicious use of these techniques can increase the accuracy of diagnosis of myeloma.

The immunoglobulin class and isotype help little to distinguish MM from MGUS. A patient with IgD benign monoclonal gammopathy has been described.[75] In a series of 128 patients diagnosed with MGUS who were followed over a 20-year period, analysis of different presenting features for predictive value of the malignant transformation suggested that the IgA type of MGUS was the only variable associated with a higher probability of such an event.[76] MM bone marrow plasma cells express an excess of the same immunoglobulin light-chain isotype as found in the serum or urine in MM. In MGUS, marrow plasma cells are sometimes polyclonal.[77] In our experience, plasma cells in almost all patients with MGUS show predominance of the involved immunoglobulin light chain with ratios > 4:1 in the case of κ and < 0.5:1 in the case of λ MGUS.

Myeloma cell DNA content aneuploidy by flow cytometry probably helps very little to distinguish MM from MGUS. Aneuploidy was detected in the bone marrow of 25 of 46 patients with MM (54 per cent) and in one of 15 patients with MGUS (7 per cent).[78] In our experience, 50 per cent of patients with MGUS are aneuploid and 67 per cent of those with MM are aneuploid.

Many new observations of differences between MGUS and MM have been made; each needs to be evaluated on its own merits and in relation to previously observed differences between MGUS and MM. Single studies involving few patients provide preliminary evidence that can be confirmed in larger, more comprehensive analyses.

SOLUBLE INTERLEUKIN–6 RECEPTOR (sIL–6R)

IL–6 is a central growth factor for myeloma cells. sIL–6R synthesized by myeloma cells and normal cells may modulate IL–6 activity. Soluble receptors have been shown to be potent immunomodulators of their respective ligands. We have previously reported a novel IL–6R mRNA from myeloma cells that exhibits a 94-nt deletion of the entire transmembrane domain from codons 356 (G–TG) to 387 (AG–G).[79] The transmembrane domain deletion results in a shift in the translational reading frame with the insertion of ten new amino acids followed by a stop codon. Sequence analysis shows the ligand-binding domain of the sIL–6R to be identical to that of the membrane-bound IL–6R up to the transmembrane domain deletion. The sIL–6R cDNA was expressed and supernates were collected from mock or sIL–6R transfected PA–1 cells after 48 hours and assayed for their ability to stimulate or suppress the growth of an IL–6-dependent cell line, ANBL–6. Soluble IL–6R alone had no effect on the growth of the ANBL–6 cells. However, growth stimulation of ANBL–6 cells by sIL–6R was potentiated in the presence of IL–6 and could be blocked by anti-IL–6 antibody. The above results suggest that, in the presence of IL–6, sIL–6R associates with gp130 leading to signal transduction and cell growth.[80]

A myeloma cell line, ANBL–6, exhibited a clonally rearranged immunoglobulin locus but was comprised of both near-diploid and near-tetraploid populations. Cytogenetic studies showed two aneuploid karyotypes with numerous shared structural abnormalities and a near-tetraploid and near-diploid subclone, probably arising during clonal evolution.[81] These studies support observations by Epstein and colleagues suggesting the possible origin

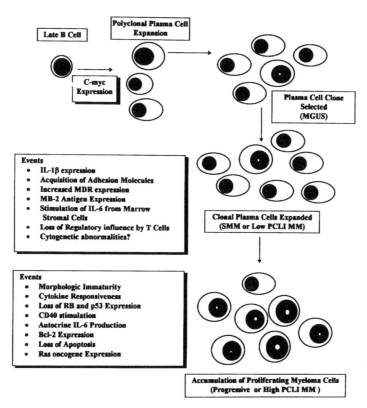

Fig. 9.3 A schema depicting events occurring during the development of MM from MGUS.

of DNA aneuploid myeloma cells from DNA diploid cells.[82]

ROLE OF CD40

CD40 may play an important role in the development of myeloma. A myeloma cell line, ANBL–6, can be induced to produce autocrine IL–6 using anti-CD40 as an artificial ligand.[83] CD40 expression has been observed on the surface of malignant plasma cell lines and plasma cells from almost all patients with myeloma but not normal plasma cells. CD40 ligand stimulation of autocrine IL–6 may be one mechanism whereby myeloma tumor cells are induced to proliferate. Further studies should increase our understanding of the role of CD40 in myeloma.

Summary

Cell proliferation changes are a key element in the progression from MGUS to MM. A model showing many of the putative mechanisms activated during transition to MM is shown in Fig. 9.3. Acquisition

of adhesion markers by neoplastic plasma cells may play a role in homing of myeloma cells to marrow stromal cells. *Mb–2* expression and loss of MDR may play a role in the development of MM, but the mechanisms are not known. IL–1 production by myeloma plasma cells is likely an early event that mediates (1) the acquisition of adhesion markers, (2) the development of lytic bone lesions, and (3) the stimulation of the central growth factor IL–6 from marrow stromal cells, including osteoblasts. Autocrine IL–6 likely follows IL–1 in the multistep process. Soluble IL–6R production by myeloma cells enhances response of myeloma cells to IL–6. IL–3 may aid the differentiation of clonal B cells to myeloma cells and their subsequent proliferation.

Oncogenes may play a critical role in the development and progression of MM. It is likely that c-*myc* maintains B-cell clones and later in the course of disease loss of *rb* or *p53* and *ras* overexpression may allow uncontrolled myeloma cell proliferation by increasing expression of autocrine IL–6. Cytogenetic changes underlying these changes have been difficult to elucidate. Simple karyotyping is inadequate because myeloma cells proliferate slowly and do not produce many metaphases. FISH techniques that identify single cell copies of known genes like *rb* are capable of demonstrating the number of such genes in interphase cells. FISH techniques are proving useful for demonstrating abnormalities in gene number in slowly replicating myeloma cells.

Circulating MM cells probably play a much greater role than previously recognized in the spread of the disease. Precursor B cells, including these circulating in the blood, need to be studied intensively to determine their role in the development of MM. Molecular, biological, cell culture systems, and animal models need to be developed to realize this goal. Various serum markers may be measured to obtain an estimate of myeloma activity, but they are inadequate in the differentiation of MGUS and MM. Newer markers such as neopterin, sIL–6R, soluble CD16 (Fc receptor), and soluble CD56 show promise and are worth studying as discriminant variables in differentiating MM from MGUS. Single studies of LAP score and serum osteocalcin need confirmatory study. Loss of apoptosis, or programmed cell death, may result in tumor cell accumulation. Loss of apoptosis may be mediated by BCL–2, IL–6, or CD40.

Simple changes in the percentage of T cells, T-cell subsets, and NK cells are not likely to play a major role in the progression of MGUS to MM. However, the observation that B-cell light-chain isotype ratios shift to favor the malignant clone as patients progress to MM suggests that a functional change in T-cell regulation of B cells operates in MGUS and may be lost in MM. The role of anti-idiotype T cells and T cells with restricted V-region receptor expression is actively being studied.

β_2-Microglobulin, TK, LDH, IL–6, CRP, and serum albumin levels are not very helpful in distinguishing MM from MGUS.

Critique

Standard criteria for differentiating MGUS from MM must be improved. The PCLI and enumeration of circulating monoclonal plasma cells are very helpful ancillary tests. A scheme for more accurate differentiation of MGUS and MM has been offered (Tables 9.2 and 9.3). Further investigation is needed to improve diagnostic accuracy. Methods such as measurement of IL–1 production by monoclonal plasma cells could have useful diagnostic value and should be examined.

In order to get a better idea where to target future investigation, it is appropriate to develop a model based on current understanding of the pathway from MGUS to MM. The model shown (Fig. 9.3) involves the secretion of cytokines by monoclonal plasma cells and marrow stromal cells. It also involves oncogenes, CD40 stimulation, circulating myeloma cells, and

myeloma precursors. Other features of this model are proliferation mediated by cytokines, cytogenetic abnormalities, loss of apoptosis, and myeloma regulatory cells.

Caution is necessary when interpreting results of studies comparing MGUS and MM. Conflicting results are caused by (1) differences in patient characteristics, e.g. stage and aggressiveness of the myeloma being compared with MGUS, (2) contamination of monoclonal plasma cells by residual normal plasma cells in MGUS, (3) differences in analytic methods and use of bioassay versus enzyme-linked immunosorbent assay (ELISA) for IL–6, (4) misinterpretation of phenotypic and molecular changes, and (5) conclusions based on too few patients. Each investigator interpreting studies needs to take these pitfalls into consideration.

Conclusion

MM and MGUS have major biological and molecular differences, including increased cell proliferation, cytokine changes, cytogenetic abnormalities, oncogene mutations, alteration in oncogene expression, and difference in regulatory cell number and function. We do not know which are the critical differences. Further knowledge may be gained through a prospective serial study in which changes are observed as patients move from MGUS or SMM to MM. We must identify the phenotypic and biological characteristics of the myeloma precursor cell because they may hold the key to a better understanding of the differences between MGUS and MM.

References

1. Kyle RA. 'Benign' monoclonal gammopathy – after 20 to 35 years of follow-up. *Mayo Clinic Proceedings* 1993; **68**: 26–36.
2. Kyle RA, Beard CM, O'Fallon WM, Kurland LT. Incidence of multiple myeloma in Olmsted County, Minnesota: 1978 through 1990, with a review of the trend since 1945. *Journal of Clinical Oncology*, 1994; **12**: 1577–83.
3. Kyle RA, Greipp PR. Smoldering multiple myeloma. *New England Journal of Medicine* 1980, **302**: 1347–9.
4. Greipp PR, Kyle RA. Clinical, morphological, and cell kinetic differences among multiple myeloma, monoclonal gammopathy of undetermined significance, and smoldering multiple myeloma. *Blood* 1983; **62**: 166–71.
5. Boccadoro M, Gavarotti P, Fossati G et al. Low plasma cell 3(H) thymidine incorporation in

monoclonal gammopathy of undetermined significance (MGUS), smouldering myeloma and remission phase myeloma: a reliable indicator of patients not requiring therapy. *British Journal of Haematology* 1984; **58**: 689–96.

6. Boccadoro M, Durie BGM, Frutiger Y et al. Lack of correlation between plasma cell thymidine labelling index and serum beta–2-microglobulin in monoclonal gammopathies. *Acta Haematologia* 1987; **78**: 239–42.

7. Greipp PR, Witzig TE, Gonchoroff NJ et al. Immunofluorescence labeling indices in myeloma and related monoclonal gammopathies. *Mayo Clinic Proceedings* 1987; **62**: 969–77.

8. Büchi G, Girotto M, Veglio M et al. Kappa/lambda ratio on peripheral blood lymphocytes and bone marrow plasma cell labeling index in monoclonal gammopathies. *Haematologica* 1990; **75**: 132–6.

9. Büchi G, Veglio M, Reviglione G et al. The discriminating analysis in the differential diagnosis of monoclonal gammopathies. *Pathologica* 1993; **85**: 175–81.

10. Leo R, Boeker M, Peest D et al. Multiparameter analyses of normal and malignant human plasma cells: CD38++, CD56+, cIg+ is the common phenotype of myeloma cells. *Annals of Hematology* 1992; **64**: 132–9.

11. Girino M, Riccardi A, Luoni R, Ucci G, Cuomo A. Monoclonal antibody Ki–67 as a marker of proliferative activity in monoclonal gammopathies. *Acta Haematologica* 1991; **85**: 26–30.

12. Ide K, Ahmann GJ, Roche PC, Greipp PR. Evaluation of a PCNA labeling index in multiple myeloma (MM). *Blood* 1992; **80** (suppl. 1): 468a, no. 1863.

13. Van Camp B, Durie BGM, Spier C et al. Plasma cells in multiple myeloma express a natural killer cell-associated antigen: CD56 (NKH–1; Leu–19). *Blood* 1990; **76**: 377–82.

14. Sonneveld P, Durie BGM, Lokhorst HM et al. Analysis of multidrug-resistance (MDR–1) glycoprotein and CD56 expression to separate monoclonal gammopathy from multiple myeloma. *British Journal of Haematology* 1993; **93**: 63–7.

15. Harada H, Kawano MM, Huang N et al. Phenotypic difference or normal plasma cells from mature myeloma cells. *Blood* 1993; **81**: 2658–63.

16. Barker HF, Hamilton MS, Ball J et al. Expression of adhesion molecules LFA–3 and N-CAM on normal and malignant human plasma cells. *British Journal of Haematology* 1992; **81**: 331–5.

17. Ahsmann EJM, Lokhorst HM, Dekker AW, Bloem AC. Lymphocyte function-associated antigen–1 expression on plasma cells correlates with tumor growth in multiple myeloma. *Blood* 1992; **79**: 2068–75.

18. Dehou MF, Schots R, Lacor P et al. Diagnostic and prognostic value of the MB2 monoclonal antibody in paraffin-embedded bone marrow sections of patients with multiple myeloma and monoclonal gammopathy of undetermined significance. *American Journal of Clinical Pathology* 1990; **94**: 287–91.

19. Solary E, Guiguet M, Zeller V et al. Radioimmunoassay for the measurement of serum IL–6 and its correlation with tumor cell mass parameters in multiple myeloma. *American Journal of Hematology* 1992; **39**: 163–71.

20. Nachbaur DM, Herold M, Maneschg A, Huber H. Serum levels of interleukin–6 in multiple myeloma and other hematological disorders: correlation with disease activity and other prognostic parameters. *Annals of Hematology* 1991; **62**: 54–8.

21. Pettersson T, Metsarinne K, Teppo AM, Fyhrquist F. Immunoreactive interleukin–6 in serum of patients with b-lymphoproliferative diseases. *Journal of Internal Medicine* 1992; **232**: 439–42.

22. Merico F, Bergui L, Gregoretti MG et al. Cytokines involved in the progression of multiple myeloma. *Clinical and Experimental Immunology* 1993; **92**: 27–31.

23. Bataille R, Chappard D, Marcelli C et al. Recruitment of new osteoblasts and osteoclasts is the earliest critical event in the pathogenesis of human multiple myeloma. *Journal of Clinical Investigation* 1991; **88**: 62–6.

24. Gaillard JP, Bataille R, Brailly H et al. Increased and highly stable levels of functional soluble interleukin–6 receptor in sera of patients with monoclonal gammopathy. *European Journal of Immunology* 1993; **23**: 820–4.

25. Yi Q, Bergenbrandt QY, Österborg A et al. T-cell stimulation induced by idiotypes on monoclonal immunoglobulins in patients with monoclonal gammopathies. *Scandinavian Journal of Immunology* 1993; **38**: 529–34.

26. Vacca A, Di Stefano R, Frassanito A et al. A disturbance of the IL–2/IL–2 receptor system parallels the activity of multiple myeloma. *Clinical and Experimental Immunology* 1991; **84**: 429–34.

27. Lichtenstein A, Berenson J, Norman D et al. Production of cytokines by bone marrow cells obtained from patients with multiple myeloma. *Blood* 1989; **74**: 1266–73.

28. Cozzolino F, Torcia M, Aldinucci D et al. Production of interleukin–1 by bone marrow myeloma cells. *Blood* 1989; **74**: 380–7.

29. Klein B, Lu ZY, Gaillard JP et al. Inhibiting IL–6 in human multiple myeloma. *Current Topics in Microbiology and Immunology* 1992; **182**: 237–44.

30. Hawley TS, Lach B, Burns BF et ak. Expression of retrovirally transduced IL–1α in IL–6 dependent B cells: a murine model of aggressive multiple myeloma. *Growth Factors* 1991; **5**: 327–38.

31. Hawley RG, Wang MH, Fong AZ, Hawley TS. Association between ICAm–1 expression and metastatic capacity of murine B-cell hybridomas. *Clinical and Experimental Metastasis* 1993; **11**: 213–26.

32. Lewinshohn DM, Nagler A, Ginzton N et al. Hematopoietic progenitor cell expression of the H-CAM (CD44) homing-associated adhesion molecule. *Blood* 1990; **75**: 89–95.

33. Drach J, Gattringer C, Huber H. Expression of the neural cell adhesion molecule (CD56) by human myeloma cells. *Clinical Experimental Immunology* 1991; **83**: 418–22.

34. Hamilton MS, Ball J, Bromidge E, Franklin IM. Surface antigen expression of human neoplastic plasma cells includes molecules associated with lymphocyte recirculation and adhesion. *British Journal of Haematology* 1991; **78**: 60–5.

35. Miyake K, Weissman IL, Greenberger JS, Kincade PW. Evidence for the role of the integrin VLA–4 in lympho-hemopoiesis. *Journal Experimental Medicine* 1991; **173**: 599–607.

36. Bagby GC. Interleukin–1 and hematopoiesis. *Blood Review* 1989; **3**: 152–61.

37. Carter A, Merchav S, Silvian DI, Tatarsky I. The role of interleukin–1 and tumor necrosis factor-alpha in human multiple myeloma. *British Journal of Haematology* 1990; **74**: 424–31.

38. Riccardi A, Ucci G, Luoni R et al. Bone marrow biopsy in monoclonal gammopathies: correlations between pathological findings and clinical data. The Cooperative Group for Study and Treatment of Multiple Myeloma. *Journal of Clinical Pathology* 1990; **43**: 469–75.

39. Greipp PR, Raymond NM, Kyle RA, O'Fallon WM. Multiple myeloma: Significance of plasmablastic subtype in morphological classification. *Blood* 1985; **65**: 305–310.

40. Danova M, Riccardi A, Ucci G et al. Ras oncogene expression and DNA content in plasma cell dyscrasias: a flow cytofluorimetric study. *British Journal of Cancer* 1990; **62**: 781–5.

41. Tsuchiya H, Epstein J, Selvanayagam P et al. Correlated flow cytometric analysis of H-ras p21 and nuclear DNA in multiple myeloma. *Blood* 1988; **72**: 796–800.

42. Neri A, Murphy JP, Cro L et al. *Ras* oncogene mutation in multiple myeloma. *Journal of Experimental Medicine* 1989; **170**: 1715–25.

43. Seremetis S, Inghirami G, Ferrero D et al. Transformation and plasmacytoid differentiation of EBV-infected human B lymphoblasts by ras oncogenes. *Science* 1989; **243**: 660–3.

44. Matozaki S, Nakagawa T, Nakao Y, Fujita G. RAS gene mutations in multiple myeloma and related monoclonal gammopathies. *Kobe Journal of Medical Sciences* 1991; **37**: 35–45.

45. Tricot G, Dao D, Gazitt Y et al. Deletion of the retinoblastoma gene (Rb–1) in multiple myeloma. *Blood* 1993; **82**: 261a.

46. Santhanam U, Ray A, Sehgal PB. Repression of the interleukin–6 gene promotor by p53 and the retinoblastoma susceptibility gene product. *Proceedings of the National Academy of Sciences* 1991; **88**: 7605–9.

47. Goto H, Shimazaki C, Ashihara E, et al. Effects of interleukin–3 and interleukin–6 on peripheral blood cells from multiple myeloma patients and their clinical significance. *Acta Haematologica* 1992; **88**: 129–35.

48. Fend F, Weyrer K, Drach J et al. Immunoglobulin gene rearrangement in plasma cell dyscrasias: detection of small clonal cell populations in peripheral blood and bone marrow. *Leukemia and Lymphoma* 1993; **10**: 223–9.

49. Bataille R, Duperray C, Zhang XG et al. CD5 B lymphocyte antigen in monoclonal gammopathy. *American Journal of Hematology* 1992; **41**: 102–6.

50. Jensen GS, Mant MJ, Belch AJ et al. Selective expression of CD45 isoforms defines CALLA+ monoclonal B-lineage cells in peripheral blood from myeloma patients as late stage B-cells. *Blood* 1991; **78**: 711–19.

51. Jensen GS, Belch AR, Kherani F et al. Restricted expression of immunoglobulin light chain mRNA and of the adhesion molecule CD11b on circulating monoclonal B lineage cells in peripheral blood of myeloma patients. *Scandinavian Journal of Immunology* 1992; **36**: 843–53.

52. Omede P, Boccadoro M, Gallone G et al. Multiple myeloma: increased circulating lymphocytes carrying plasma cell-associated antigens as an indicator of poor survival. *Blood* 1990; **76**: 1375–9.

53. Oritani K, Katagiri S, Tominaga N et al. Aberrant expression of immunoglobulin light chain isotype in B lymphocytes from patients with monoclonal gammopathies: isotypic discordance and clonal B-cell excess. *British Journal of Haematology* 1990; **75**: 10–15.

54. Witzig TE, Dhodapkar MV, Kyle RA, Greipp PR. Quantitation of circulating peripheral blood plasma cells and their relationship with disease activity in patients with multiple myeloma. *Cancer* 1993; **72**: 108–13.

55. Billadeau D, Quam L, Thomas W et al. Detection and quantitation of malignant cells in the peripheral blood of multiple myeloma patients. *Blood* 1992; **80**: 1818–24.

56. Billadeau D, Greipp PR, Ahmann G et al. Detection of B cells clonally related to the tumor population in multiple myeloma and MGUS. *Current Topics in Microbiology and Immunology*, 1994; **194**;: 9–16.

57. Österborg A, Steinitz M, Lewin N et al. Establishment of idiotype bearing B-lymphocyte clones from a patient with monoclonal gammopathy. *Blood* 1991; **78**: 2642–9.

58. Berenson JR, Lichtenstein A, Hart S et al. Expression of shared idiotypes of paraproteins from patients with multiple myeloma and monoclonal gammopathy of undetermined significance. *Blood* 1990; **75**: 2107–11.

59. Janson CH, Grunewald J, Osterborg A et al. Predominant T cell recept V gene usage in patients with abnormal clones of B cells. *Blood* 1991; **77**: 1776–80.

60. Famularo G, D'Ambrosio A, Quinitieri F et al. Natural killer cell frequency and function in patients with monoclonal gammopathies. *Journal of Clinical and Laboratory Immunology* 1992; **37**: 99–109.

61. Massaia M, Bianchi A, Dianzani U et al. Defective interleukin–2 induction of lymphokine-activated killer (LAK) activity in peripheral blood T lymphocytes of patients with monoclonal gammopathies. *Clinical and Experimental Immunology* 1990; **79**: 100–4.

62. Lefrere JJ, Debbia M, Lambin P. Prospective

follow-up on monoclonal gammopathies in HIV-infected individuals. *British Journal of Haematology* 1993; **84**: 151–5.

63. Durie BGM, Mason DY, Giles F et al. Expression of the BCL–2 oncogene protein in multiple myeloma. *Blood* 1990; **76**: 347a.

64. Williams GT. Programmed cell death. Apoptosis and oncogenesis. *Cell* 1991; **65**: 1097–8.

65. Ucci G, Riccardi A, Luoni R, Ascari E. Presenting features of monoclonal gammopthies: an analysis of 684 newly diagnosed cases. Cooperative Group for the Study and Treatment of Multiple Myeloma. *Journal of Internal Medicine* 1993; **234**: 165–73.

66. Williams AT, Shearer MJ, Oyeyi J et al. Serum osteocalcin in the management of myeloma. *European Journal of Cancer* 1992; **29A**(1): 140–2.

67. Boccadoro M, Battaglio S, Omede P et al. Increased serum neopterin concentration as indicator of disease severity and poor survival in multiple myeloma. *European Journal of Haematology* 1991; **47**: 305–9.

68. Majumdar G, Hunt M, Singh AK. Use of leucocyte alkaline phosphatase (LAP) score in differentiating malignant from benign paraproteinaemias. *Journal of Clinical Pathology* 1991; **44**: 606–7.

69. Majumdar G, Heard SE, Singh AK. Use of cytoplasmic 5′nucleotidase for differentiating malignant from benign monoclonal gammopathies. *Journal of Clinical Pathology* 1990; **43**: 891–2.

70. Luoni R, Ucci G, Riccardi A et al. Serum thymidine kinase in monoclonal gammopathies. A prospective study. *Cancer* 1992; **69**: 1368–72.

71. Back H, Jagenburg R, Rödjer S, Westin J. Serum deoxythymidine kinase: no help in the diagnosis and prognosis of monoclonal gammopathy. *British Journal of Haematology* 1993; **84**: 746–8.

72. Dubost JJ, Ristori JM, Soubrier M et al. Acute phase proteins in monoclonal gammopathies. *Pathologie Biologie* 1991; **39**: 769–73.

73. Elias J, Dauth J, Senekal JC et al. Serum beta–2-microglobulin in the differential diagnosis of monoclonal gammopathies. *South African Medical Journal* 1991; **79**: 650–3.

74. Feggi LM, Scutellari PN, Prandini N et al. Bone marrow imaging in plasma cell dyscrasias: review of 130 cases. *Journal of Nuclear Biology and Medicine* 1992; **36**: 303–8.

75. O'Connor ML, Rice DT, Buss DH, Muss HB. Immunoglobulin D benign monoclonal gammopathy. A case report. *Cancer* 1991; **68**: 611–16.

76. Blade J, Lopez-Guillermo A, Rozman C et al. Malignant transformation and life expectancy in monoclonal gammopathy of undetermined significance. *British Journal of Haematology* 1992; **81**: 391–4.

77. Majumdar G, Grace RJ, Singh AK, Slater NG. The value of the bone marrow plasma cell cytoplasmic light chain ratio in differentiating between multiple myeloma and monoclonal gammopathy of undetermined significance. *Leukemia and Lymphoma* 1992; **8**: 491–3.

78. Tienharra A, Pelliniemi TT. Flow cytometric DNA analysis and clinical correlations in multiple myeloma. *American Journal of Clinical Pathology* 1992; **97**: 322–30.

79. Lust JA, Donovan KA, Kline MP et al. Isolation of an mRNA encoding a soluble form of the human interleukin–6 receptor. *Cytokine* 1992; **4**(2): 96–100.

80. Lust JA, Jelinek DF, Donovan KA et al. Sequence, expression, and function of an mRNA encoding a soluble form of the human interleukin–6 receptor (sIL–6R). *Current Topics in Microbiology and Immunology* 1994; **194**: 199–206.

81. Jelinek DF, Ahmann GJ, Greipp PR et al. Coexistence of aneuploid subclones within a myeloma cell line that exhibits clonal immunoglobulin gene rearrangement: clinical implications. *Cancer Research* 1993; **53**: 5320–37.

82. Epstein J, Barlogie B, Katzmann J, Alexanian R. Phenotypic heterogeneity in aneuploid multiple myeloma indicates pre-B cell involvement. *Blood* 1988; **71**: 861–5.

83. Westendorf JJ, Ahmann GJ, Armitage RJ et al. CD40 expression in malignant plasma cells: Role in stimulation of autocrine IL–6 secretion by a human myeloma cell line. *Journal of Immunology* 1994; **152**: 117–128.

CHAPTER 10

Clinical features and staging

ANDERS ÖSTERBORG, HÅKAN MELLSTEDT

Clinical features	98	Clinical staging	103
Diagnostic criteria	101	References	105
Prognostic factors	102		

Clinical features

ASYMPTOMATIC MULTIPLE MYELOMA

In the past, the initial clinical symptoms of multiple myeloma were commonly those of advanced disease. However, the widespread use of routine health examinations in recent years and a better general health care system have led to improved case ascertainment resulting in more patients being diagnosed with multiple myeloma during an early, asymptomatic phase of the disease. A monoclonal immunoglobulin fraction (M-component or M-protein) in serum may be suspected following the finding of an increased erythrocyte sedimentation rate (ESR), or a persistent, unexplained proteinuria may be the primary finding, indicating the possibility of Bence Jones proteinuria.

Electrophoresis of urine and serum is recommended in all cases of undetermined proteinuria, especially as intravenous radiological contrast agents used for pyelography may cause damage to the kidneys in patients with Bence Jones proteinuria.[1] Serum protein electrophoresis is widely used for routine examination of many clinical conditions and the initial sign might just be that of an unexpected serum monoclonal immunoglobulin fraction. The frequent use of routine serum electrophoresis analyses may be the main explanation for the increased incidence of multiple myeloma and thus does not necessarily reflect a true rise in the incidence rate.[2]

GENERAL SYMPTOMS

As for most malignant diseases, patients presenting with advanced multiple myeloma may have a case history comprising 6–12 months of general weakness, weight loss, anorexia, and fatigue. In very advanced disease, these symptoms may proceed to acute nausea/vomiting or confusion, indicating the development of hypercalcemia or uremia.

ANEMIA

Anemia, usually of the normocellular, normochromic type, is found in most myeloma patients at diagnosis although the hemoglobin concentration may still be within the normal range at an early, asymptomatic stage of the disease. Anemia may result from a combination of several factors, such as heavy infiltration in the bone marrow of malignant cells, renal impairment, or a deficient endogenous production of erythropoietin.[3] Reduction of the hemoglobin concentration but without change in the red blood cell volume may sometimes occur. This false anemia is due to hemodilution, induced by expansion of the plasma volume from a high concentration of the M-protein.

SKELETAL SYMPTOMS

Bone pain occurs in approximately 75 per cent of patients and about 50 per cent have radiologically detectable myeloma-related skeleton lesions at diagnosis. Patients may have a history of recurrent and gradually increasing back pain for more than 6 months, sometimes with normal radiographs (absence of osteolytic lesions). If myeloma is suspected from the other laboratory findings, magnetic resonance imaging (MRI) of the spine may be recommended due to its superior sensitivity in the detection of bone or marrow disease.[4] The superiority of MRI compared to conventional X-rays is illustrated in Fig. 10.1.

Patients may also present with a sudden onset of severe back, extremity, or rib pains, which may be due to pathologic fracture. Paraplegia develops due to spinal cord compression by the tumor in about 10 per cent of patients. This is a clinical condition that requires immediate surgical decompression and stabilization by laminectomy followed by local radiotherapy (2 Gy × 20 or 3 Gy × 10). In general, radiotherapy may be very useful and effective to control pain from localized myeloma lesions but also to prevent pathological fractures in cases with advanced, high-risk osteolytic lesions.

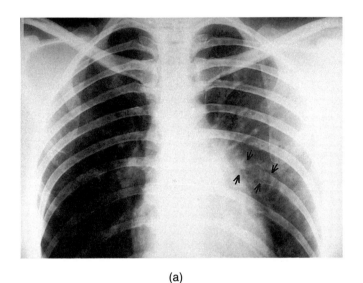

(a)

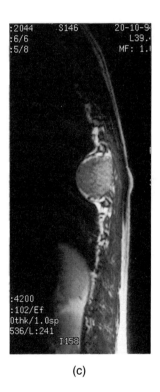

(c)

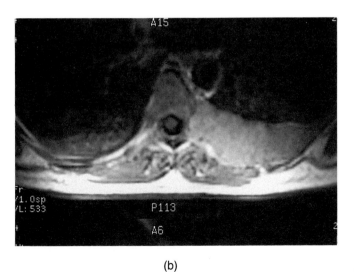

(b)

Fig. 10.1 Conventional X-ray of the ribs from a patient with solitary plasmacytoma of the 8th left rib. The osteolytic lesion is indicated by arrows. The magnetic resonance imaging (MRI) of this area (b and c) demonstrated that the tumor was not only restricted to the involved rib but formed a large bulky tumor which also infiltrated the vertebra.

LOCALIZED PLASMACYTOMAS

Myeloma may occasionally present as a single bone marrow tumor, usually in combination with a localized osteolytic lesion. These patients have an otherwise normal radiograph and less than 5 per cent bone marrow plasma cells. Such cases, comprising < 10 per cent of all plasma cell tumors are referred to as 'solitary plasmacytomas' and should be treated with curative intention by local radiotherapy (usually 2 Gy × 20). Median survival exceeds 10 years, but at that time not more than 15 per cent of the patients remain disease-free.[5] Thus, most patients with solitary bone plasmacytomas will develop multiple myeloma requiring systemic treatment, but even from that time-point the prognosis seems to be better than for those presenting with disseminated multiple myeloma.[6]

Soft tissue localized plasmacytoma termed 'extra-medullary plasmacytoma' seems to have a much better prognosis. More than 80 per cent are localized to the oral cavity or upper respiratory tract. In this patient category 90 per cent are cured by surgical resection followed by radiotherapy.[7,8]

RENAL DYSFUNCTION

Approximately 20 per cent of myeloma patients present with Bence Jones proteinuria alone ('light-chain disease', 'Bence Jones myeloma'). In addition, a further 40–50 per cent secrete Bence Jones protein together with the serum M-component (IgG, IgA, IgD). The presence of Bence Jones protein in the urine is strongly associated with the risk of developing renal failure. More than half of all patients with myeloma develop renal dysfunction, and it is the second most common cause of death (after infection). Monoclonal light chains are filtered, reabsorbed and catabolized by the tubular epithelial cells of the nephron. This process may induce irreversible cell damage resulting in a decreased ability to concentrate the urine. Increased loss of electrolytes and other filtered substances may occur.[9] The end-stage of this process is characterized by a progressively decreased glomerular filtration rate, terminating in uremia. However, not all patients with Bence Jones proteinuria develop renal impairment. Patients excreting κ light chains tend to have a lower risk of developing renal failure than those excreting λ chains, which may partly explain the longer survival of patients with Bence Jones myeloma of the κ type compared with patients with λ type.[10] Patients may continuously secrete large amounts of Bence Jones proteins without developing renal damage. On the other hand, patients with minute amounts of proteinuria might experience kidney failure. The etiology of renal failure in myeloma with regard to Bence Jones protein excretion is considered to be due to chemico-physical properties of the light chain rather than to the amount of the excreted protein.[11] Moreover, particularly during the first chemotherapy course an extensive tumor lysis with release of nephrotoxic products may occur which might cause acute renal failure. Careful pretreatment with hydration and forced diuresis is recommended in all patients with Bence Jones protein during the first treatment cycle, especially when intensive induction treatment regimens are used. Several factors other than Bence Jones proteinuria, such as hypercalcemia/hypercalciuria, hyperuricemia, amyloidosis, and dehydration, also contribute to renal impairment.

SUSCEPTIBILITY TO INFECTIONS

The first sign of myeloma may be an acute bacterial infection, most commonly a pneumonia but also urinary tract infection. Medical case history may reveal repeated infectious episodes during the last 12–24 months. The most common agents are *Streptococcus pneumoniae* and *Haemophilus influenzae*. The risk of viral infections other than herpes zoster is not enhanced. The major reason for the susceptibility to repeated infections is subnormal serum concentrations of polyclonal (background) immunoglobulins which are present in 90 per cent of the myeloma patients at the time of diagnosis. The low CD4/CD8 T-lymphocyte ratio found in patients with advanced disease may also contribute.[12]

HYPERVISCOSITY

Hyperviscosity syndrome (HVS), which is one of the hallmarks of Waldenström's IgM macroglobulinemia, is not commonly observed in patients with myeloma. However, since the incidence of myeloma is about 10-fold higher than that of macroglobulinemia, HVS is, in practice, more frequently encountered in patients with myeloma.[13] The most common mechanism of HVS in myeloma seems to be an extremely high serum level of non-aggregated monoclonal IgG (> 120 g/l). The IgG$_3$ subtype seems to be more frequently associated with HVS than other subtypes. IgA M-proteins and IgG M-components may also form aggregates (dimers–tetramers), which behave rheologically as macromolecules and may, at low serum concentrations (< 40 g/l), give rise to HVS.[13] General symptoms associated with HVS are weakness, fatigue, anorexia, and dyspnea. The most common focal symptom is diffuse bleedings due to defective platelet function from immunoglobulin coating of the platelets with the M-component. A disturbed microcirculation may result in a gradual or sudden visual impairment, and unspecific neuro-psychiatric manifestations. Cardiovascular complications may occur, usually hyperkinetic congestive heart failure. The treatment of choice for acute complications is plasmapheresis followed by chemotherapy to achieve a long-term effect on the M-protein-producing tumor cell clone.

POLYNEUROPATHY

Symptomatic peripheral neuropathy is observed in 5–15 per cent of myeloma patients, but as many as 50 per cent of the patients may have subclinical neuropathy.[14] Symptoms may precede the diagnosis or develop gradually during the disease. Notably, 10 per cent of all patients with peripheral neuropathy of unknown origin has a serum M-component.[15] The clinical picture is heterogeneous with both motor and sensory neuropathy, resulting in muscular weakness as well as distal numbness and painful dysethesia. Of special interest is the frequent association between peripheral neuropathy of osteosclerotic type myeloma.[15] The reason for this association is not clear, but since many of these patients may have signs of a multisystem disorder (POEMS syndrome) an immunological mechanism might be anticipated.

AMYLOIDOSIS

Amyloidosis develops in about 10 per cent of myeloma patients. Symptoms may occur gradually during months to years and do not differ from those of primary amyloidosis. Large amounts of the variable portion or the complete monoclonal light chain are deposited in the tissues.[16] In contrast to the usual distribution in myeloma with a predominance for κ chains, λ light chains are more frequent than κ chains in cases of amyloidosis. General symptoms include weakness, loss of weight, edema, and dyspnea. Purpura and macroglossia may be present. Congestive heart failure and nephrotic syndrome might develop, but renal insufficiency leading to uremia is usually a late-occurring phenomenon. The presence of amyloidosis in patients with myeloma is associated with a poor prognosis.[17]

HYPERCALCEMIA

An increase in serum calcium concentration is found in about a third of patients at diagnosis. This finding is usually associated with advanced disease and in particular with extensive osteolytic bone lesions. Typical symptoms include anorexia, nausea/vomit-

ing, polydipsia/polyuria, dehydration, confusion, and eventually coma. Damage to the kidneys may occur and the renal excretion of calcium may be further impaired by dehydration. Early detection of hypercalcemia is important. The primary therapy should include hydration, forced diuresis and corticosteroids but in advanced or refractory cases also bisphosphonates and/or calcitonin are recommended.

Diagnostic criteria

The term 'myeloma' (i.e. marrow tumor) implies that the presenting and predominant clinical manifestations of the disease are those related to marrow infiltration and bone destruction by malignant plasma cells. Most patients present with more than 25 per cent plasma cells in the bone marrow, but in some cases only a slight elevation of the number of plasma cells is noted. The diagnosis of multiple myeloma relies on three main criteria:

1. More than 10 per cent plasma cells must be present on bone marrow smears. Marrow infiltration may be focal rather than diffuse and repeated bone marrow aspirations may be necessary.
2. The presence of an M-component in serum and/or urine, together with subnormal concentrations of at least one non-monoclonal immunoglobulin class (IgG, IgM, or IgA) is required. The probability of multiple myeloma is high if the M-component concentration is more than 30 g/l (IgG) or more than 10 g/l (IgA).
3. Osteolytic and/or osteoporotic bone lesions compatible with myeloma should be present.

The diagnosis of myeloma is most likely confirmed if two of these three criteria are fulfilled. However, rare cases may not fulfil these criteria.

SMOLDERING MULTIPLE MYELOMA

This term is used to define a subgroup of myeloma patients with early asymptomatic disease that may remain stable without therapy for several years.[18] Diagnostic criteria for smoldering myeloma include bone marrow plasmacytosis of 10–30 per cent plasma cells, an M-component concentration of < 70 g/l (IgG) or < 50 g/l (IgA), absence of bone lesions and no symptoms or associated disease features, including Karnofsky index < 70 per cent, hemoglobin concentration > 100 g/l, serum creatinine concentration < 170 μmol/l, normal serum calcium concentration, and no infections.[19] The plasma cells invariably have a very low proliferation rate as evident by labeling indices (LI). These patients should not be treated unless progression of laboratory variables or symptoms of myeloma occur.[18]

DIFFERENCES BETWEEN MYELOMA AND MGUS

The most important differential diagnosis has to be made between myeloma and MGUS.[20] Table 10.1 shows the major differential diagnostic criteria between these two disorders. However, there is a large overlap between early Stage I myeloma and MGUS. If the diagnosis is uncertain, the patient should be observed off therapy and the M-component concentration monitored at regular intervals (3–6 months) for increase in the M-protein concentration. The diagnosis of multiple myeloma usually become apparent within 1–2 years. However, patients with an M-component compatible with MGUS should be followed indefinitely as 25 per cent of these patients will develop a malignant disease.[21] Delaying commencement of therapy until symptoms occur is justifiable as there is no evidence showing that early treatment of Stage I myeloma improves the outcome of the disease.[22]

Prognostic factors

The survival of patients with myeloma ranges from a few months to several years, indicating the need for factors predicting survival in individual patients. Identificaton of prognostic factors is important to provide optimal initial assessments and follow-up of patients. An increased knowledge of such factors would enable clinicians to categorize patients into high or low risk groups in order to determine who

needs aggressive treatment and who should receive conservative therapy. Moreover, identification of prognostic factors is important to be able to compare patient groups in clinical trials.

For a long time it has been clear that a number of clinical and laboratory features in myeloma have prognostic impact on survival. Among conventional prognostic variables, anemia and renal impairment have the strongest adverse effects on survival.[23–26] Other traditional pretreatment variables that may affect survival are poor performance status, hypercalcemia, hypoalbuminemia, and the presence of lytic bone lesions.[25,27,28] Controversy exists with regard to the prognostic impact of age. Some studies indicate that age may be an independent prognostic factor[29,30] but others have failed to show an influence of age on survival.[28,31]

The prognostic significance of response to treatment, defined as a 50 or 75 per cent reduction of the serum M-component concentration, has been questioned in several studies.[28,30,32,33] However, in a recent study[34] comprising 243 patients, objective response (> 50 per cent) to initial therapy was of utmost importance in predicting survival. Interestingly, also a partial response, defined as < 50 per cent reduction of the M-component concentration but with normalization of hypercalcemia, hypoalbuminemia and improvement of performance status by at least 2 grades, had a positive impact on survival. Rapid responders have previously been reported to experience an inferior prognosis than slow responders,[26,35] a finding that was not verified in this large study.[34]

The uncorrected level of serum β_2-microglobulin $(S\beta_2\text{-M})$ is one of the most important prognostic factors in multiple myeloma, giving a reliable fit for survival prediction.[33,36,37] In a study of 612 patients $S\beta_2$-M was found to be the most powerful prognostic factor available in myeloma.[38] It may, however, be a less reliable parameter for follow-up.[39] For this purpose serial M-component determinations are still

Table 10.1 Major differences in diagnostic criteria between multiple myeloma and monoclonal gammopathy of undetermined significance (MGUS)

Multiple myeloma	MGUS
I. Bone marrow plasmacytosis with > 10 % plasma cells	I. Bone marrow plasma cells < 10 %
II. Monoclonal immunoglobulin spike on serum electrophoresis > 30 g/l (G peaks) or > 10 g/l (A peaks): ≥ 1.0 g/24 h of κ or λ light chain excretion on urine electrophoresis (+ amyloidosis)	II. M-component level: IgG ≤ 30 g/l IgA ≤ 10 g/l Bence Jones protein ≤1.0 g/24 h
III. Lytic bone lesions	III. No bone lesions
IV. Symptoms and signs	IV. No symptoms

Adapted from ref. 20.

recommended. Serum levels of thymidine kinase may also be of value during follow-up and seem to add prognostic information.[39–41]

Labeling index (LI) indicates the percentage of myeloma cells that incorporates [³H-]thymidine. Several studies have emphasized the major importance of LI for the prognosis of myeloma.[29,42,43] A pretreatment LI of < 1 per cent was associated with a long survival, whereas LI > 3 per cent conferred a very poor prognosis. In a multivariate analysis only LI and $S\beta_2$-M levels were found to give independent prognostic information.[42]

Interleukin-6 (IL-6) is a growth factor for myeloma cells and the serum level of IL-6 was identified as a significant predictor of survival.[44] High levels of C-reactive protein (CRP), probably induced by IL-6, also indicated a poor prognosis.[45] Another interesting molecule is the IL-6 receptor which, in its soluble form (sIL-6R), might facilitate myeloma cell proliferation by attaching IL-6 to the plasma cell membrane. Increased levels of sIL-6R were also associated with a short survival[46] and independent prognostic information was demonstrated in a multivariate analysis comprising 514 patients.[47]

The search for new prognostic factors has identified several important variables that may not only give a more reliable prognosis than conventional parameters but may also more accurately reflect the biology of the disease.

Clinical staging

Prognostic factors have been combined in order to create clinical staging systems that may be applied prospectively in chemotherapy trials. Adequate staging systems are a prerequisite for optimal treatment of all types of human cancers.

In 1975, Durie and Salmon presented a clinical

Table 10.2 Durie and Salmon staging system

Criteria	Measured myeloma cell mass (cells $\times 10^{12}/m^2$)
Stage I: All of the following: 　Hemoglobin value > 100 g/l 　Serum calcium value normal or < 2.60 mmol/l 　Bone X-ray, normal bone structure 　　(scale 0) or solitary bone plasmocytoma only 　Low M-component production rates 　　IgG value < 50 g/l 　　IgA value < 30 g/l 　　Urine light chain M-component on electrophoresis < 4 g/24 h	< 0.6 (low)
Stage II. Fitting neither stage I nor stage III	0.6–1.20 (intermediate)
Stage III: One or more of the following: 　Hemoglobin value < 85 g/l 　Serum calcium value ⩾ 3.00 mmol/l 　Advanced lytic bone lesions (scale 3) 　High M-component production rates 　　IgG value > 70 g/l 　　IgA value > 50 g/l 　　Urine light chain M-component on electrophoresis > 12 g/24 h	> 1.20 (high)
Subclassification (A or B) 　A Relatively normal renal function 　　(serum creatinine value < 170 μmol/l) 　B Abnormal renal function 　　(serum creatinine value ⩾ 170 μmol/l)	
Examples: Stage IA Low cell mass with normal renal function 　　　　　Stage IIIB High cell mass with abnormal renal function	

Adapted from ref. 19.

staging system for myeloma[19,48] which is still the most widely used. The system was based on measurement of the total body myeloma cell number. Clinical parameters were correlated with the estimated tumor cell mass using stepwise, multivariate regression analysis. Patients were grouped into three stages (I–III) based on the hemoglobin and serum calcium values, the M-component concentration, and the extent of lytic bone lesions (Table 10.2). Subdivisions with regard to normal (A) or abnormal (B) renal function were also made. Although the percentage of bone marrow plasma cells was an important prognostic factor, this parameter was deleted from the final version of the staging system because other parameters provided equal prognostic information. Moreover, bone marrow aspiration does not always provide adequate samples.

Statistically significant differences in survival were noted between the three clinical stages (Fig. 10.2). The validity of this staging system has been confirmed.[24] In a study comprising 1356 patients from 11 trials, Stage I patients had a median survival time of 48 months, Stage II 32 months and Stage III patients 20 months.[26] In other studies, however, no significant difference in survival was noted, especially between clinical Stages II and III.[27,49,50]

The addition of the platelet count might improve the accuracy of the Durie and Salmon staging system, by separating patients with clinical Stages II and III into a larger subgroup with a normal platelet count and a median survival of 48 months and a smaller high-risk group of thrombocytopenic patients (< 150 × 10[9] platelets/l) with median survival of only 9 months.[51]

Other staging systems have been proposed by the Eastern Cooperative Oncology Group (ECOG) divid-ing patients into a 'good risk' and a 'poor risk' group based on the levels of urea, calcium, white blood cells, and platelets,[52] and by Carbone et al.[53] using hemoglobin, blood urea, and serum calcium concentrations and performance status. Merlini et al.[54] suggested a three-stage system based on hemoglobin, serum creatinine, serum calcium, the M-component concentrations, and the percentage of plasma cells in the bone marrow. In the staging system of the British Medical Research Council (MRC) patients were classified into Stage A, B, or C dependent on the hemoglobin and blood urea concentrations in combination with clinical performance status.[55] All these alternative staging systems are fairly good in identifying high- and low-risk groups of patients.[49] Some studies indicate that the MRC staging system might have a somewhat better discriminating power than the others.[27,56,57] None of them, however, has demonstrated any major advantages compared to the Durie and Salmon staging system and they have not been widely accepted.

A disadvantage of all staging systems is that they require arbitrarily defined cut-off points. It is thus important to ensure that the systems divide the patients into groups of similar size to make them clinically useful. Unfortunately this is not always the case, as exemplified by the Durie and Salmon system which allocates most of the patients to the high-risk group, and thereby fails to identify patients with an intermediate risk. Moreover, the different systems seem inadequate in identifying long-term survivors, as a considerable difference in survival exists for patients with Durie and Salmon Stage I disease.

It may therefore be concluded that the clinical staging systems predict the total tumor cell mass

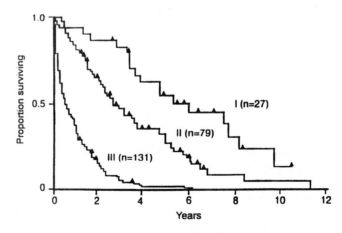

Fig. 10.2 Life-table analysis of patients grouped by clinical stage according to the Durie and Salmon clinical staging system. Significant differences exist between survival of the three groups (*p* < 0.0001). ▲, living patients. (Adapted from ref. 24.)

Table 10.3 Specification of patients with multiple myeloma using Sβ_2-M plus labeling index (LI) at diagnosis

Grade of malignancy	Descriminative presenting[a]	Expected median[b] survival (months) Example
Low[c]	Low Sβ_2-M and low LI	48 months
Intermediate	Low Sβ_2-M but high LI	29 months
High[d]	High Sβ_2-M	12 months

Adapted from ref. 58.
[a] The most useful cut-offs for LI ranged from 0.4 to 1 per cent.
[b] Example uses Sβ_2-M cut-off of < 4 mg/l and 4 mg/l plus LI of 0.4 per cent.
[c] Includes patients with solitary plasmacytoma, smoldering myeloma and indolent myeloma.
[d] Includes patients with plasma cell leukemia and those with biclonal/hypodiploid tumors (DNA content) regardless of Sβ_2-M (which can be low) and labeling index.

relatively well, but none of them are clearly superior to single-risk factors such as hemoglobin and creatinine concentrations in determining prognosis.[49] This is evident from a study by Gobbi et al.,[56] which demonstrated that the predictive ability of each staging system was improved by adding Sβ_2-M, especially if used as a continuous rather than a binary variable. Moreover, the staging systems are not very efficient in indicating the proliferative activity of the tumor. Thus, there is a need for new and better staging systems that take not only the tumor cell load but also the biology of the tumor into account. The most promising variables seem to be the plasma cell labeling index (LI) for determination of proliferative activity, and Sβ_2-M as the most efficient single variable in determining the tumor burden. In a study by Greipp et al.,[42] multivariate analysis showed that no other parameters added prognostic information to that obtained from LI and Sβ_2-M. The Durie and Salmon staging system was not even of statistically significant importance.[42]

By combining Sβ_2-M and LI, Durie and Bataille[58] grouped patients into three prognostic categories: low-grade disease (low Sβ_2-M and low LI), intermediate-grade disease (low Sβ_2-M and high LI) and high-grade disease (high Sβ_2-M) (Table 10.3). The median survival times for these categories were 48, 29, and 12 months, respectively. A major disadvantage is, however, that LI is not yet available in most hematopathological laboratories, despite the fact that it is a simple and reproducible method. The powerful prognostic information obtained from LI clearly underlines that the test should be in routine use in hematologic centers. Instead of LI, the monoclonal antibody against bromo-deoxy-uridine (BRDU) may be used.[59] LI might also be replaced by another strong independent prognostic factor, the serum albumin concentration, which in combination with Sβ_2-M enabled the identification of low-, intermediate-, and poor-risk patients with survival times of 55, 19, and 4 months, respectively.[19] The most recently identified prognostic factors, CRP[45] and sIL-6R,[47] need further

evaluation to clarify their contribution to myeloma staging.

In conclusion, there is a need to continue the search for biological prognostic factors to create a new international myeloma staging system based preferentially on biologic features of the disease.

References

1. Perillie PE, Conn HO. Acute renal failure after intravenous pyelography in plasma cell myeloma. *Journal of the American Medical Association* 1958; **167**: 2186–9.
2. Kyle RA, Beard CM, O'Fallon WM, Kurland LT. Incidence of multiple myeloma in Olmsted County, Minnesota: 1978 through 1990, with a review of the trend since 1945. *Journal of Clinical Oncology* 1994; **12**: 1577–83.
3. Ludwig H, Fritz E, Leitgeb C et al. Erythropoietin treatment for chronic anemia of selected hematological malignancies and solid tumors. *Annals of Oncology* 1993; **4**: 161–7.
4. Dimopoulos MA, Moulopoulos A, Smith T et al. Risk of disease progression in asymptomatic multiple myeloma. *American Journal of Medicine* 1993; **94**: 57–61.
5. Bataille R. Localized plasmacytomas. *Clinics in Haematology* 1982; **11**: 113–22.
6. Alexanian R. Localized and indolent myeloma. *Blood* 1980; **56**: 521–5.
7. Soesan M, Paccagnella A, Chiarion-Sileni V et al. Extramedullary plasmacytoma. Clinical behaviour and response to treatment. *Annals of Oncology* 1992; **3**: 51–7.
8. Schabel SI, Rogers CI, Rittenberg GM, Bubanj R. Extra-medullary plasmacytoma. *Radiology* 1978; **128**: 625–8.
9. Maldonado JE, Velosa JA, Kyle RA et al. Fanconi syndrome in adults: A manifestation of a latent form of myeloma. *American Journal of Medicine* 1975; **58**: 354–64.
10. Shustik C, Bergsagel DE, Pruzanski W. κ and λ light

chain disease: Survival rates and clinical manifestations. *Blood* 1976; **48**: 41–51.

11. Coward RA, Delamore IW, Mallick NP, Robinson EL. The importance of urinary immunoglobulin light chain isoelectric point (pI) in nephrotoxicity in multiple myeloma. *Clinical Science* 1984; **65**: 229–32.

12. Mellstedt H, Holm G, Pettersson D et al. T cells in monoclonal gammopathies. *Scandinavian Journal of Haematology* 1982; **29**: 57–64.

13. Somer T. Rheology of paraproteinaemias and the plasma hyperviscosity syndrome. *Clinical Haematology* 1987; **1**: 695–723.

14. Meier C. Polyneuropathy in paraproteinaemia. *Journal of Neurology* 1985; **232**: 204–14.

15. Kelly JJ, Kyle RA, Miles JM et al. The spectrum of peripheral neuropathy in myeloma. *Neurology* 1981; **31**: 24–31.

16. Glenner GG, Ein D, Eanes ED et al. Creation of 'amyloid' fibrils from Bence-Jones proteins *in vitro*. *Science* 1971; **174**: 712–14.

17. Kyle RA. Amyloidosis. *Clinics in Haematology* 1982; **11**: 151–80.

18. Kyle RA, Greipp PR. Smoldering multiple myeloma. *New England Journal of Medicine* 1980; **302**: 1347–9.

19. Durie BGM. Staging and kinetics of multiple myeloma. *Seminars in Oncology* 1986; **13**: 300–9.

20. Salmon SE, Cassady JR. Plasma cell neoplasma. In: DeVita VT, Hellman S, Rosenberg SA eds *Cancer: Principles and practice of oncology*. Philadelphia: J.B. Lippincott, 1989: 1853–95.

21. Kyle RA. 'Benign' monoclonal gammapathy – after 20 to 35 years of follow-up. *Mayo Clinic Proceedings* 1993; **68**: 26–36.

22. Hjort M, Hellqvist L, Holmberg E et al. Initial versus deferred melphalan-prednisone therapy for asymptomatic multiple myeloma stage I. A randomized study. *European Journal of Haematology* 1993; **50**: 95–102.

23. Imamura Y, Takatsuki K. Ten-year survival and prognostic factors in multiple myeloma. *British Journal of Haematology* 1994; **87**: 832–4.

24. Woodruff RK, Wadsworth J, Malpas JS, Tobias JS. Clinical staging in multiple myeloma. *British Journal of Haematology* 1979; **42**: 199–205.

25. Hannisdal E, Kildahl-Andersen O, Grottum KA, Lamvik J. Prognostic factors in multiple myeloma in a population-based trial. *European Journal of Haematology* 1990; **45**: 198–202.

26. Hansen OP, Galton DAG. Classification and prognostic variables in myelomatosis. *Scandinavian Journal of Haematology* 1985; **35**: 10–19.

27. San Miguel JF, Sanchez J, Gonzalez M. Prognostic factors and classification in multiple myeloma. *British Journal of Cancer* 1989; **59**: 113–18.

28. Palmer M, Belch A, Hanson J, Brox L. Reassessment of the relationship between M-protein decrement and survival in multiple myeloma. *British Journal of Cancer* 1989; **59**: 110–12.

29. Greipp PR, Katzmann JA, O'Fallon WM, Kyle RA. Value of β2-microglobulin level and plasma cell labeling indices as prognostic factors in patients with newly diagnosed myeloma. *Blood* 1988; **72**: 219–23.

30. Tsuchiya J, Murakami H, Kanoh T et al. Ten-year survival and prognostic factors in multiple myeloma. *British Journal of Haematology* 1994; **87**: 832–4.

31. Cohen HJ, Bartolucci A. Age and the treatment of multiple myeloma. *American Journal of Medicine* 1985; **79**: 316–24.

32. Marmont F, Levis A, Falda M, Resegotti L. Lack of correlation between objective response and death rate in multiple myeloma patients treated with oral melphalan and prednisone. *Annals of Oncology* 1991; **2**: 191–5.

33. Baldine L, Radaelli F, Chiorboli O et al. No correlation between response and survival in patients with multiple myeloma treated with vincristine, melphalan, cyclophosphamide and prednisone. *Cancer* 1991; **68**: 62–7.

34. Bladé J, Lopez-Guillermo A, Bosch F et al. Impact of response to treatment on survival in multiple myeloma: Results in a series of 243 patients. *British Journal of Haematology* 1994; **88**: 117–21.

35. Boccadoro M, Marmont F, Tribalto M et al. Early responder myeloma: Kinetic studies identify a patient subgroup characterized by very poor prognosis. *Journal of Clinical Oncology* 1989; **7**: 119–25.

36. Bataille R, Grenier J, Sany J. Beta-2-microglobulin in myeloma: optimal use for staging, prognosis, and treatment – a prospective study of 160 patients. *Blood* 1984; **63**: 468–76.

37. Cuzick J, Cooper EH, MacLennan ICM. The prognostic value of serum β2 microglobulin compared with other presentation features in myelomatosis. *British Journal of Cancer* 1985; **52**: 1–6.

38. Durie BGM, Stock-Novack D, Salmon SE et al. Prognostic value of pretreatment serum β2 microglobulin in myeloma: A Southwest Oncology Group Study. *Blood* 1990; **75**: 823–30.

39. Boccadoro M, Omede P, Frieri R et al. Multiple myeloma: Beta-2-microglobulin is not a useful follow-up parameter. *Acta Haematologica* 1989; **82**: 122–5.

40. Simonsson B, Källander CF, Brenning G et al. Evaluation of serum deoxythymidine kinase as a marker in multiple myeloma. *British Journal of Haematology* 1985; **61**: 215–24.

41. Brown RD, Ioannidis RA, Joshua DE, Kronenberg H. Serum thymidine kinase as a marker of disease activity in patients with multiple myeloma. *Australian and New Zealand Journal of Medicine* 1989; **19**: 226–32.

42. Greipp PR, Lust JA, O'Fallon WM et al. Plasma cell labeling index and β2-microglobulin predict survival independent of thymidine kinase and C-reactive protein in multiple myeloma. *Blood* 1993; **81**: 3382–7.

43. Durie BGM, Salmon SE, Moon TE. Pretreatment tumor mass, cell kinetics and prognosis in multiple myeloma. *Blood* 1980; **55**: 364–72.

44. Reibnegger G, Krainer M, Herold M et al. Predictive value of interleukin-6 and neopterin in patients with multiple myeloma. *Cancer Research* 1991; **51**: 6250–3.

45. Bataille R, Boccadoro M, Klein B et al. C-Reactive protein and β2-microglobulin produce a simple and powerful myeloma staging system. *Blood* 1992; **80**: 733–7.

46. Pulkki K, Pelliniemi T-T, Irjala K et al. High levels of soluble interleukin-6 receptor (sIL-6R) and immunoreactive interleukin-6 (IL-6) predict poor prognosis in multiple myeloma (MM). Abstract. *Blood* 1994; (suppl. 1): 385.

47. Greipp PR, Gaillard JP, Klein B et al. Independent prognostic value for plasma cell labeling index (PCLI), immunofluorescence microscopy plasma cell percent (IMPCP), beta-2-microglobulin (β2M), soluble interleukin-6 receptor (sIL-6R), and C-reactive protein (CRP) in myeloma trial E9487. Abstract. *Blood* 1994; **84** (suppl. 1): 385.

48. Durie BGM, Salmon SE. A clinical staging system for multiple myeloma. *Cancer* 1975; **36**: 842–54.

49. Gassmann W, Pralle H, Haferlach T et al. Staging systems for multiple myeloma: A comparison. *British Journal of Haematology* 1985; **59**: 703–11.

50. Pennec Y, Mottier D, Youinou P et al. Critical study of staging in multiple myeloma. *Scandinavian Journal of Haematology* 1983; **30**: 183–90.

51. Cavo M, Galieni P, Zuffa E et al. Prognostic variables and clinical staging in multiple myeloma. *Blood* 1989; **74**: 1774–80.

52. Costa G, Engle RL, Schilling A et al. Melphalan and prednisone: An effective combination for the treatment of multiple myeloma. *American Journal of Medicine* 1973; **54**: 589–99.

53. Carbone PP, Kellerhouse LE, Gehan EA. Plasmacytic myeloma. A study of the relationship of survival to various clinical manifestations and anomalous protein type in 112 patients. *American Journal of Medicine* 1967; **42**: 937–48.

54. Merlini G, Waldenström JG, Jayakar SD. A new improved clinical staging system for multiple myeloma based on analysis of 123 treated patients. *Blood* 1980; **55**: 1011–19.

55. Medical Research Council's working part on leukaemia in adults. Prognostic features in the third MRC myelomatosis trial. *British Journal of Cancer* 1980; **42**: 831–40.

56. Gobbi PG, Bertolini D, Grignani G et al. A plea to overcome the concept of 'staging' and related inadequacy in multiple myeloma. *European Journal of Haematology* 1991; **46**: 177–81.

57. Bladé J, Rozman C, Cervantes F et al. A new prognostic system for multiple myeloma based on easily available parameters. *British Journal of Haematology* 1989; **72**: 507–11.

58. Durie BGM, Bataille R. Therapeutic implications of myeloma staging. *European Journal of Haematology* 1989; **43** (suppl. 51): 111–16.

59. Greipp PR, Witzig TE, Gonchoroff NJ et al. Immunofluorescence labeling indices in myeloma and related monoclonal gammopathies. *Mayo Clinic Proceedings* 1987; **62**: 969–77.

CHAPTER 11

Principles of chemotherapy and radiotherapy

DIANA SAMSON

Introduction	108	The role of steroids	118
Indications for treatment	108	Maintenance therapy	119
Aims of treatment and criteria for evaluation	110	High-dose therapy other than transplantation	119
Single alkylating agents with or without predniso(lo)ne	112	Intermediate-dose melphalan	120
		Drug resistance	120
Combination chemotherapy	113	New drugs	121
Is combination chemotherapy more effective than single alkylating agents with prednis(ol)one?	115	Radiotherapy	122
		Conclusions	123
VAD and similar regimens	117	References	123

Introduction

Multiple myeloma remains a difficult disease to treat, because of its marked resistance to chemotherapy. With conventional treatment not all patients respond, and even in responding patients complete remissions are rare and, with the extremely rare exception,[1] relapse is inevitable. There are a number of new approaches to treatment which have improved the outlook for myeloma patients, including allogeneic and autologous transplantation and interferon-α, either in induction or as maintenance. These are discussed in Chapters 13, 15 and 16. Conventional dose chemotherapy, however, remains the mainstay of treatment, for some patients as the only therapy, for others as a preliminary to one of these new alternative approaches. Radiotherapy used in various ways is an important adjunct to chemotherapy.

The average survival in myeloma is of the order of 2.5–3.5 years, but there is a wide variation in prognosis depending on factors such as hemoglobin level,

serum albumin, and renal function.[2–4] These have been combined in the widely used Durie–Salmon staging system (*see* Chapter 10).[3] The most important single prognostic factor is, however, the level of serum β_2-microglobulin,[5] which reflects both tumor mass and renal function. Younger patients and those with poor prognostic factors may be considered for more aggressive treatment approaches, whereas the elderly and younger patients with good prognostic factors may do as well, if not better, with simple treatment.

Indications for treatment

Because myeloma is not curable with current therapy, treatment is not necessarily indicated immediately after diagnosis. Increasing numbers of patients are being found by chance investigations to have multiple myeloma, without any relevant symptoms. It is important to be sure that a patient does have multiple myeloma rather than MGUS or solitary myeloma (*see* Chapters 9 and 10). Even where the diagnostic criteria for multiple myeloma are fulfilled, in some asymptomatic patients the disease will remain non-progressive over a period of time, sometimes many years, without treatment. The terms smoldering myeloma or indolent myeloma have been used to describe these patients.[6,7] The consensus view is that such patients should not be treated, since treatment is not required to alleviate symptoms, is not curative, and has significant short-term and long-term side-effects, including the development of secondary leukemia.[8] However it is impossible to know at diagnosis whether patients will fall into the smoldering or indolent group or will progress.

Dimopoulos et al.[9] looked at the factors affecting the risk of disease progression in 95 patients with asymptomatic multiple myeloma seen at the MD Anderson Hospital over many years, who were initially untreated. Not all were Stage I as some had a lytic bone lesion on standard X-rays. At the time of reporting, progression had occurred in 74 patients, after a median of 26 months, while 19 patients were still progression-free at a median follow-up time of 23 months. Factors which significantly predicted for earlier progression were the presence of a lytic bone lesion at presentation ($p < 0.01$; all patients progressed within 8 months), serum paraprotein > 50 g/l ($p < 0.01$), and light chain excretion > 0.5 g/24 h ($p =$ 0.02). These characteristics were used to define three risk groups as follows: low risk – no risk factors; intermediate risk – no lytic lesion but one other factor; and high risk – lytic lesion and/or both other factors. The presence and extent of abnormality revealed by MRI of the spine was also found to correlate with earlier progression. Treatment when given appeared equally effective as in comparable patients treated at the time of diagnosis, with response rates of 50–60 per cent in the different risk groups.

There has also been one randomized study comparing initial versus deferred treatment in asymptomatic Stage I patients.[10] This study excluded patients with any lytic lesion; such patients were all treated at diagnosis. Fifty patients were randomized to receive MP (for explanation of abbreviations, see Table 11.1) either at diagnosis or at the time of disease progression. The median time from diagnosis to the start of treatment in the latter group was 12 months, with a range of 2 months to 3 years, with one patient still untreated at over 6 years. The reasons for starting therapy were increasing M-protein in 8 cases, anemia in 5, and symptomatic bone disease in 9. (Skeletal X-rays were not performed as part of routine follow-up, only if there were relevant symptoms or other signs of progression.) Median survival was 52 months in both treatment groups. It was concluded that while deferred treatment was possible, it conferred no over-

Table 11.1 Chemotherapy regimens used in multiple myeloma

ABCM	Adriamycin, BCNU, cyclophosphamide, melphalan
BCP	BCNU, cyclophosphamide, prednis(ol)one
C-weekly	cyclophosphamide weekly
CWAP	cyclophosphamide weekly plus alternate day prednis(ol)one
EDAP	etoposide, dexamethasone, cytosine arabinoside, cisplatin
MCBP	melphalan, cyclophosphamide, BCNU, prednis(ol)one
MOCCA	methyl prednisolone, vincristine, cyclophosphamide, CCNU, melphalan
MOD	mitoxantrone (Novatrone®), vincristine, dexamethasone
MP	melphalan and prednis(ol)one
NOP	mitoxantrone (Novatrone®), vincristine, prednis(ol)one
VAD	vincristine, Adriamycin, dexamethasone
VAMP	vincristine, Adriamycin, methyl prednisolone
C-VAMP	VAMP plus C-weekly
VBAD	vincristine, BCNU, Adriamycin, dexamethasone
VBAM-Dex	vincristine, BCNU, Adriamycin, melphalan, dexamethasone
VBAP	vincristine, BCNU, Adriamycin, prednis(ol)one
VBAPP	vincristine, BCNU, Adriamycin, procarbazine, prednis(ol)one
VBMCP (M2 protocol)	vincristine, BCNU, melphalan, cyclophosphamide, prednis(ol)one
VMCP	vincristine, melphalan, cyclophosphamide, prednis(ol)one
VMCPP	vincristine, melphalan, cyclophosphamide, procarbazine, prednis(ol)one
Z-DEX	idarubicin (Zavedos®) plus dexamethasone

all survival benefit and there was a risk of disease progression before start of treatment.

These studies do not negate the observation that some patients remain stable for many years without treatment, but such patients can only be identified by careful and regular follow-up. Patients with any evidence of bone disease on radiological examination should be treated at diagnosis. In other patients follow-up should include regular skeletal X-rays as well as M-protein measurements, since bone destruction is usually irreversible.

Aims of treatment and criteria for evaluation

The aims of treatment are to relieve symptoms and to prolong life without causing unacceptable side-effects. What side-effects are acceptable will vary in different patient groups. In a young patient the risk of allogeneic bone marrow transplant (BMT) may be justified by the chance of long-term relapse-free survival, while in an elderly patient the possible survival benefit of combination chemotherapy may outweigh the increased toxicity.

The intrinsic variability of prognosis in myeloma is extremely important when attempting to compare the results of different treatment approaches. Results from non-randomized studies must be interpreted with caution. Even where randomized studies are concerned, randomization has not usually been stratified for prognostic factors and the most important prognostic factor, serum β-2-microglobulin, has only

been available in more recent studies. Particularly in small studies, differences in outcome between different treatment arms may occur because of patient heterogeneity, and it is not surprising that many studies addressing the same question have yielded different results.

Criteria by which treatment regimens are usually evaluated include the proportion of patients achieving an objective response (i.e. a given degree of tumor reduction), the duration of the response, and survival. There are two widely used criteria for the assessment of response, those of the Myeloma Task Force (MTF) of the National Cancer Institute[11] which are based on reduction in paraprotein concentration, and those of the South West Oncology Group (SWOG),[12] which are based on the calculated rate of M-protein synthesis (Table 11.2). This calculation depends on the serum level and fractional catabolic rate of the particular immunoglobulin isotype and subclass. The reason for this rather complicated analysis is that in some immunoglobulin subclasses the fractional catabolic rate declines markedly once levels come down to below 30 g/l, so that below this level changes in protein concentration alone will underestimate tumor reduction. However the SWOG criteria have not been widely adopted by other groups, and most studies have used the MTF criteria for objective response. In the Medical Research Council (MRC) myelomatosis trials, stabilization of disease with lack of progression (plateau) is used to assess response to treatment; this does not require any specific degree of reduction of M-protein. In this chapter, response data quoted are based on MTF criteria, except where otherwise stated.

Although response is therefore usually defined as a reduction in paraprotein of at least 50 per cent,

Table 11.2 Criteria for objective response in myeloma as defined by the Myeloma Task Force (MTF), South West Oncology Group (SWOG), and the Medical Research Council (MRC)

MTF	SWOG	MRC criteria for plateau
One or more of the following: reduction of 50 % in serum paraprotein level reduction of 50 % in light chain excretion if originally > 1.0 g/24 h or a fall to less than 0.1 g/24 h if originally < 1.0 g/24 h reduction of 50 % in plasmacytoma size radiographic evidence of skeletal healing	One or both of the following, sustained for at least 2 months: reduction of 75 % in calculated rate of synthesis of serum paraprotein decrease of light chain excretion to less than 10 % of initial value and < 0.2 g/2 h plus in addition: Hb maintained above 9 g/dl albumin maintained above 30 g/l calcium level normal	Stable or undetectable serum paraprotein and/or light chain excretion for 6 months (Trials IV and V) or 3 months (Trial VI) Few or no symptoms attributable to myeloma Transfusion-independent

Table 11.3 Effect of objective response on survival in previously untreated myeloma

First author	Treatment	No. evaluable	Median survival by MTF criteria		Median survival by SWOG criteria	
			Responders (months)	Non-responders (months)	Responders (months)	Non-responders (months)
Palmer[14]	MP	164				
	Stage II	67	43.8	40.3 ($p = 0.29$)	48.3	39.0 ($p = 0.12$)
	Stage III	71	34.0	21.7 ($p = 0.01$)	35.5	24.4 ($p = 0.04$)
Marmont[16]	MP	76	32.6	29.3 ($p = 0.43$)		
	Stage II	33	46.3	33.2 ($p = 0.76$)		
	Stage III	43	32.2	15.6 ($p = 0.03$)		
Baldini[15]	VMCP	80	26.5	29	30	27

patients who respond slowly and who do not reach this degree of tumor reduction may still have long survival. In fact it has become apparent that there is little, if any, correlation between the degree of response and subsequent survival. Alexanian et al.,[12] in one of the first studies of combination chemotherapy for myeloma, reported that the survival of patients treated with combination chemotherapy was directly correlated with the degree of reduction of M-protein synthesis. However Harley et al.,[13] in another early study of combination chemotherapy, noted that although there was an increased response rate with combination therapy this did not confer a survival benefit. Several subsequent studies have also reported a lack of correlation of response rate and survival.[14–16]

Palmer et al.[14] pointed out that in analyzing differences between responders and non-responders, it is important not to include early deaths as non-responders, as this will a priori shorten the survival time of the non-responder group. It is also important that follow-up is long enough (the 'guarantee time') to allow response to be reached, because some patients respond slowly and may be misclassified as non-responders, and also long enough to allow accurate assessment of survival. They looked at a series of 173 patients treated with MP, most of whom had died by the time of the analysis, and censored those patients who died early. The degree of tumor response in patients treated with MP did not correlate with survival in Stage II patients. There was a weak correlation in Stage III patients. In addition, there was no difference in outcome according to whether MTF or SWOG criteria were used to define response (Table 11.3). Similar observations were made in another study looking at patients treated with MP.[16]

Baldini et al.[15] observed a similar lack of correlation between objective response and survival in 85 patients treated with VMCP. Response was evaluable in 80 patients; early deaths were excluded. According to Myeloma Task Force criteria there were 55 responders and 25 non-responders; median survival was 26.5 months in the responders and 29 months in the non-responders. Based on SWOG criteria, there were 25 responders and 55 non-responders; median survival was 30 months in responders and 27 months in non-responders. Thus not only was there no survival difference between responders and non-responders, however defined, but there was also no survival difference between responders by Myeloma Task Force criteria and responders by SWOG criteria.

Thus response rates as currently defined are of limited value in comparing different treatment regimens or in selecting patients for different forms of post-induction therapy. There are a number of possible reasons why response is not correlated with survival. Indeed a greater degree of response may be associated with a higher cell turnover rate and hence confer a poorer prognosis. The important factor in determining relapse is probably the biological characteristics of the residual cells remaining after treatment rather than the number of plasma cells. Chemotherapy may result in recruitment of cells into cycle.[17] After alkylating agent therapy, the labeling index of plasma cells may actually increase.[18–22] Initially this was thought to explain the failure of cell mass to decrease any further when plateau was reached, but more recent analysis suggests that, although labeling index may increase during the initial phase of therapy, when stable plateau is reached the cells are kinetically quiescent.[23] However, kinetically inactive cells can still give rise to subsequent relapse. Patients responding to VAMP were found to have increased numbers of clonogenic myeloma cells in the marrow despite a marked reduction in plamacytosis; this applied even to patients who were in complete remission.[24] Peest et al.[25] observed a shorter survival after VBMCP than high-dose MP alone despite an equal degree of tumor reduction and speculated that the combined chemotherapy exerted a pressure for the emergence of resistant clones.

Rapidity of response may be a poor prognostic factor, depending on the chemotherapy in question. Hobbs first reported that rapid response to alkylating agent therapy was associated with more rapid relapse and shorter survival[26] and this was confirmed by Belch et al.,[27] the IVth MRC myelomatosis trial[28] and Marmont et al.[16] These authors suggested that this was a manifestation of the increased chemosensitivity of more rapidly proliferating tumors. Boccadoro et al.[29] looked to see if there was a correlation between plasma cell labeling index at diagnosis, rapidity of response to therapy (MP or VMCP/VBAP) and survival. Patients achieving 50 per cent tumor reduction within 3 months were called early responders (ERM), and those reaching the same criteria after 3 months were called slow responders (SRM). Survival was not statistically different between ERM and SRM. However, plasma cell labeling index (LI) was significantly higher in the ERM group overall, and if LI was used to divide ERM into two subgroups, the subgroup with a high LI had a median survival of 16 months compared with 46.9 months for ERM with a low LI. Thus an early response may be due to a high LI, in which case it will confer a poor prognosis, but if LI is low, the rapid response may be due to an intrinsically more sensitive tumor. These results suggest that the rate of response cannot be used as an independent prognostic factor. The rate of response is probably irrelevant when regimens such as VAD are used, since rapid response is almost universal and in fact patients who fail to respond after two courses usually prove refractory to continuation of the same regimen.

Response duration is the second measure of outcome which can be evaluated. It has most usually been used in evaluating maintenance therapy, e.g. with interferon. Response duration does not, however, necessarily correlate with survival because of variable response to treatment at relapse.

Survival is the ultimate measure of outcome, but this also is not without problems of analysis. First, in a disease affecting the elderly, a number of deaths will occur from unrelated causes, and survival analyses do not always censor these deaths. Peest et al.[30] showed that median overall survival was one year shorter than median tumor-related survival in a series of 320 patients. Another problem in using survival as the end-point to assess first-line therapy is the effectiveness of therapy given at relapse.

Single alkylating agents with or without prednis(ol)one

Before the introduction of melphalan and prednisolone (MP) the average survival of myeloma patients was only a few months.[31] Alexanian et al.[32] first published the results of treatment with oral melphalan, using different dosage regimens with or without prednisolone, and demonstrated prolongation of survival to between 17 and 24 months. Oral M ± P has been shown in numerous subsequent studies to prolong survival to between 2 and 3 years.[33–35] Both continuous and intermittent treatment with melphalan are equally effective, but the intermittent schedule causes less myelosuppression and requires less dose modification and has become the accepted method of giving melphalan. An equivalent dose given intravenously was not found to improve response rate or survival.[36] In Alexanian's original study the best results were achieved with a combination of intermittent melphalan (0.25 mg/kg/day) together with prednisone (2 mg/kg/day) both for 4 days every 6 weeks. Objective response rate was 70 per cent (SWOG criteria), as compared with 35 per cent for intermittent melphalan alone and 19 per cent for continuous daily melphalan; median survival was 24 months in the prednisone group compared with 17–18 months with melphalan alone. Most subsequent studies using oral melphalan have also used predniso(lo)ne, although there are other randomized studies which do not support the conclusion that the addition of steroid to melphalan or combination chemotherapy improves long-term survival;[37–39] this question is discussed further later in this chapter. There is a little evidence to suggest that dose intensity of oral melphalan is correlated with outcome. Fernberg et al.[35] observed a correlation with cumulative melphalan dose and response. Myelosuppression was similar in responders and non-responders, so that poor absorption could not be the only reason for lack of response.

Oral cyclophosphamide given at a dose of 2–4 mg/kg/day continuously was also found to prolong survival compared with placebo,[40–41] and cyclophosphamide was also found to be effective in melphalan-resistant patients.[42] Several studies show melphalan and oral cyclophosphamide to be equally effective.[37,43,44] The third MRC myeloma trial showed no difference between MP and i.v. cyclophosphamide (600 mg/m^2 3-weekly).[45] The use of intravenous cyclophosphamide was further developed in the CWAP regimen – weekly i.v. cyclophosphamide (150–300 mg/m^2) with alternate day prednisolone. This was found to be more effective than cyclophosphamide given every 4 weeks and produced responses in approximately 50 per cent of patients with advanced disease, including patients who were resistant to melphalan and also some who had failed on oral cyclophosphamide therapy.[46,47] Weekly cyclophosphamide is considerably less myelotoxic than melphalan, and was used without prednisolone (C-weekly) in the MRC fifth myeloma trial for patients with cytopenia precluding combination therapy (ABCM) or oral melphalan. It is interesting that patients changing from ABCM to C-weekly because of myelosuppression actually survived at least as long if not longer than non-cytopenic patients continuing on ABCM.[39] There is, however, no formal study of C-weekly as a first-line therapy. An oral equivalent of the i.v. C-weekly schedule is cyclophosphamide 200–400 mg/m^2 weekly.

Overall, 50–60 per cent of patients will respond to this type of therapy, usually over a period of 3–6 months, and will reach a stable plateau phase, during which the paraprotein level does not continue to fall but remains steady. Complete remission, i.e. disappearance of the paraprotein with a normal number of plasma cells in the bone marrow, is exceptional. Treatment is usually stopped when a stable plateau is reached, since giving further chemotherapy does not prolong the duration of the remission and may favor the development of drug resistance.[48] The median duration of remission is around 18–24 months, and median survival duration varies between 19 and 45 months in different series (Table 11.4). The latter figure has improved a little in more recent series, probably because of more effective salvage therapy at relapse.

Combination chemotherapy

Vincristine, Adriamycin and the nitrosoureas, together with melphalan and cyclophosphamide, have been the mainstay of combination chemotherapy schedules. In an early SWOG study patients receiving vincristine had survival times longer than those in any previous SWOG study.[57] Vincristine alone given after initial chemotherapy produced a further reduction in tumor mass in some responding patients,[21] possibly because cells were recruited into cell cycle when the tumor mass was reduced by prior chemotherapy. A Cancer and Leukaemia Group B (CALGB) study compared melphalan, BCNU and CCNU as initial single agent therapy, and found the three drugs to be equally effective in terms of response rate, myelotoxicity and survival.[58] Alberts et al. reported the effectiveness of Adriamycin and combined Adriamycin/BCNU in relapsed patients resistant to melphalan.[59,60] Several multidrug regimens have been developed incorporating some or all

of these agents. The most widely used have been the VBMCP regimen (a minor modification of the M-2 protocol) used by the Eastern Cooperative Oncology Group (ECOG) and the VMCP/VBAP protocol developed by the Southwest Oncology Group (SWOG). The relative contribution of the various components to overall efficacy remains unclear, particularly in the case of vincristine, since the IVth MRC myeloma trial failed to show any advantage for bolus injections of vincristine in addition to MP.[61]

THE VBMCP (M-2) PROTOCOL

The use of the M-2 protocol in myeloma was first published by Case et al.[62] who reported results in 73 patients. Over 80 per cent of previously untreated and 13/26 previously treated patients obtained objective responses (50 per cent tumor reduction). Projected median survival in the previously untreated patients was over 3 years, which compared favorably with historical controls treated with MP in whom median survival was 22 months. The ECOG subsequently carried out a large randomized study comparing VBMCP with MP.[63,64] This confirmed higher response rate and longer remission duration and although there was no long-term survival benefit the VBMCP regimen has continued to be the basis for ECOG studies and has also been used by other groups. MOCCA is a similar regimen incorporating methyl prednisolone, vincristine, cyclophosphamide, CCNU, and melphalan.[65,66]

VMCP AND VMCP/VBAP

The VMCP/VBAP regimen, using the same drugs as VBMCP plus Adriamycin, was a development from early studies at the M.D. Anderson Hospital in which the following combinations were evaluated: VMCP, VCAP, VBAP, CAP, and VCP.[57] Response rates (SWOG criteria) were 55–64 per cent for all the four-drug combinations, but only 14 per cent for VCP and 45 per cent for CAP. Thus inclusion of both vincristine and Adriamycin gave a higher response rate. The next step was to evaluate alternating combinations including all the available drugs known to be active in myeloma – VMCP/VCAP and VMCP/VBAP. The response rates for both these alternating regimens were the same, and were not significantly different from those achieved in the earlier study using any of the three four-drug regimens.[67] However the alternating combination including BCNU, i.e. VMCP/VBAP, was adopted for several subsequent studies by the Southwest Oncology Group (SWOG) and has become the most widely

Table 11.4 Randomized studies comparing combination chemotherapy (CCT) with MP and other simple regimens (unless stated, comparison arm is MP)

First author	Year	Group/ country	Combin- ation	No. evaluated	Objective response[a] CCT/ MP(%)	Median survival CCT/MP (months)	Comments
Abramson[49]	1982	ECOG	BCP	188	50/43	25/19	Survival difference not significant
Cohen[50]	1984	SECSG	BCP	373	49/52	36/36	
Pavlovsky[51]	1988	Argentina	CCNU/CP	150	40/40	30/38	
Harley[13]	1979	CALGB	MCBP	250	68/56[b]	Not stated	Survival advantage in poor-risk patients but no significant difference overall
Bergsagel[8]	1979	NCI Canada	MCBP	301	39/40[b]	31/28	
Cooper[52]	1986	CALGB	MCBP +/−A	615	43/40	29/34	Adding A did not improve results, nor did giving alkylating agents sequentially
Tribalto[53]	1985	Italy	VMCP	133	46/35[b]	45/30	Survival difference not significant
Peest[25]	1988	Germany	VMCP	320	33/33[b]	48/not yet reached	Survival difference $p < 0.02$ in favor of MP
The M-2 Regimen (VBMCP) and variants							
Montalban[76]	1984	Spain	VBMCP	39	77/59	24/41	Survival difference not significant
Hansen[54]	1985	Denmark	VBMCP	96		30/21	
Kildahl- Andersen[55]	1988	Norway	VBMCP	67	74/67	33/32	
Oken[64]	1993	ECOG	VBMCP	479	72/51	—	Preliminary analysis 2-yr survival 57 % both arms
Palva[77]	1987	Finland	MOCCA	130	75/54	41/45	
Pavlovsky[51]	1988	Argentina	V/CCNU/ MCP	260	44/33	44/42	
Adriamycin-containing regimens							
Alexanian[67]	1984	MDAH	VMCP/ VCAP VMCP/ VBAB	106	55/53[b] 60/53[b]	27/38	More good risk features in MP group
Salmon[82]	1983	SWOG 7704/05	VMCP/ VCAP or VMCP/ VBAP	237	54/32[b]	43/23	$p = 0.004$ for survival difference
Durie[83]	1986	SWOG 7927/28	VMCP/ VBAP	200	50/28[b]	48/29	Comparison arm VCP not MP p for survival = 0.08
Osterborg[84]	1989	Central Sweden	VMCP/ VBAP	86	52/61	19/22	No significant difference survival; all Stage III
Boccadoro[81]	1991	Italy	VMCP/ VBAP	304	77/64	37/32	No significant difference survival even in poor-risk subgroups
Hjorth[56]	1990	Western Sweden	VMCP VMCP/ VBAP	54 106	56/69 57/58	33/46 24/26	VMCP used in Stage II patients and VMCP/VBAP in Stage III patients
Blade[80]	1993	Spain	VMCP/ VBAP	449	45/32	32/26	No significant difference except in IgA myeloma: median survival 40 versus 20 months
Peest[78]	1990	Germany	VBAM-Dex	138	70/51	Not yet evaluable	Stage III only. M in MP given i.v. Response > 25 % tumor reduction
Maclennan[39]	1992	MRC	ABCM	630	61/49[c]	32/24	Comparison arm M alone $p < 0.0003$ for survival

[a] Myeloma Task Force criteria unless otherwise stated.
[b] SWOG criteria.
[c] Plateau.

used of the standard combination chemotherapy regimens.

ABCM

The ABCM regimen used by the MRC in the Myeloma V and VI studies is similar in many ways to VMCP/VBAP, since it includes four of the same drugs and alternates drug combinations at 3-weekly intervals (AB alternating with CM). It differs in that both vincristine and prednisone are omitted, since neither had been shown to improve survival in previous MRC studies.[28,37,61] Data from cross trial analysis comparing results of the MRC trials using ABCM (V and VI) and SWOG studies suggests that ABCM and VMCP/VBAP are similarly effective.[68]

OTHER STANDARD DOSE COMBINATIONS

There are a number of other similar standard dose combination chemotherapy regimens, including VBAM-Dex[69] and VBAD.[70] More recently, new combinations have been evaluated in relapsed patients, including drugs such as cytosine,[71,72] etoposide,[72–74] platinum,[72] teniposide,[75] epirubicin[65] and ifosfamide.[65,73] Etoposide in particular appears to be useful since Barlogie et al.[72] showed that response to a combination of etoposide, dexamethasone, ara-C, and cisplatin (EDAP) was significantly more effective than DAP alone in relapsed patients. The EDAP combination is now being used by this group as consolidation therapy after VAD induction and prior to high-dose therapy with stem cell rescue.

Is combination chemotherapy more effective than single alkylating agents with prednis(ol)one?

In spite of over twenty randomized studies comparing various combination chemotherapy regimens with MP, this remains a controversial question. The results of these studies are shown in Table 11.4. Several have shown a higher objective response rate with combination chemotherapy but in the majority of studies this has not translated into a significant improvement in survival.[13,64,76–81] This illustrates again the fact that objective response does not necessarily prolong survival. The majority of the studies

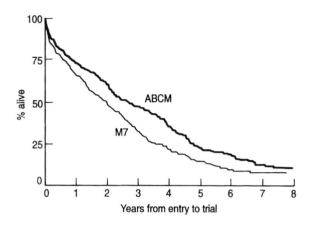

Fig 11 1. Survival from entry in the Medical Research Council's Myeloma V trial, in which 314 patients were randomized to M7 (oral melphalan) and 316 to ABCM; *p* = 0.0003. (Reproduced from ref. 38, Maclennan et al. *Lancet* 1992 339:200–5, with permission. © The Lancet Ltd.)

showed no significant difference in survival between the two types of therapy, possibly because the number of patients was too small or possibly because there is no real difference. Only two studies, the SWOG study 7704/05 and the MRC Vth Myeloma Trial, have shown a significant survival benefit for the combination therapy arm. In SWOG study 7704/05, VMCP/VBAP produced a significant improvement in high response rate and survival compared with MP, maintained at a follow-up of 5 years.[82,83] Other studies of the same regimen have failed to confirm these observations, however (*see* Table 11.4).

The only study to demonstrate an unequivocal survival benefit for combination chemotherapy is the MRC Vth Myeloma Study comparing ABCM with melphalan alone (M7).[39] This is the largest randomized study so far reported and produced a highly significant benefit for survival in patients treated with ABCM (Fig. 11.1). This was not due to the omission of prednisone from the melphalan arm, since cross-trial analysis with previous MRC studies showed that results of M7 were similar to those of the melphalan and prednisone or MVP (melphalan, vincristine, prednisone). Significant survival benefit was observed in all prognostic subgroups, stratified either by β_2-microglobulin level or by the Cuzick index.[4] The magnitude of the survival advantage is, however, relatively modest in clinical terms, with median survival being prolonged by only a few months, and less than 25 per cent of patients in the ABCM arm surviving at 5 years.

Contrary results were reported from another large study from the Eastern Cooperative Oncology Group (ECOG), in which 479 patients were randomized to receive either VBMCP or MP. Preliminary analysis showed 2-year survival to be identical at 57 per cent for both arms, in spite of a higher response rate with VBMCP (72 versus 51, $p < 0.001$) and longer response duration (24 versus 18 months, $p = 0.007$).[64] Further data from this study will be of great interest.

It is unlikely that there are important differences in effectiveness between the different combination chemotherapy regimens used in all these studies. The reasons for the failure to observe a significant benefit for combination therapy in most of the studies could be either that there is no benefit, that the benefit is so small that a very large study is required to detect it, that the difference only becomes apparent after long follow-up, or that the benefit is confined to certain subgroups, in which case a large study would again be required to detect it.

The MRC study shows that even when a highly significant difference is detected, the magnitude of the difference is relatively small, and would have been missed by a smaller study. Both this study and the Italian study from Boccadoro et al.[81] suggest that differences are more readily apparent with longer follow-up. In the latter study, there was no significant difference in median survival, but the data at 5 years suggested a small long-term advantage for the combination arm.

The effect, if any, of standard prognostic factors on the difference in outcome between MP and combination chemotherapy is unclear. In the MRC Vth Myeloma Trial, the benefit of ABCM was consistent at all levels of β_2-microglobulin. Other studies have suggested that poor-risk patients benefit from the combination therapy even though there is no difference in the patient groups as a whole.[13] A Swedish study confined to Stage III patients,[84] however, showed no advantage of combination therapy over MP. In the Italian study of VMCP/VBAP versus MP,[81] separate analysis of two poor-prognosis subgroups, patients with β_2-microglobulin over 6 mg/l and those in Stage III at diagnosis, showed no significant benefit for combination therapy, but by 5 years the survival curves in the group with high β_2-microglobulin start to diverge, with a trend in favor of the combination. In a study from the Spanish group PETHEMA, Blade et al.[80] observed that though there was no overall survival benefit from combination chemotherapy, in patients with IgA myeloma survival was significantly longer (median 38 versus 20 months, $p < 0.005$).

It is possible that the conflicting results of these numerous studies could be resolved by meta-analysis of the published data. Gregory et al.[85] analyzed the results of eighteen published studies comparing combination chemotherapy with MP, most of which are shown in Table 11.3. The MRC data on ABCM were

not included, because the comparison arm was melphalan alone rather than MP. The authors looked at the differences in the percentage of patients surviving at 2 years, since minimum follow-up was usually more than 2 years, and calculated the number of deaths that had occurred by 2 years in each treatment arm. From this data the number of observed deaths in the combination chemotherapy arm was compared with the number of deaths which would have been expected if the death rate was the same as in the control MP arm, giving a typical odds ratio (TOR). This is less than 1.0 if there are fewer than expected deaths in the combination chemotherapy arm and conversely greater than 1.0 if there are more deaths in the combination chemotherapy arm than in the MP arm (*see* Fig. 11.2). The figures are then summed over the whole set of trials to produce an overall result. The variance of the individual results is also derived and summed to give an overall variance and standard deviation for the overall result. The overall odds ratio was just over 1.0, i.e. there was no evident difference in efficacy between combination therapy and MP. However the authors noted that the 2-year survival in the MP arm varied widely in the different studies, from 48 to 87 per cent, and that in those studies where survival on MP was shorter than average, there was a difference in favor of combination chemotherapy,

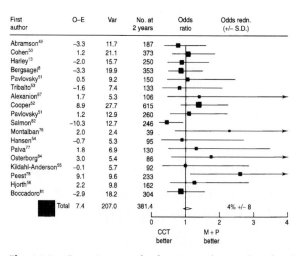

Fig. 11.2 Overview results for 18 randomized trials of MP versus combination chemotherapy in myeloma. O–E signifies the difference between observed and expected deaths. Black boxes represent the variance for individual trials. Lines represent 99 per cent confidence intervals for the odds ratio of each study, i.e. the probability that there is a difference in favor of one or other therapy. The diamond gives the 95 per cent confidence interval for all results combined. (Reproduced from ref. 76, Gregory, Richards, Malpas, *Journal of Clinical Oncology* 1992; 10:334–42, with permission.)

whereas the reverse was true in studies with high MP survival rate. This led them to postulate that in studies selecting poorer risk patients (as defined by worse than average survival on MP), combination chemotherapy had an advantage, and vice versa. However it can be shown that exactly the opposite conclusion would be reached if the authors had defined poor risk patients on the basis of worse than average survival on combination chemotherapy, rather than on MP. Furthermore, as discussed above, data from individual studies do not support the conclusion that poor-risk patients (as defined by standard criteria) do better on combination therapy. Perhaps the time has now come to leave this long-standing controversy, with the conclusion that the survival advantage of combination chemotherapy is at best small, and probably confined to poor-risk patients.

VAD and similar regimens

The first of these regimens to be introduced was the VAD regimen, developed by Barlogie et al.[86] It is a combination of vincristine (V), Adriamycin (A), and dexamethasone (D), but differs from conventional regimens in that the vincristine and Adriamycin are given not as bolus injections but by continuous infusion over 4 days. Modifications of the regimen include VAMP, where methyl prednisolone replaces dexamethasone, and MOD, where mitoxantrone is substituted for Adriamycin. The underlying rationale is that because myeloma cells are slowly dividing, usually with under 1 per cent in S-phase at any time, drugs which act only against cycling cells are likely to kill more cells if given over a longer period. The VAD regimen was found to be more effective than any previously described regimen in relapsed patients, with a response rate of approximately 40 per cent and a survival of over 1 year in responding patients.[86] These results were confirmed in other studies.[87–90] Subsequently studies in previously untreated patients[88,91,92] showed that over 80 per cent of these patients respond, with 10–20 per cent achieving complete remission (CR). Similar results are achieved using high-dose methyl prednisolone instead of dexamethasone.[93] Unfortunately these remissions are not durable, lasting on average only 18 months, even in patients who achieve CR.[91] In an attempt to improve the duration of the remissions observed with VAD, one group has given MP in between VAD cycles,[94] and another has used either concurrent or maintenance interferon;[95] neither of these approaches appear to result in more durable responses.

For many patients, therefore, VAD is not superior to other chemotherapy combinations, and in the only published study comparing VAD with other first-line therapy there was no benefit of VAD versus VBMCP in terms of response rate or survival.[96] Nevertheless, the VAD regimen has advantages in certain situations. Since none of the drugs are renally-excreted, it can be given without dosage modification in patients in severe renal failure, including those on dialysis.[91,97] It produces almost no myelosuppression and so is particularly useful in patients presenting with neutropenia or thrombocytopenia. It is also an ideal initial cytoreductive therapy for younger patients in whom it is planned to proceed to stem cell harvest. The rate of response is dramatic, with most patients achieving 90 per cent of their maximum response within 6 weeks, an advantage in patients who require rapid tumor reduction, e.g. those with rapidly progressive bone disease. The cardiotoxicity of Adriamycin is reduced by giving it as a continuous infusion.[98] The main disadvantages are the necessity for a central venous line for administration, and the high incidence of steroid-related side-effects.

The method of administration does appear in practice to be important for the effectiveness of the VAD regimen. In a non-randomized study where the VA infusion was given over 2 hours rather than over 24 hours, the response rate was less than expected.[99]

MOD, NOP AND NOP-BOLUS

In these regimens mitoxantrone is used instead of Adriamycin. Potential advantages include less cardiotoxicity and less alopecia. MOD is a true variant of VAD in that both mitoxantrone and vincristine are given by continuous infusion. A randomized study of MOD versus VAD in relapsed patients showed VAD and MOD to be equally effective.[100] MOD is therefore a useful regimen in patients who would be suitable for a VAD-type regimen but in whom there is reduced cardiac function or in whom the accepted total Adriamycin dosage has already been reached. NOP is a variant where the vincristine is given by infusion but the mitoxantrone (N) is given by bolus injection, and in NOP-bolus both drugs are given as a bolus. In non-randomized studies of NOP and NOP-bolus in relapsed patients there was a 20–25 per cent response rate, rather lower than with other studies using VAD.[101,102] A randomized study of NOP-bolus versus MP in newly diagnosed patients suggested that NOP was inferior as primary treatment, with no survival advantage and a number of toxic deaths.[103]

CONTINUOUS-INFUSION CYCLOPHOSPHAMIDE

Leoni et al.[104] have used a 7-day continuous infusion of cyclophosphamide (200 mg/m^2/day) on the basis that activity could be enhanced when given in this manner. Cyclophosphamide was combined with teniposide 30 mg/m^2 i.v. on days 1–2 and dexamethasone 40 mg i.v. for 7 days. Response was observed in 50 per cent of forty-three patients with refractory disease, including patients refractory to prior therapy containing cyclophosphamide, and the median survival of all patients was 20 months.

The role of steroids

Steroids are active against myeloma cells while sparing normal bone marrow, and hence form a useful part of the therapeutic repertoire. The addition of steroids in moderate doses to conventional chemotherapy regimens can increase the speed of response and reduce myelosuppression, although it is not clear whether there is a long-term survival benefit. In the first report of Alexanian et al. in 1969,[32] comparing intermittent oral melphalan, continuous oral melphalan, and intermittent melphalan with prednisolone (1 mg/kg alternate days throughout, or 2 mg/kg/day for 4 days every 6 weeks) the inclusion of prednisolone was found to increase response rate (61 per cent versus 17 per cent for continuous melphalan and 32 per cent for intermittent melphalan, SWOG criteria) and improve survival (median survival 24 months versus 17–18 months for either melphalan schedule). The addition of prednisolone to conventional alkylating agent regimens has subsequently been almost universal. However the second MRC Myelomatosis Trial,[37] comparing melphalan, melphalan/prednisolone, and cyclophosphamide, found that the addition of prednisolone did not improve survival; median survival was identical at 20 months for both melphalan and melphalan/prednisolone. The Finnish Leukaemia Group also failed to demonstrate a benefit of conventional dose steroid (methyl prednisolone 0.8 mg/kg/day on days 1–7) added to a combination regimen comprising VCR, CCNU, cyclophosphamide and melphalan (MOCCA versus COLA).[38] The response rate was 72 per cent with steroid and 62 per cent without (ns) while survival was 56 versus 61 months (ns). In addition there was an 8 per cent early death rate in the steroid arm and 0 per cent in the other arm. Preliminary results from the MRC VIth Myeloma Trial showed no survival benefit of adding pulsed prednisolone (60 mg/m^2 for 4 days) during initial therapy with ABCM.[39] In contrast, the preliminary results of a SWOG study[105] comparing VMCP/VBAP with or without addition of prednisone showed a benefit of the additional steroid on both response rate and survival. Response rate was 49 per cent with additional prednisone and 36 per cent in the control arm, while median survival was 42 months compared with 23 months ($p = 0.04$).

Thus the evidence is unclear as to whether there is any benefit of steroids in conventional doses, although there is no doubt that in individual cases the addition of conventional dose steroid can reduce myelosuppression, allowing chemotherapy to be more easily administered, and can also improve well-being. Care must be taken, however, about increased risk of infection.

More recently, following the success of the VAD regimen, there has been renewed interest in the role of steroids, and it has become clear that when used at high dosage, steroids alone can produce rapid and extensive responses both in untreated and relapsed patients. Alexanian and co-workers had already demonstrated the effectiveness of pulsed prednisolone therapy in refractory patients.[106] In seventy patients resistant to melphalan and prednisolone, 25 per cent responded to intermittent high-dose prednisone (60 mg/m^2 daily for 5 days) and 47 per cent responded when vincristine and Adriamycin were added – forming the basis of the VAD regimen (no survival data are given). It is likely that much of the effect of VAD is due to the steroid component. Alexanian et al.[107] treated forty-nine refractory patients (relapsed on treatment or primary refractory) with high-dose dexamethasone (Dex) alone, giving three 4-day pulses of 40 mg/day every 36 days, and compared their results with a historical group of thirty-nine refractory patients treated with the VAD regimen. In the patients who were refractory to initial treatment there was little difference in response: 27 per cent and 32 per cent for Dex and VAD respectively (SWOG criteria). In relapsed patients, however, the response rate to Dex was only 21 per cent versus 65 per cent with VAD. Thus responses were achieved with dexamathasone alone, but the additional vincristine and Adriamycin more than doubled the response in relapsed patients. However, the median survival was the same in patients treated with Dex or with VAD – 12 months overall and 22 months in responders. Alexanian et al. have more recently reported a study using Dex as primary treatment in 112 patients.[108] Using the same Dex schedule as before, response rate was 43 per cent compared with 53 per cent for a historical control group treated with VAD. The tumor halving time was the same in both groups (median 0.5 months), suggesting that the steroid accounts for most of the cytoreduction in VAD. All responders received IFN maintenance. Projected survival was similar to the VAD group but toxicity was significantly less, only 4 per cent serious complications.

Friedenberg et al.[109] also used single-agent high-dose Dex in relapsed/refractory patients, using a more intensive schedule of 40 mg daily for 4 days per week every week for 8 weeks. Those who responded were then maintained on the same dose at 2-week intervals. There were 40 per cent objective responses, but 55 per cent had moderate to severe side-effects including CNS effects, gastrointestinal bleeding, and infection, with one treatment-related death. Median survival was shorter than in Alexanian's study – 31 weeks in the responders. Norfolk and Child[110] reported the use of pulsed oral prednisolone 60 mg/m^2 for 5 days every 14 days in 17 patients with relapsed or refractory myeloma. There were 10 responses and 8 patients reached stable plateau. The median survival for all patients was 19–20 months.

High-dose steroids, particularly high-dose dexamethasone, are therefore very effective at inducing rapid responses with a marked degree of cytoreduction. Approximately 40 per cent of *de novo* patients and 20–25 per cent of relapsed/refractory patients will achieve objective responses to single-agent dexamethasone. It has the major advantage of lack of myelosuppression, and is not associated with secondary leukemia, but infection is a significant risk. However, the long-term remission duration is not likely to be better than that observed with VAD, i.e. about 18 months in previously untreated patients. It is thus an ideal agent to use alone or in combination to achieve rapid cytoreduction as an initial treatment, but would need to be followed by some form of chemotherapy and/or interferon to act on residual myeloma cells.

Maintenance therapy

In the majority of responding patients, paraprotein level tends to reach a plateau and does not fall any further even if treatment is continued. This may result from the low turnover rate of the myeloma cell population when a stable plateau is reached.[23] The question of whether continuing chemotherapy at this stage offers any advantage in terms of preventing relapse has been addressed in a number of studies. An early SWOG study[48] showed no benefit for maintenance with either MP or BCNU plus prednisolone given 6-weekly after an initial six cycles of MP-based induction therapy. Comparison with a control group given no maintenance showed no difference in response duration or survival between the three arms. The incidence of pneumonia and herpes zoster was, however, increased in the maintenance groups. The third MRC Myelomatosis Trial[61] randomized patients in

plateau after induction with either MP or cyclophosphamide to receive no further treatment or maintenance with intermittent induction courses alternating with azathioprine and vincristine. Survival was slightly better in the maintenance arm but the difference was not significant. In the MRC IVth Myeloma Trial, patients reaching plateau after MP with or without vincristine were randomized to continue the same therapy for a further year or to stop treatment. In contrast to the previous study, at the time of reporting there was a small but not significant survival advantage for those receiving no maintenance. Long-term follow-up of the IVth trial[28] confirmed no significant difference in survival.

Belch et al.[27] randomized patients who had reached stable plateau (at least 4 months) with MP induction to continue MP to relapse or to stop treatment. Time to relapse was significantly shorter in the no maintenance group (median 23 versus 31 months) but many of these patients responded to restarting MP and the time to final progression on MP was actually longer in the no-maintenance arm (39 versus 31 months). Median survival was not significantly different. They concluded that maintenance offered no advantage to patients who had reached stable plateau phase. A small Norwegian study[111] also reported no differences in patients randomized to maintenance or no maintenance with continued induction therapy which was either MP or VBMCP.

None of these studies therefore indicate any benefit from continuing chemotherapy once stable plateau phase is reached. The role of interferon maintenance in prolonging response is discussed in Chapter 14.

High-dose therapy other than transplantation

With conventional dose chemotherapy, complete remissions are rare. By analogy with the treatment of leukemia, it can be assumed that the first step to prolonging relapse-free survival (and ultimately to achieving cure in a proportion of patients) would be to achieve more frequent complete remissions. The first attempt to induce complete remission with high-dose therapy was with the use of high-dose melphalan, pioneered by McElwain and colleagues at the Royal Marsden Hospital. A single administration of melphalan at a dose of 140 mg/m^2 resulted in an encouraging 25 per cent complete remission rate in previously untreated patients, in spite of a significant treatment-related mortality.[112] However with longer follow-up it became evident that these remissions were not durable, and the median duration of remission was only 18 months, even in patients who had achieved CR. (These results are very similar to those achieved with the VAD regimen,[91] but at the expense of greater toxicity.)

Similar data on the use of high-dose melphalan in previously untreated patients were reported by three other groups.[91,113,114] All reported a high response rate but a short median response duration (7–12 months) and a significant treatment-related mortality (15–29 per cent). The toxicity of high-dose melphalan can be reduced significantly by the use of growth factors provided there is adequate bone marrow reserve.[115] However, high-dose melphalan is not an ideal regimen for use as induction therapy in young patients who may go on to high-dose therapy with stem cell rescue, because it severely compromises the ability to harvest adequate cells from marrow or peripheral blood.[116] It is useful, however, for the occasional younger patient who is refractory to other regimens.

Although high-dose melphalan ultimately proved disappointing as a first-line treatment, this work nevertheless provided an impetus for the subsequent development of even more intensive therapy combined with bone marrow or peripheral blood stem cell rescue.[93] High-dose melphalan is still frequently used as part of transplantation schedules, either with total body irradiation, or as a single agent at very high dosage. High-dose melphalan and other high-dose therapies such as high-dose cyclophosphamide or high-dose cyclophosphamide and etoposide with growth factor support are also being increasingly used in relapsed and refractory patients.[117,118] With the development of new growth factors, high-dose therapy with growth factor support may in future be employed early in the course of the disease, as an alternative to high-dose therapy with re-infusion of autologous stem cells, avoiding the possible risk of relapse from reinfused myeloma cells.

Intermediate-dose melphalan

In the last few years several authors have investigated the use of intravenous melphalan at intermediate dosage. So far this approach has only been used in patients with relapsed or refractory disease. Tsakanikas et al.[119] used a single dose of melphalan 50–70 mg/m^2 to treat eighteen patients with advanced disease resistant to VAD. There was a 50 per cent objective response rate with a median response duration of 6 months and median survival of 11.5 months.

Two other groups have used serial courses of intravenous melphalan at a dose approximately equivalent to the dose which would be given orally in standard melphalan therapy. Petrucci et al.[120] treated thirty-four patients with relapsed or refrac-

tory myeloma using a dose of 25 mg/m^2 melphalan plus prednisolone every 4–6 weeks for 12 courses. Treatment was administered on an outpatient basis. Thirty-five per cent of patients responded and the median duration of survival in these patients was not reached at 28 months of follow-up. The median duration of response was 16 months. A Swedish study[121] used a similar regime of 30 mg/m^2 every 4–6 weeks for up to 12 courses. Six of 8 patients reached objective response after 1–5 courses. Although there was moderate myelotoxicity in both these studies, there were no treatment-related deaths. This is an approach which may therefore prove useful as a less toxic alternative to high-dose melphalan in patients refractory to standard induction therapy.

Drug resistance

Resistance to chemotherapy is a common problem in myeloma. At least 30 per cent of patients are resistant to primary chemotherapy. Over 50 per cent of those who fail to respond to initial alkylating agents will respond to VAD, but very few of those who are resistant to VAD or relapse thereafter will respond to any other therapy. In many cases drug resistance is associated with expression of the *mdr-1* phenotype. The product of the *mdr-1* gene is a 170 kDa glycoprotein, p-glycoprotein or PGP, which is over-expressed in the cell membrane and which functions as a drug efflux pump. It confers resistence to a number of different chemotherapeutic agents including Adriamycin, vincristine, and etoposide. The extent of overexpression of PGP in myeloma is correlated with resistance to chemotherapy.[122–124] The extent of PGP expression is also closely correlated with prior treatment, and is correlated not with duration of the disease, but with treatment using Adriamycin and/or vincristine.[124] Whereas only 6 per cent of untreated patients expressed PGP, this proportion rose to 50 per cent in patients who had received a cumulative VCR dose of > 20 mg, 83 per cent in those who had received a cumulative Adriamycin dose of > 340 mg, and 100 per cent in cases where both these dose levels were exceeded. Induction of the *mdr* phenotype is therefore a significant potential disadvantage of using Adriamycin and/or vincristine as part of initial chemotherapy.

Several drugs, including verapamil and quinidine, can reverse PGP function *in vitro*, probably by competitive binding, but in general the doses required for *in vitro* effect cannot be achieved *in vivo* without significant toxicity. Dalton et al. in 1989[123] reported preliminary results using a continuous infusion of verapamil to reverse resistance to VAD. Partial responses were observed in 2 of 7 patients, of whom 5 were PGP-positive. This same group subsequently reported results of verapamil–VAD in 22 VAD-resistant patients.[125] There were 5 partial responses (23 per cent), including 4 of 10 patients shown to be PGP-positive; in contrast none of 5 *mdr*-negative patients responded. The median verapamil concentration achieved *in vivo* was 295 ng/ml; hypotension was the limiting factor in dose escalation. The level used to demonstrate reversibility of drug resistance *in vitro* was 1000 ng/ml and lower doses had not been tested.

The demonstration that cyclosporine could reverse drug-resistance *in vitro* at levels of 800–1000 ng/ml led to clinical studies in leukemia and other malignancies. Preliminary data in myeloma suggest that cyclosporine can reverse resistance due to the *mdr* phenotype.[126] Twenty-one patients with relapsed or refractory myeloma were treated with VAD–cyclosporine, using three different dose levels of cyclosporine: 5, 7.5, and 10 mg/kg/day, given by continuous infusion throughout VAD therapy. Serum levels of cyclosporine adequate to reverse drug resistance *in vitro* were reached in all patients receiving 7.5 or 10 mg/kg/day. Ten patients responded (48 per cent) including 7 of 15 who had been previously resistant to VAD. Toxic effects were mild and reversible. The presence of the *mdr* phenotype was assessed in 15 patients and was closely correlated with outcome: of the 12 patients in whom the *mdr* phenotype was demonstrated, 7 (58 per cent) responded, whereas none of the 3 *mdr*-negative patients responded. Other mechanisms of drug resistance must therefore exist which are not due to the *mdr* phenotype and which are not reversible by cyclosporine. It is thus worthwhile to test for the presence of PGP before using agents that modulate the *mdr* phenotype.

New drugs

PURINE ANALOGS

It is disappointing that none of the new purine analogs has shown any evidence of activity against multiple myeloma. No objective responses were observed in thirty-two relapsed patients treated with fludararabine, either as a loading dose with continuous infusion[127] or as daily injections for 5 days.[128] Chlorodeoxyadenosine (CDA) was also found to be ineffective in myeloma, even in previously untreated patients, with none of ten patients showing any decrease in paraprotein or bone marrow plasmacytosis,[129] even though 50 per cent of patients with resistant Waldenström's disease respond to CDA.[130] *In vitro* studies using the DiSC assay (differential stain-

ing cytotoxicity assay) also indicated a major difference between myeloma and other lymphoid and myeloid malignancies, with minimum inhibitory concentration of CDA at least 40-fold higher in myeloma than any other cell type tested.[131] The activity of CDA depends on the relative levels of 5′nucleotidase deoxycytidine kinase, levels of these enzymes could be different in plasma cells as compared with lymphocytes, but this has not been tested; furthermore earlier lymphoid cells are clearly involved in myeloma. The reason for the complete lack of activity in myeloma thus remains unclear.

The third member of this drug group, deoxycoformycin (DCF), acts by inhibition of ADA deaminase, and the level of this enzyme is as high in plasma cells as in T or non-T lymphocytes, thus it might be expected that DCF would have some activity in myeloma. Belch et al.[132] used DCF in a dose of 5 mg/m²/day for 3 days every 2 weeks in thirteen patients with advanced disease. Six died early; of 7 evaluable for response 2 had an objective response (50 per cent reduction in paraprotein) and 2 others had a significant reduction in soft tissue disease. The dose used was higher than is now generally used, and three patients had neurotoxicity. In another study using the more conventional dose of 4 mg/m² every 2 weeks, Grever et al.[133] observed no responses in 14 evaluable patients, in spite of moderate toxicity.

IDARUBICIN

This new orally active anthracycline offers an attractive alternative to Adriamycin for the treatment of myeloma. Its long half-life *in vivo* could confer a similar advantage as infused Adriamycin. There are also data suggesting that idarubicin may be less affected than daunorubicin by the *mdr* phenotype.[134] Three studies using idarubicin either as a single agent or with prednisolone[135–137] indicate a response rate of over 40 per cent in relapsed and *de novo* patients. Preliminary data using idarubicin in combination with dexamethasone (Z-Dex) indicate a high response rate with little toxicity other than that due to the dexamethasone.[138] The Z-Dex protocol is now being used to achieve initial cytoreduction in *de novo* patients prior to intensive consolidation and then autologous transplantation. Idarubicin and dexamethasone in combination with CCNU for relapsed patients is currently being evaluated by the Riverside Haematology Group (UK).

INVESTIGATIONAL DRUGS

A number of new drugs have been studied, predominantly in relapsed or refractory patients. These include amsacrine,[139] aclacinomycin,[140] bisantrene,[141] amonafide,[142] and carboquone.[143] No responses were observed with bisantrene or amonafide, and only very rarely with aclacinomycin or amsacrine (1/43 and 2/74 patients). Carboquone was studied in newly diagnosed patients and appeared more promising; in a schedule using carboquone and prednisolone, response was observed in 44 per cent of 18 patients. One case of POEMS syndrome responding to tamoxifen has been reported.[144]

Radiotherapy

Multiple myeloma is a very radiosensitive tumor, as shown by the ability of radiotherapy alone to cure a significant proportion of patients with solitary myeloma of bone; relapse in these patients is more likely to reflect the fact that disease may be more widespread than is detectable by standard techniques. The limitation of radiotherapy as a treatment modality is myelosuppression. Until recently the use of radiotherapy was confined mainly to treating local areas, but recently its role has expanded to include whole-body treatment, either in the form of double hemi-body irradiation or as part of high-dose preparative regimens for bone marrow or peripheral blood transplantation.

LOCAL RADIOTHERAPY

Local radiotherapy is a very effective means of relieving bone pain,[145–147] and pain usually improves within a few days of starting treatment. This is more rapid than can be achieved with any form of chemotherapy, including VAD-type regimens. Treatment may be given as a single fraction (usually 8 Gy) or as a fractionated course (usually 15–20 Gy in 7–10 fractions or 30–35 Gy in 10–15 fractions).[148] A dose of 8–10 Gy is usually sufficient to provide pain relief and above this dose there is no evidence of a dose–response curve in relation to quality and duration of symptomatic control.[147] Higher doses (30–35 Gy), however, may be needed for a tumoricidal effect.[149] A randomized study of single fraction versus fractionated radiotherapy showed no difference in rapidity of onset or duration of pain relief.[150] In cases of fracture or impending fracture of a long bone, surgical fixation is required but radiation may be given immediately wound healing is complete. Fractionated radiotherapy is preferable for the relief of spinal cord compression (30–35 Gy) because there is less risk of radiation-induced edema; dexamethasone should also be given (e.g. 4 mg four times a day) immediately this complication is diagnosed and con-

tinued to cover the period of radiation. MRI scanning (or, if unavailable, CT with contrast) is extremely valuable in defining the site(s) of compression, which are often multiple.

DOUBLE HEMI-BODY IRRADIATION (DHBI) AND TOTAL BODY IRRADIATION (TBI)

Bergsagel in 1971[151] suggested that a 3-fold greater degree of tumor reduction could be achieved with 7–10 Gy of total body radiation (TBI) (as used in BMT conditioning schedules) than was possible with standard MP therapy, and that this could have a significant impact on survival. He suggested that as an alternative to TBI with bone marrow support, the radiation could be given in two doses, allowing recovery of hematopoiesis between treatment fractions. This suggestion was taken up by a number of workers, and it appears in practice that hematopoietic cells from the unirradiated sites can indeed repopulate the irradiated sites, over a period of 4–6 weeks, and the other half of the body can then be safely treated. Normally the upper half of the body is treated first and then the lower part, each usually being treated in a single fraction. Different centers vary in the exact technique employed, including dose, dose rate, exact field limits (i.e. whether the head and the lower legs are included) and the areas shielded. These details have been well covered in a review by Rowell and Tobias.[148] In addition to predictable myelosuppression, nausea, vomiting, mucositis, and gastrointestinal toxicity are common. Prolonged myelosuppression can occur after either the first treatment fraction (preventing completion of DHBI) or after the second treatment. Prolonged loss of taste and dryness of the mouth may occur, also renal impairment. Radiation pneumonitis is the most serious potential complication and may occur up to 6 months after treatment. The risk of pneumonitis is correlated with radiation dose and dose rate, and the risk is higher in patients who have previously received nitrosoureas and alkylating agents.

A number of authors have reported the results of DHBI in patients with relapsed and refractory myeloma.[152–158] Marked symptomatic relief is achieved in the majority of patients who have bone pain. It is difficult to assess response rate and survival from these studies since the patients treated are very heterogeneous, but objective responses are reported in 25–60 per cent of patients. A wide range of values for median survival have been reported, from 1.5 to 29 months, but for patients who respond, median survival is of the order of one year. These results are similar to those of second-line chemotherapy, but myelosuppression is significantly greater with DHBI. This form of therapy is therefore most useful in relapsed patients with generalized bone pain.

Two studies have addressed the question of the value of DHBI as consolidation in patients who have responded to initial induction chemotherapy. In one of these studies, which was non-randomized, sixty-three patients responding to VMCP received DHBI followed by a planned further eight cycles of VMCP.[159] The addition of DHBI did not confer any major benefit in terms of increasing the number of objective responses and did not appear to prolong survival, since median survival was approximately 30 months. Eight patients developed long-standing neutropenia or thrombocytopenia. The SWOG group randomized 180 patients responding to VMCP/VBAP to receive either DHBI or one further year of VMCP.[160] Relapse-free survival and overall survival were superior in the VMCP arm than the DHBI arm (26 versus 20 months and 36 versus 28 months). Tobias et al.[161] pointed out, however, that the radiation doses in this study were lower than generally used. Also commenting on the results of the SWOG study, a French group[162] noted that they had obtained more encouraging results using DHBI as initial treatment in eighteen newly diagnosed patients with Stage III disease. They observed objective response in 9 of 11 evaluable patients, 2 of whom entered CR, and overall median survival was 25 months. In the absence of any randomized trials, however, it is not possible to recommend DHBI as initial therapy.

One of the factors limiting the long-term effectiveness of DHBI may be reseeding of the marrow with myeloma cells from the non-irradiated areas. This is avoided, of course, by giving the radiotherapy in one treatment or treatment course and restoring hemopoiesis with infused donor or autologous stem cells. Many conditioning regimens and transplantation still include TBI, as discussed in Chapters 15 and 16.

Conclusions

It is unlikely that further advances will be achieved by further modifications in conventional dose chemotherapy, using currently available drugs. It is disappointing that none of the new purine analogs have shown any activity in myeloma. In the absence of any new active drugs, attention is currently focused on the role of high-dose therapies and biological response modifiers, particularly interferon. The role of initial chemotherapy is changing from the only treatment given at this stage towards being the first in a series of therapeutic modalities. The choice of initial therapy will vary in different patient groups depending on prognostic factors, age, the presence of renal failure, and the effect of treatment on ability to harvest marrow or peripheral blood progenitor cells. The unwanted effects of treatment need to be considered, including the development of drug resistance and the risk of secondary leukemia. For younger patients the increasing use of high-dose therapy with transplantation is likely to lead to the choice of induction treatment towards VAD-type regimens rather than combinations with alkylating agents. For the elderly, standard MP therapy is probably still the most appropriate form of initial therapy. In those younger patients with good risk features, where high-dose therapy may not be considered appropriate, MP may also still be preferable, since the results of MP are as good as those of combination chemotherapy, and MP avoids the risk of inducing the *mdr* phenotype and compromising treatment at relapse. At present, therefore, the use of standard combination chemotherapy does not appear to be the first treatment choice in any particular patient group. It may be now more appropriate to reserve such regimens for use in the event of non-response or relapse.

References

1. van Hoeven KH, Reed LJ, Factor SM. Autopsy-documented cure of mutliple myeloma 14 years after M2 chemotherapy. *Cancer* 1990; **60**: 1472–4.
2. Bataille R, Durie BGM, Greinier J, Sany J. Prognostic factors and staging in multiple myeloma: a reappraisal. *Journal of Clinical Oncology* 1986; **4**: 80–7.
3. Durie BGM, Salmon SE. A clinical staging system for multiple myeloma. *Cancer* 1975; **36**: 842–54.
4. Cuzick J, Galton DAG, Peto R, for the MRC Working Party on Leukaemia in Adults. Prognostic factors in the third MRC myelomatosis trial. *British Journal of Cancer* 1980; **43**: 831–40.
5. Cuzick J, De Stavola BL, Cooper EH et al. Long term prognostic value of serum B2-microglobulin in myelomatosis. *British Journal of Haematology* 1990; **75**: 506–10.
6. Kyle RA, Greipp PR. Smoldering multiple myeloma. *New England Journal of Medicine* 1980; **302**: 1347–9.
7. Alexanian R. Localized and indolent myeloma. *Blood* 1980; **56**: 521–5.
8. Bergsagel DE, Bailey AJ, Langley GR et al. The chemotherapy of plasma-cell myeloma and the incidence of acute leukemia. *New England Journal of Medicine* 1979; **301**: 743–8.
9. Dimopoulos A, Moulopoulos MA, Smith T et al. Risk of disease progression in asymptomatic multiple myeloma. *American Journal of Medicine* 1993; **94**: 57–61.
10. Hjorth M, Hallquist L, Holmberg E et al. Initial

versus deferred melphalan–prednisone therapy for asymptomatic multiple myeloma stage I: a randomized study of the Myeloma Group of Western Sweden. *European Journal of Haematology* 1993; **50**: 95–102.

11. Chronic Leukemia and Myeloma Task Force of the National Cancer Institutes. I. Proposed guidelines for protocol studies. II. Plasma cell myeloma. *Cell Chemotherapy Reports* 1973; **4**: 145–58.

12. Alexanian R, Bonnet J, Gehan E et al. Combination chemotherapy for multiple myeloma. *Cancer* 1972; **30**: 382–9.

13. Harley JB, Pajak TF, MacIntyre OR et al. Improved survival of increased-risk myeloma patients on triple-alkylating-agent-therapy: a study of the CALGB. *Blood* 1979; **54**: 13–21.

14. Palmer M, Belch A, Brox L et al. Are the current criteria for response useful in the management of multiple myeloma? *Journal of Clinical Oncology* 1987; **5**: 1373–7.

15. Baldini L, Radaelli F, Chiorboli O et al. No correlation between response and survival in patients with multiple myeloma treated with vincristine, melphalan, cyclophosphamide and prednisone. *Cancer* 1991; **68**: 62–7.

16. Marmont F, Levis A, Falda M, Resegotti L. Lack of correlation between objective response and death rate in multiple myeloma patients treated with oral melphalan and prednisone. *Annals of Oncology* 1991; **2**: 191–5.

17. Karp JE, Humphrey RL, Burke PJ. Timed sequential chemotherapy of cytoxan-refractory multiple myeloma with cytoxan and adriamycin based on induced tumour proliferation. *Blood* 1981; **57**: 468–75.

18. Alberts DS, Golde DW. Perturbation of DNA synthesis in multiple myeloma cells following cell cycle non-specific chemotherapy. *Cancer Research* 1974; **34**: 2911–14.

19. Drewinko B, Brown BW, Humphrey R, Alexanian R. Effect of chemotherapy on the labelling index of myeloma cells. *Cancer* 1974; **34**: 526–31.

20. Salmon SE, Smith BA. Induction of tumour susceptibility to cycle-active agents in IgG multiple myeloma. *Clinical Research* 1972; **20**: 572 (abstract).

21. Salmon SE. Expansion of the growth fraction in multiple myeloma with alkylating agents. *Blood* 1975; **45**: 119–29.

22. Pileri A, Bernengo MG, Boccadoro M et al. Early recruitment in human myeloma population after cytostatic treatment. *Haematologica* 1976; **61**: 184–93.

23. Durie BGM, Russell DH, Salmon SE. Reappraisal of plateau phase in myeloma. *Lancet* 1980; **2**: 65–8.

24. Bell JGB, Millar JA, Maitland JA et al. Increased clonogenic tumour cells in bone marrow after VAMP therapy. *Lancet* 1988; **2**: 931–3.

25. Peest D, Deicher H, Coldewey R et al. Induction and maintenance therapy in multiple myeloma: a multicenter trial of MP versus VMCP. *European Journal of Cancer and Clinical Oncology* 1988; **24**: 1061–7.

26. Hobbs JR. Growth rate and responses to treatment in human myelomatosis. *British Journal of Haematology* 1969; **16**: 607–17.

27. Belch A, Shelley W, Bergsagel DE et al. A randomized trial of maintenance versus no maintenance melphalan and prednisolone in responding multiple myeloma patients. *British Journal of Cancer* 1988; **57**: 94–9.

28. MacLennan ICM, Kelly K, Crockson RA et al. Results of the MRC myelomatosis trials for patients entered since 1980. *Hematological Oncology* 1988; **6**: 145–58.

29. Boccadoro M, Marmont F, Tribalto M et al. Early responder myeloma: kinetic studies identify a patient subgroup characterized by very poor prognosis. *Journal of Clinical Oncology* 1989; **7**: 119–25.

30. Peest D, Coldewey R, Deicher H. Overall vs tumor-related survival in multiple myeloma (letter). *European Journal of Cancer and Clinical Oncology* 1991; **27**: 672.

31. Osgood E. The survival time of patients with plasmacytic myeloma. *Cancer Chemotherapy Reports* 1960; **9**: 1–10.

32. Alexanian R, Haut A, Khan A et al. Treatment for multiple myeloma: combination chemotherapy with different melphalan dose regimes. *Journal of the American Medical Association* 1969; **208**: 1680–5.

33. McArthur JR, Athens JW, Wintrobe MM, Cartwright GE. Melphalan and myeloma. Experience with a low dose continuous regimen. *Annals of Internal Medicine* 1970; **1970**: 665–70.

34. Mellstedt H, Bjorkholm M, Holm G. Intermittent melphalan and prednisolone therapy in plasma cell myeloma. *Acta Medica Scandinavica* 1977; **202**: 5–9.

35. Fernberg JG, Johansson B, Lewensohn R, Mellstedt H. Oral dosage of melphalan and response to treatment in multiple myeloma. *European Journal of Cancer* 1990; **26**: 393–6.

36. Osterborg A, Ahre A, Bjorkholm M et al. Oral versus intravenous melphalan and prednisone treatment in multiple myeloma stage II. A randomized study from the Myeloma Group of Central Sweden. *Acta Oncologica* 1990; **29**: 727–31.

37. MRC Working Party on Leukaemia in Adults. Report on the second myelomatosis trial after 5 years of follow-up. *British Journal of Cancer* 1980; **42**: 813–22.

38. Palva IP, Ala-Harja K, Almquist A et al. Corticosteroid therapy is not beneficial in multiple-drug combination chemotherapy for multiple myeloma. Finnish Leukaemia Group. *European Journal of Haematology* 1993; **52**: 98–101.

39. MacLennan ICM, Chapman C, Dunn J, Kelly K, for the MRC Working Party on Leukaemia in Adults. Combined chemotherapy with ABCM versus melphalan for treatment of myelomatosis. *Lancet* 1992; **339**: 200–5.

40. Rivers SL, Whittington RM, Patno ME. Comparison of effect of cyclophosphamide and a placebo in treatment of multiple myeloma. *Cancer Chemotherapy Reports* 1963; **29**: 115–19.

41. Korst DR, Clifford GO, Fowler WM et al. Multiple myeloma: II. analysis of cyclophosphamide therapy in 165 patients. *Journal of the American Medical Association* 1964; **189**: 758–62.

42. Bergsagel DE, Cowan DH, Hasselback R. Plasma cell myeloma: response of melphalan-resistant patients to high-dose intermittent cyclophosphamide. *Journal of the Canadian Medical Association* 1972; **107**: 851–5.

43. Rivers SL, Patno ME. Cyclophosphamide vs melphalan in the treatment of plasma cell myeloma. *Journal of the American Medical Association* 1969; **207**: 1328–34.

44. MRC Working Party on Leukaemia in Adults. Myelomatosis: comparison of melphalan and cyclophosphamide therapy. *British Medical Journal* 1971; **1**: 640–1.

45. MRC Working Party on Leukaemia in Adults. Treatment comparisons in the third MRC myelomatosis trial. *British Journal of Cancer* 1980; **42**: 823–30.

46. Brandes LJ, Israels LG. Treatment of advanced plasma cell myeloma with weekly cyclophosphamide and alternate-day prednisone. *Cancer Treatment Reports* 1982; **66**: 1413–15.

47. Brandes LJ, Israels LG. Weekly low-dose cyclophosphamide and alternate-day prednisone: an effective low-toxicity regimen for multiple myeloma. *European Journal of Haematology* 1987; **39**: 362–8.

48. SWOG Southwest Oncology Group. Remission maintenance therapy for multiple myeloma. *Archives of Internal Medicine* 1975; **135**: 147–52.

49. Abramson N, Lurie P, Mietlowski WL et al. Phase III study of intermittent carmustine (BCNU), cyclophosphamide, and prednisone versus intermittent melphalon and prednisone in myeloma. *Cancer Treatment Reports* 1982; **66**: 1273–7.

50. Cohen HJ, Silberman HR, Tornyos K et al. Comparison of two long-term chemotherapy regimens, with or without agents to modify skeletal repair, in multiple myeloma. *Blood* 1984; **63**: 639–48.

51. Pavlovsky S, Corrado C, Santurelli MT et al. An update of two randomized trials in previously untreated multiple myeloma comparing melphalan and prenisone versus three- and five-drug combinations: An Argentine group for the treatment of acute leukemia study. *Journal of Clinical Oncology* 1988; **6**: 769–75.

52. Cooper MR, McIntyre OR, Propert KJ et al. Single, sequential, and multiple alkylating agent therapy for multiple myeloma: A CALGB study. *Journal of Clinical Oncology* 1986; **4**: 1331–9.

53. Tribalto M, Amadori S, Cantonetti M et al. Treatment of multiple myeloma: A randomized study of three different regimens. *Leukemia Research* 1985; **9**: 1043–9.

54. Hansen OP, Clausen NAT, Drivsholm A et al. Phase III study of intermittent 5-drug regimen (VBCMP) versus intermittent 3-drug regimen (VMP) versus intermittent melphalan and prednisone (MP) in myelomatosis. *Scandinavian Journal of Haematology* 1985; **35**: 518–24.

55. Kildahl-Andersen O, Bjark P, Bondevik A et al. Multiple myeloma in central and northern Norway 1981–1982; A follow-up study of a randomized clinical trial of 5-drug combination therapy versus standard therapy. *European Journal of Haematology* 1988; **41**: 47–51.

56. Hjorth M, Hellquist L, Holmberg E et al. Initial treatment in multiple myeloma: No advantage of multidrug chemotherapy over melphalan-prednisone. *British Journal of Haematology* 1990; **74**: 185–91.

57. Alexanian R, Salmon S, Bonnet J et al. Combination therapy for multiple myeloma. *Cancer* 1977; **40**: 2675–771.

58. Cornwell GG, Pajak TF, Kochwa S et al. Comparison of oral melphalan, CCNU, and BCNU with and without vincristine in the treatment of multiple myeloma. Cancer and Leukaemia Group B experience. *Cancer* 1982; **50**: 1669–75.

59. Alberts DS, Durie BGM, Salmon SE. Doxorubicin/BCNU chemotherapy for multiple myeloma in relapse. *Lancet* 1975; **1**: 926–8.

60. Alberts DS, Salmon SE. Adriamycin in the treatment of alkylator-resistant multiple myeloma. A pilot study. *Cancer Chemotherapy Reports* 1975; **59**: 345–50.

61. MRC Working Party on Leukaemia in Adults. Objective evaluation of the role of vincristine in induction and maintenance therapy for myelomatosis. *British Journal of Cancer* 1985; **52**: 52–158.

62. Case DCJ, Lee BJ, Clarkson BD. Improved survival times in multiple myeloma treated with melphalan, prednisolone, vincristine, cyclophosphamide and BCNU – M2 protocol. *American Journal of Medicine* 1977; **63**: 897–903.

63. Oken MM, Tsiatis A, Abramson N, Glick J. Evaluation of intensive (VBMCP) vs standard (MP) therapy for multiple myeloma. *ASCO Proceedings* 1987; **6**: A 802.

64. Oken MM. Standard primary treatment of multiple myeloma: recent ECOG studies. Abstracts of the IVth International Workshop on Multiple Myeloma. Rochester, 1993: 70–1.

65. Palva IP, Ahrenberg P, Ala-Harja K et al. Intensive chemotherapy with combinations containing anthracyclines for refractory and relapsing multiple myeloma. *European Journal of Haematology* 1990; **44**: 121–4.

66. Finnish Leukaemia Group. Combination chemotherapy MOCCA in resistant and relapsing multiple myeloma. *European Journal of Haematology* 1992; **48**: 37–40.

67. Alexanian R, Dreicer R. Chemotherapy for multiple myeloma. *Cancer* 1984; **53**: 583–8.

68. Kelly K, Durie B, Maclennan ICM. Prognostic factors and staging system for multiple myeloma: comparisons between the MRC studies in the United Kingdom and the Southwest Oncology Group studies in the United States. *Hematological Oncology* 1988; **6**: 131–40.

69. Peest D, Schmoll HJ, Schedel I et al. VBAMDex chemotherapy in advanced multiple myeloma. *European Journal of Haematology* 1988; **40**: 245–9.

70. Blade J, San Miguel J, Sanz-Sanz MA et al. Treatment of melphalan-resistant multiple myeloma with vincristine, BCNU, doxorubicin and high-dose dexamethasone. *European Journal of Cancer* 1993; **29A**: 57–60.

71. Dodwell DJ, McGill IG. The treatment of poor-prognosis myelomatosis with a cytosine arabinoside containing regimen. *Hematology and Oncology* 1989; **7**: 295–6.

72. Barlogie B, Velasquez WS, Alexanian R, Cabanillas F. Etoposide, dexamethasone, cytarabine and cisplatin in vincristine, doxorubicin and dexamethasone-refractory multiple myeloma. *Journal of Clinical Oncology* 1989; **7**: 1514–17.

73. Ikeda K, Abe N, Morioka A et al. Etoposide and ifosfamide for MP (melphalan and presnisolone) and VAD (vincristine, Adriamycin and dexamethasone)-resistant plasma-cell myeloma: a case report. *Japanese Journal of Medicine* 1990; **29**: 516–18.

74. Ohrling M. Bjorkholm M, Osterborg A et al. Etoposide, doxorubicin, cyclophosphamide and high-dose betamethasone as outpatient salvage therapy for multiple myeloma. *European Journal of Haematology* 1993; **51**: 45–9.

75. Leoni F, Ciolli S, Salti F, Teodori P, Ferrini PR. Teniposide, dexamethasone and continuous infusion cyclophosphamide in advanced refractory multiple myeloma. *British Journal of Haematology* 1991; **77**: 180–4.

76. Montalban C, Zapatero Z, Blanco L et al. Tratamiento del mieloma multiple en estudios II y III. Estudio comparativo del protocol M-2 y el de melfalan-prednisona. *Sangre* 1984; **29**: 993–9.

77. Palva IP, Ahrenberg P, Ala-Harja K et al. Treatment of multiple myeloma with an intensive 5-drug combination or intermittent melphalan and prednisone: A randomized multicentre trial. *European Journal of Haematology* 1987; **38**: 50–4.

78. Peest D, Deicher H, Coldewey R et al. Melphalan and prednisone (MP) versus vincristine, BCNU, adriamycin, melphalan and dexamethasone (VBAMDex) induction chemotherapy and interferon maintenance treatment in multiple myeloma; current results of a multicenter trial. *Onkologie* 1990; **13**: 458–60.

79. Blade J, San-Miguel J, Alcala A et al. A randomized multicentric study comparing alternating combination chemotherapy (VMCP/VBAP) and melphalan-prednisolone in multiple myeloma. *Blut* 1990; **60**: 319–22.

80. Blade J, San Miguel JF, Alcala A et al. Alternating combination VMCP/VBAP chemotherapy versus melphalan/prednisone in the treatment of multiple myeloma: a randomized multicentric study of 487 patients. *Journal of Clinical Oncology* 1993; **11**: 1165–71.

81. Boccadoro M, Marmont F, Tribalto M et al. Multiple myeloma: VMCP/VBAP alternating combination chemotherapy is not superior to melphalan and prednisolone even in high-risk patients. *Journal of Clinical Oncology* 1991; **9**: 444–8.

82. Salmon SE, Haut A, Bonnet JD et al. Alternating combination chemotherapy and levamisole improves survival in multiple myeloma: a Southwest Oncology Group study. *Journal of Clinical Oncology* 1983; **1**: 453–61.

83. Durie BGM, Dixon DO, Carter S et al. Improved survival with combination chemotherapy induction for multiple myeloma: A Southwestern Oncology Group Study. *Journal of Clinical Oncology* 1986; **4**: 1227–37.

84. Osterborg A, Ahre A, Bjorkholm M et al. Alternating combination chemotherapy (VMCP/VBAP) is not superior to melphalan/prednisone in the treatment of multiple myeloma patients stage III – A randomized study from the MGCS. *European Journal of Haematology* 1989; **43**: 54–62.

85. Gregory WM, Richards MA, Malpas JS. Combination chemotherapy versus melphalan and prednisolone in the treatment of multiple myeloma: an overview of published trials. *Journal of Clinical Oncology* 1992; **10**: 334–42.

86. Barlogie B, Smith L, Alexanian R. Effective treatment of advanced multiple myeloma resistant to alkylating agents. *New England Journal of Medicine* 1984; **310**: 1353–6.

87. Scheithauer W, Cortelezzi A, Kutzmits R et al. VAD protocol for treatment of advanced refractory multiple myeloma. *Blut* 1987; **55**: 245–52.

88. Anderson H, Scarffe JH, Lambert M et al. VAD chemotherapy – toxicity and efficacy – in patients with multiple myeloma and other lymphoid malignancies. *Hematological Oncology* 1987; **5**: 213–22.

89. Lokhorst HM, Meuwissen OJAT, Bast EJEG, Dekker AW. VAD chemotherapy for refractory multiple myeloma. *British Journal of Haematology* 1989; **71**: 25–30.

90. Stenzinger W, Blomker A, Hiddeman W, van de Loo J. Treatment of refractory multiple myeloma with the vincristine–adriamycin–dexamethasone (VAD) regimen. *Blut* 1990; **61**: 55–9.

91. Samson D, Gaminara E, Newland AC et al. Infusion of vincristine and doxorubicin with oral dexamethasone as first-line therapy for multiple myeloma. *Lancet* 1989; **2**: 882–5.

92. Alexanian R, Barlogie B, Tucker S. VAD-based regimens as primary treatment for multiple myeloma. *American Journal of Hematology* 1990; **33**: 86–9.

93. Gore ME, Selby PJ, Viner C et al. Intensive treatment of multiple myeloma and criteria for complete remission. *Lancet* 1989; **2**: 879–82.

94. Lejeune C, Sotto JJ, Fuzibet JG et al. Alternating combinations of alkylating agents and vincristine, doxorubicin and dexamethasone in multiple myeloma (letter). *Journal of Clinical Oncology* 1991; **9**: 1090–1.

95. Bird J, Samson D, Newland AC et al. VAD with concurrent IFN or VAD followed by maintenance IFN in newly-diagnosed myeloma: updated results of a randomized study. *British Journal of Haematology* 1993; **84** (suppl. 1):44.

96. Monconduit M, Menard JF, Michaux JL et al. VAD or VBMCP in severe multiple myeloma. *British Journal of Haematology* 1992; **80**: 199–204.

97. Aitchison RG, Reilly IA, Morgan AG, Russell NH. Vincristine, adriamycin and high-dose steroids in myeloma complicated by renal failure. *British Journal of Cancer* 1990; **61**: 765–6.

98. Lieverse RJ, Ossenkoppele GJ. Prevention of doxorubicin-induced congestive cardiac failure by continuous intravenous infusion in multiple myeloma; a case report and review of the literature. *Netherlands Journal of Medicine* 1991; **38**: 33–4.

99. Browman GP, Belch A, Skillings J et al. Modified Adriamycin-vincristine-dexamethasone (m-VAD) in primary refractory and relapsed plasma cell myeloma: an NCI (Canada) pilot study. *British Journal of Haematology* 1992; **82**: 555–9.

100. Phillips JK, Pearce R, Davies JM et al. Comparison of mitoxantrone, vincristine and dexamethasone (MOD) and adriamycin, vincristine, dexamethasone (VAD) in the treatment of relapsed and refractory multiple myeloma. Abstracts of the 24th Congress of International Society for Haematology, London, Blackwell: 1992; 72.

101. Gimsing P, Bjerrum O, Brandt E et al. Refractory myelomatosis treated with mitoxantrone in combination with vincristine and prednisone (NOP-regimen): a phase II study. *British Journal of Haematology* 1991; **77**: 73–9.

102. Wisloff F, Gimsing P, Hedenus M et al. Bolus therapy with mitoxantrone and vincristine in combination with high-dose prednisone (NOP-bolus) in resistant multiple myeloma. *European Journal of Haematology* 1992; **48**: 70–4.

103. Keldsen N, Bjerrum OW, Dahl IMS et al. Multiple myeloma treated with mitoxantrone in combination with vincristine and prednisolone (NOP regimen) versus melphalan and prednisolone: a phase III study. *European Journal of Haematology* 1993; **51**: 80–5.

104. Leoni F, Ciolli S, Caporale R, Salti F, Ferrini PR. Continuous infusion cyclophosphamide in combination with teniposide and dexamethasone in refractory myeloma. *Leukemia and Lymphoma* 1992; **7**: 481–7.

105. Salmon SE, Crowley J. Impact of glucocorticoids (GC) and interferon (IFN) on outcome of multiple myeloma. *ASCO Proceedings* 1992; **11**: 316 (abstract).

106. Alexanian R, Yap BS, Bodey GP. Prednisone pulse therapy for refractory myeloma. *Blood* 1983; **62**: 572–7.

107. Alexanian R, Barlogie B, Dixon DO. High dose glucocorticoid treatment for resistant multiple myeloma. *Annals of Internal Medicine* 1986; **105**: 8–11.

108. Alexanian R, Dimopoulos MA, Delasalle K, Barlogie B. Primary dexamethasone treatment of multiple myeloma. *Blood* 1992; **80**: 887–90.

109. Friedenberg WR, Kyle RA, Knospe WH et al. High-dose dexamethasone for refractory relapsing multiple myeloma. *American Journal of Hematology* 1991; **36**: 171–5.

110. Norfolk DR, Child JA. Pulsed high dose oral prednisolone in relapsed or refractory multiple myeloma. *Hematological Oncology* 1989; **7**: 61–8.

111. Kildahl-Andersen O, Bjark P, Bondevik A et al. Multiple myeloma in central Norway 1981–1982: a randomized clinical trial of 5-drug combination therapy versus standard therapy. *Scandinavian Journal of Haematology* 1986; **35**: 518–24.

112. Selby PJ, McElwain TJ, Nandi AC et al. Multiple myeloma treated with high-dose intravenous mephalan. *British Journal of Haematology* 1987; **66**: 55–62.

113. Lokhorst HM, Meuwissen OJAT, Verdonck LF, Dekker AW. High-risk multiple myeloma treated with high-dose melphalan. *Journal of Clinical Oncology* 1992; **10**: 47–51.

114. Case DC, Coleman M, Gottlieb A, McCarroll K. Phase I–II trial of high-dose melphalan in previously untreated stage III multiple myeloma: Cancer and Leukaemia Group B Study 8512. *Cancer Investigations* 1992; **10**: 11–17.

115. Barlogie B, Jagannath S, Dixon DO et al. High-dose melphalan and granulocyte-macrophage colony-stimulating factor for refractory multiple myeloma. *Blood* 1990; **76**: 677–80.

116. Harousseau JL, Milpied N, Laporte JP et al. Double-intensive therapy in high-risk multiple myeloma. *Blood* 1992; **79**: 2827–33.

117. Lenhard RE, Oken MM, Barnes JM et al. High dose cyclophosphamide: an effective treatment for advanced myeloma. *Cancer* 1984; **53**: 1456–60.

118. Dimopoulos MA, Delasalle KB, Champlin R, Alexanian R. Cyclophosphamide and etoposide therapy with GM-CSF for VAD-resistant multiple myeloma. *British Journal of Haematology* 1993; **83**: 240–4.

119. Tsakanikas S, Papanastasiou K, Stamatelou M, Maniatis A. Intermediate dose of intravenous melphalan in advanced multiple myeloma. *Oncology* 1991; **48**: 369–71.

120. Petrucci M, Avvisati G, Tribalto M et al. Intermediate-dose (25 mg/m^2) intravenous melphalan for patients with multiple myeloma in relapse or refractory to standard treatment. *European Journal of Haematology* 1989; **42**: 233–7.

121. Back HRL, Rodjer S, Westin J. Single-dose intravenous melphalan in advanced multiple myeloma. *Acta Haematologica* 1990; **83**: 183–6.

122. Epstein J, Xiao H, Koba B. P-glycoprotein expression in myeloma is associated with resistance to VAD. *Blood* 1989; **74**: 913–17.

123. Dalton WS, Grogan TM, Meltzer PS et al. Drug-resistance in multiple myeloma and non-Hodgkin's lymphoma: detection of p-glycoprotein and potential circumvention by addition of verapamil to chemotherapy. *Journal of Clinical Oncology* 1989; **7**: 415–24.

124. Grogan TM, Spier CM, Salmon SE et al. p-Glycoprotein expression in human plasma cell myeloma: correlation with prior chemotherapy. *Blood* 1993; **81**: 490–5.

125. Salmon SE, Dalton WS, Grogan TM et al. Multidrug resistant myeloma: laboratory and clinical effects of verapamil as a sensitiser. *Blood* 1991; **78**: 44–50.

126. Sonneveld PDB, Lokhorst HM, Marie J-P et al. Modulation of multidrug-resistant multiple myeloma by cyclosporin. *Lancet* 1992; **340**: 255–9.

127. Lichtman SM, Mittelman A, Budman DR et al. Phase II trial of fludarabine phosphate in multiple myeloma using a loading dose and continuous infusion schedule. *Leukemia and Lymphoma* 1991; **6**: 61–3.

128. Kraut EH, Crowley JJ, Grever MR et al. Phase II study of fludarabine phosphate in multiple myeloma. *Investigational New Drugs* 1990; **8**: 199–200.

129. Dimopoulos MA, Kantarjian HM, Estey EH, Alexanian R. 2-Chlorodeoxyadenosine in the treatment of multiple myeloma (letter). *Blood* 1992; **79**: 1626.

130. Dimopoulos MA, Kantarjian H, Estey EH, Alexanian R. Treatment of Waldenstrom's macroglobulinaemia with 2-chlorodeoxyadenosine. *Annals of Internal Medicine* 1993; **118**: 195–8.

131. Nagourney RA, Evans SS, Messenger JC et al. 2 Chlorodeoxyadenosine activity and cross resistance patterns in primary cultures of human hematologic neoplasms. *British Journal of Cancer* 1992; **67**: 10–14.

132. Belch AR, Henderson JF, Brox LW. Treatment of multiple myeloma with deoxycoformycin. *Cancer Chemotherapy and Pharmacology* 1985; **14**: 49–52.

133. Grever MR, Crowley J, Salmon S et al. Phase II investigation of pentostatin in multiple myeloma: a Southwest Oncology Group study. *Journal of the National Cancer Institute* 1990; **82**: 1778–9.

134. Berman E, McBride M. Comparative cellular pharmacology of daunorubicin and idarubicin in human multidrug-resistant leukaemia cells. *Blood* 1992; **79**: 3267–73.

135. Chisesi T, Capnist G, De Dominicis E, Dini E. A phase II study of idarubicin (4-methoxydaunorubicin) in advanced myeloma. *European Journal of Cancer and Clinical Oncology* 1988; **24**: 681–4.

136. Alberts AS, Falkson G, Rapoport BL, Uys A. A phase II study of idarubicin and prednisone in multiple myeloma. *Tumori* 1990; **70**:465–6.

137. Eridani S, Slater NGP, Singh AK, Pearson TC. Intravenous and oral demethoxydaunorubicin (NSC 256–439) in the treatment of acute leukaemia and lymphoma: a pilot study. *Blut* 1985; **50**: 369–72.

138. Franklin IM, Sharp S, Murphy J, Tansey P. Z-DEX therapy for myeloma. *British Journal of Haematology* 1994; **86**: (suppl. 1): 77 (abstract).

139. Greipp PR, Colemen M, Anderson K, McIntyre OR. Phase II trial of amsacrine in patients with multiple myeloma. *Medical Pediatric Oncology* 1989; **17**: 76–8.

140. Karanes C, Crowley J, Sawkar L et al. Aclacinomycin-A in the treatment of multiple myeloma: a Southwest Oncology Group study. *Investigational New Drugs* 1990; **8**; 101–4.

141. Durie BG, Crowley J, Coltman CAJ et al. Phase II evaluation of bisantrene in refractory multiple myeloma. *Investigational New Drugs* 1991; **9**: 329–31.

142. Hanson KH, Crowley J, Salmon SE et al. Evaluation of amonafide in refractory and relapsing multiple myeloma. *Anti-Cancer Drugs* 1991; **2**: 247–50.

143. Akashi M, Sakamoto S, Ohla M et al. Treatment of multiple myeloma with carboquone-prednisolone. *European Journal of Haematology* 1989; **42**: 265–9.

144. Barrier JH, Le Noan H, Mussini JM, Brisseau JM. Stabilization of a severe case of POEMS syndrome after tamoxifen administration (letter). *Journal of Neurology Neurosurgery and Psychiatry* 1989; **52**: 286.

145. Bosch SCH, Frias Z. Radiotherapy in the treatment of multiple myeloma. *Journal of Radiation, Oncology, Biology and Physics* 1988; **15**: 1363–9.

146. Adamietz IA, Schober C, Schulte RW et al. Palliative radiotherapy in plasma cell myeloma. *Radiotherapy and Oncology* 1991; **20**: 111–16.

147. Leigh BR, Kurtts TA, Mack CF et al. Radiation therapy for the palliation of multiple myeloma. *Journal of Radiation, Oncology, Biology and Physics* 1993; **25**: 801–4.

148. Rowell NP, Tobias JS. The role of radiotherapy in the management of multiple myeloma. *Blood Reviews* 1991; **5**: 84–9.

149. Norin T. Roentgen treatment of myeloma with special consideration to the dosage. *Acta Radiologica* 1957; **47**: 46–54.

150. Price P, Hoskin PJ, Easton D et al. Prospective randomized trial of single and multifraction radiotherapy schedules in the treatment of painful bony metastases. *Radiotherapy and Oncology* 1986; **6**: 247–55.

151. Bergsagel DE. Total body irradiation for myelomatosis. *British Medical Journal* 1971; **2**: 325.

152. Jaffe JP, Bosch A, Raich PC. Sequential hemibody radiotherapy in advanced multiple myeloma. *Cancer* 1979; **43**: 124–8.

153. Rowland CG, Garrett MJ, Crowley J et al. Half-body irradiation in plasma cell myeloma. *Clinical Radiology* 1983; **34**: 507–10.

154. Rostom AY, O'Cathail SM, Folkes A. Systemic irradiation in multiple myeloma: a report on nineteen cases. *British Journal of Haematology* 1984; **58**: 423–31.

155. Tobias JS, Richards JDM, Blackman GM et al. Hemibody irradiation in multiple myeloma. *Radiotherapy and Oncology* 1985; **3**: 11–16.

156. Rostom AY. A review of the place of radiotherapy in myeloma with the emphasis on whole body irradiation. *Hematology and Oncology* 1986; **6**: 193–8.

157. Singer CR, Tobias JS, Giles F et al. Hemibody irradiation – an effective second-line therapy in drug-resistant multiple myeloma. *Cancer* 1989; **63**: 2446–51.

158. Giles FJ, Richards JDM, Tobias EJ et al. Prospective randomized study of double hemi-body irradiation with and without subsequent maintenance recombinant alpha 2b interferon on survival in patients with relapsed multiple myeloma. *European Journal of Cancer* 1992; **28A**(8–9):1392–5.

159. MacKenzie MR, Wold H, Gandara GD et al. Consolidation hemibody radiotherapy following induction combination chemotherapy in high-tumor-burden multiple myeloma. *Journal of Clinical Oncology* 1992; **10**: 1769–74.

160. Salmon SE, Tesh D, Crowley J et al. Chemotherapy is superior to sequential hemibody irradiation for remission consolidation in myeloma: a Southwest Oncology Group study. *Journal of Clinical Oncology* 1990; **8**: 1575–84.

161. Tobias JS, Rowell NP, Richards JD. Half-body irradiation in multiple myeloma (letter; comment). *Journal of Clinical Oncology* 1991; **9**: 705–7.

162. Troussard X, Leporrier M. Place of double half-body irradiation in the treatment of multiple myeloma? (letter). *Journal of Clinical Oncology* 1991; **9**: 2233–5.

CHAPTER 12

Treatment of refractory multiple myeloma with cytotoxic therapy and radiation therapy

FRANCIS J GILES

Introduction	130	**Systemic radiation therapy**	137
Classification of refractory disease	131	**Cytotoxic drug resistance**	139
Cytotoxic therapy	133	**Conclusions**	141
Myeloablative cytotoxic therapy	136	**References**	141

Introduction

Induction therapy with standard regimens induces an average objective response (OR) rate (Southwestern Oncology Group Criteria, SWOG) of 40–50 per cent in patients with previously untreated multiple myeloma (MM).[1,2] The great majority of those who initially achieve remission eventually relapse, with less than 20 per cent of patients being in ongoing remission at 5 years from time of initial therapy.[1] The majority of plateau-phase patients will ultimately have evident overt progression of disease. All patients stopping therapy must be closely monitored for the possibility of relapse. A complete battery of tests is necessary at 6 monthly intervals with an annual skeletal survey. Serum paraprotein levels may be insensitive for the detection of early relapse in some patients. Some 10–15 per cent of patients relapse with 'Bence Jones' escape, in which the serum paraprotein levels may drop rather than increase at the time of development of relapse. Relapse may develop in an extramedullary site. It is dangerous to rely on single parameters for detection of relapse and full, frequent clinical and laboratory reviews are necessary.

At least 30 per cent of MM patients fail to respond adequately to induction chemotherapy.[3,4] Patients may exhibit clinical deterioration, increasing paraprotein levels, increasing marrow suppression, refractory hypercalcemia or worsening renal failure while undergoing induction therapy. These features may occur in any combination and the diagnosis of 'primary resistant disease' may be difficult to make. No standard definition of this entity has been agreed upon. Patients who do not achieve an objective response to induction therapy are sometimes described as 'non-responders' – this group includes patients who are truly primarily resistant to front-line therapy, patients whose disease is in plateau phase at time of disease presentation, and patients who are slow responders to their induction therapy. The general term 'refractory' is often applied to patients who have either primary resistant or relapsed disease. The term 'advanced' is often used to describe patients with the latter condition. The lack of agreed terminology is a major stumbling block to data review and the logical planning of clinical studies. A classification of relapsed and resistant disease is proposed later in this chapter.

Prolongation of median survival beyond 1 year in studies of therapy for refractory MM patients is rarely achieved.[3–104] Unless MM patients die of other intervening causes, they eventually develop drug resistance and die of their disease. This has led to attempts to improve treatment by the investigation of mechanisms of cytotoxic drug resistance, the development of methods to identify drug-resistant MM cells, and attempts to either prevent or, once established, to circumvent drug resistance.[105–109] The current data on cytotoxic and/or systemic radiation therapy will be analyzed with an emphasis on the potential role of cytokines and cytotoxic drug resistance reversal agents to help improve upon our current poor results. For discussion purposes, treatment options for patients with refractory MM are arbitrarily divided into data on standard-dose cytotoxic regimens; myeloablative therapy with autologous stem cell rescue; systemic radiation therapy and cytotoxic drug resistance modulation therapy.

Classification of refractory disease

Refractory MM patients may be divided into two major groups: primarily resistant, and secondarily resistant or relapsing patients. Primarily resistant patients comprise those who either progress on initial therapy or those who initially remain stable or fail to achieve an objective response and then have evident disease progression while still receiving induction therapy. Those patients whose disease progresses while they are still receiving initial therapy represent the most unfavorable prognostic subgroup of all. Distinction between the subgroups of primary resistant patients is of great importance because of recent data which have highlighted the efficacy of myeloablative therapy in patients with true primary resistant disease.[110–117] Some patients who neither respond to induction therapy nor progress, i.e. remain stable, have a relatively good prognosis and should be clearly distinguished from the above subgroups of resistant patients. The inclusion of stable response

patients in an overall 'non-responding' category is inappropriate and misleading.

Secondarily resistant or relapsing patients include those who had an initial response and subsequently relapsed either during or off therapy. The term 'relapsing' should be reserved for patients who have completed and responded to induction cytotoxic therapy and subsequently exhibit disease progression. Interferon-α maintenance therapy has been shown to prolong plateau phase in a number of prospective randomized studies and should now be regarded as standard of care.[118–122] Thus patients may now relapse from unmaintained or maintained remissions of variable duration. Classification of refractory MM patients must thus address the duration of remission preceding relapse and whether maintenance therapy is given or not. Figure 12.1 presents a suggested classification of refractory MM patients which could form a basis for comparison of data from differing clinical studies and also allows stratification of patients for entry on to clinical protocols.

As is becoming increasingly evident from analysis of maintenance interferon data, attention must be directed to the differing criteria of response to MM therapy which are currently employed when reviewing any body of therapeutic data, including that of refractory disease. The two generally used sets of response criteria are those of the Chronic Leukemia–Myeloma Task Force of the National Cancer Institute (MTF) and those of the SWOG.[123,124] Although there are some shared parameters of response – increase in hemoglobin, decrease in serum calcium, recalcification of lytic lesions, etc. – the primary index of response in both is the measurement of serum and/or urinary paraprotein concentration. The critical difference between the MTF and SWOG criteria is that the latter requires at least a 75 per cent reduction in paraprotein synthesis (which takes into account the catabolism of IgG1, 2, and 4 and changes in plasma volume), whereas the former demands a > 50 per cent reduction in the serum paraprotein concentration. These criteria are primarily used to assess response to initial induction therapy and are equivalently irrelevant as prognostic discriminants in this circumstance.[125] These criteria may well be also of little value in predicting survival for patients with refractory disease. However, we are now moving into an era of increasing usage of maintenance immunotherapy for patients who do respond to second-line therapy and thus the specific criteria of response used may be of increasing relevance.

Reductions in M-protein levels do not seem to truly reflect changes in overall tumor mass. Analysis of current response criteria and their correlation with survival duration indicate that the magnitude of regression as assessed by M-protein reduction does

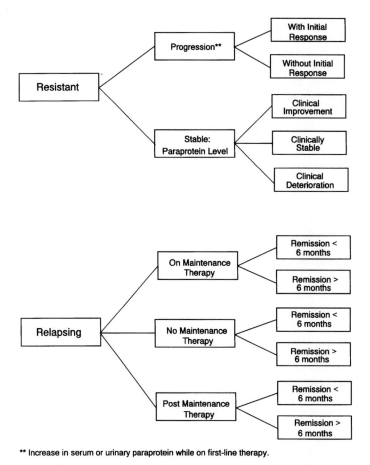

** Increase in serum or urinary paraprotein while on first-line therapy.

Fig. 12.1   Classification of refractory multiple myeloma.

not correlate in a quantitative way with the anticipated response or survival duration.[125,126] Thus a 75 per cent regression is not necessarily intrinsically better than a 25 per cent regression because all of the responses achieved with conventional therapy are partial responses. Cytotoxic therapy has improved survival by reducing early deaths from MM complications and by slowing down tumor progression in some patients. The tumor bulk at presentation, intrinsic disease kinetics and drug sensitivity in individual patients are the major determinants of survival.

Initial regression percentage cannot be used as a parameter to indicate the relative efficacy or nonefficacy of a particular induction approach in MM unless pathological complete response (CR) has occurred. Data exist to both support and refute the concept that patients who respond very quickly to induction therapy have a worse prognosis than those with a slower response. Very high initial cell kill rates are not necessarily associated with enhanced overall survival.[127] Multidrug regimens consistently give higher response rates than melphalan/prednisone

(MP), but this does not translate into prolonged survival. True pathological complete responses must become the accepted early end-point on therapeutic studies, with plateau-phase duration and overall survival being ultimate study end-points. In a disorder like MM where an elderly population is being studied, the natural mortality rate needs to be considered.

Many factors confound comparison of therapeutic studies in MM. Some studies are reported in terms of the responses achieved without the provision of survival data. 'Early' deaths are variably included or excluded in reports. Stage I patients are still included in some studies. The quality and availability of support services, particularly dialysis, is variable. Overall survival (tumor-related and tumor-unrelated deaths both counted) is reported in some studies, tumor-related survival in others. The marked influence of active and passive exclusions on the results of clinical studies in MM therapy has been highlighted recently.[128] Varying exclusion criteria, particularly on grounds of advanced age, severe renal failure and poor performance scores, may have a profound impact on survival data.

Cytotoxic therapy

Over 40 Phase II studies testing the efficacy of single agents in refractory MM have been reported. Most single agents have an objective response rate of < 0 per cent, including: acronine,[48] amsacrine (*m*-AMSA),[49] bleomycin,[46] chlorozotocin,[50] diaziquone,[52] etoposide,[25] lomustine (CCNU),[44] mitomycin,[42] mitoxantrone,[16] piperazinedione,[45] predmustine,[55] procarbazine,[43] pyrazofurin,[47] and vindesine.[72] Only twelve drugs have been reproducibly shown to induce an objective response rate > 10 per cent in patients with refractory MM. These are interferon-α,[83] busulfan,[104] cyclophosphamide,[11] 2'-deoxycoformycin,[60] doxorubicin,[56] epirubicin,[62] glucocorticoids,[40] melphalan,[8] peptichemio,[61] poly(I,C)-LC,[58] teniposide (VM–26),[57] and vincristine.[59]

The overall MTF response rate to single agent high-dose or pulsed glucocorticoids is approximately 40 per cent both for resistant and relapsing patients.[23,29,40] Lacking myelosuppression, corticosteroids are suitable for patients with poor bone marrow reserve. No data exist which would justify the use of a single agent as standard therapy for refractory MM except where marrow or other organ failure limit one's ability to treat appropriately.

While alkylator combinations are inadequate for primary resistance patients, for some patients with unmaintained prolonged (> 6 months) remissions, retreatment with the same regimen used at time of induction represents an reasonable form of therapy. However, considering the poor eventual survival that results from this approach, treatment of relapsed patients on novel protocols probably represents a better approach. Where alkylator regimens are used as salvage regimens, current data would indicate MTF response rates of < 10 per cent for primary resistant patients and some 20 per cent for relapsing patients. In many studies, single agents have been combined with high-dose or pulsed glucocorticoids for evaluation of response. Responses resulting from these combinations are usually equivalent to that seen with glucocorticoid treatment alone. The addition of alkylators or vinca alkaloids to glucocorticoids primarily enhances toxicity without significantly increasing response rate or survival and should be avoided.[12,14]

A notable exception to this principle is the data generated by the Finnish Leukemia Group on the M–2 variant MOCCA regimen.[93] This regimen consists of vincristine 0.03 mg/kg i.v. on day 1, cyclophosphamide 10 mg/kg i.v. on day 1, lomustine 40 mg p.o. on day 1, melphalan 0.25 mg/kg p.o. on days 1–4 and methylprednisolone 0.8 mg/kg p.o. on days 1–7, repeating the course every 35 days. In a Phase II study, eighty patients with refractory MM, 34 per cent with primary resistance, 66 per cent with relapsed disease, were treated with MOCCA as a second-line regimen. An objective response (MTF criteria) was achieved in 49 per cent of patients, 52 per cent in primary resistant patients and 47 per cent in relapsed patients. The median duration of response was 22 months. Severe adverse events, mainly sepsis, occurred in 30 per cent of patients, and were fatal in 9 per cent of patients. The response rate with MOCCA for all patients who had relapsed off-treatment was 92 per cent, in marked contrast to the 34 per cent response rate for patients relapsing during maintenance chemotherapy ($p < 0.001$). These data emphasize the need for patient stratification as suggested in the above classification schema.

The median survival of all patients on this study was 31 months from the start of the MOCCA regimen.[93] The median survival was 40 months for primarily resistant patients and 21 months for relapsing patients. The survival advantage gained by the response was approximately 2 years for all responders compared with non-responders, the medians being 37 and 17 months, respectively. Hemorrhagic cystitis, a complication of prolonged cyclophosphamide therapy, contributed to two fatalities on this study. Current overall response rates attributed to the non-MOCCA M–2 variant regimens is approximately 25 per cent,[35,36,38] considerably less than the 49 per cent recorded by the Finnish group. The authors stress that the different response rates are illustrative of the difficulties in comparing various strategies for second-line chemotherapy in MM. None the less, the data on the MOCCA regimen are strikingly better than those generated using seemingly similar regimens and this regimen warrants further clinical study.

Significant progress in the therapy of refractory MM with cytotoxic agents really began with the introduction of the vincristine, doxorubicin, and dexamethasone (VAD) regimen.[5] Many studies of doxorubicin-containing regimens in refractory MM had been carried out prior to this and these showed that resistant patients had a < 15 per cent MTF response rate, and that < 30 per cent of relapsing patients responded. Non-VAD doxorubicin-containing regimens have no major advantage over high-dose or pulsed glucocorticoids alone in terms of toxicity, cost, and efficacy in refractory MM. They have no clear role unless a significant contraindication to high-dose or prolonged corticosteroid therapy exists or VAD chemotherapy is impossible.

Based on initial data that a combination of vincristine, doxorubicin, and pulsed prednisone induced substantial responses in refractory MM,[40] the M.D. Anderson Cancer Center group studied even higher doses of glucocorticoids combined with a continuous infusion of vinca alkaloid and doxorubicin in an attempt to eliminate more slowly proliferating plasma cells.[5] The VAD regimen has an MTF response rate in resistant and relapsing and resistant MM patients of 50 per cent and 75 per cent respectively. This group went on to compare the response to VAD with that of high-dose dexamethasone alone, both in resistant and relapsing patients. Both regimens had similar response rates for resistant patients, but the VAD regimen seemed much more effective in relapsing patients. For relapsing patients, the VAD regimen is one of the most active combinations reported to date, it may have activity in patients who have previously failed a doxorubicin combination. The VAD regimen confers a median survival of approximately 12 months on refractory patients in studies published to date.[5,6,13,34,81,82,96,99,105] The most common serious adverse effects include infection, paralytic ileus, depression, psychosis, thrombosis, and peptic ulceration. Although moderate to severe toxicity occurs in approximately one-third of patients treated, VAD or its mitoxantrone-[10,28,92,97,98] or methylprednisolone[7,75]-containing variants constitutes the non-protocol treatment of choice for relapsing MM patients, including those refractory to preceding doxorubicin-containing regimens.

Attempts to modify the VAD regimen by giving bolus vincristine and/or doxorubicin have failed to improve upon the originally described regimen.[13,99] The NCI (Canada) have reported on a modified VAD regimen (m-VAD)[13] used in primary refractory and relapsed MM patients, in which intravenous vincristine (0.4 mg/day) and Adriamycin (9 mg/m^2 per day) infusions were administered over 2 hours on days 1–4 of each 28-day cycle. In addition, only two 4-day courses of dexamethasone (40 mg/day) were given during each cycle. All patients entered on study had received MP as induction therapy. A total of 206 courses of treatment were delivered to 44 patients evaluable for toxicity. Thirty-four patients received at least three courses of therapy, 3 patients received two courses, and 7 patients received only one course. There were no complete responses. Of 41 evaluable patients who completed at least one course of therapy 27 per cent had an objective response (MTF criteria) with a response rate of 19 per cent for primary resistant patients, and 32 per cent for those with relapsed disease. Median duration of response was 4 months. All responding patients relapsed from m-VAD therapy by 20 months and at 1 year, 30 per cent of patients were alive. Median survival for all 44 patients was 7.6 months (5.5 months for primary

resistant patients, 10 months for relapsed). Five deaths were judged as treatment-related and occurred during marrow cytopenia. These results are inferior to those reported for the original VAD regimen for which response rates were 73 per cent and 43 per cent in patients with relapsed disease and primary resistant disease, respectively.

The Spanish PETHEMA group studied another variant of the VAD regimen, VBAD,[33] in which patients received vincristine 1 mg i.v., BCNU and doxorubicin 30 mg/m^2 each i.v. on day 1, and dexamethasone 25 mg/m^2 on days 1–4, 9–12, and 17–20 in the first course and on days 1–4 in subsequent courses. Cycles were repeated at 28-day intervals. Responding patients continued to receive VBAD treatment until disease progression was observed. When a total dose of doxorubicin of 540 mg/m^2 was reached, this agent was discontinued and further therapy was restricted to vincristine, BCNU, and dexamethasone. Sixty-five patients, of whom 50 per cent had primary resistant MM and 50 per cent relapsed disease, entered on a Phase II study, from which data were available on sixty evaluable patients. All patients had received prior therapy with MP alone. All patients received H–2 blockade, and antibiotic prophylaxis with trimethoprim-sulfametoxazole during early courses of therapy. A mean number of nine cycles per patients (range 1–30) of VBAD or VBD cycles was administered. An objective response rate (MTF criteria) of 22 per cent and minor response rate of 15 per cent was recorded. Response rates were significantly higher in primarily resistant patients than in relapsing patients (48 per cent versus 24 per cent, $p < 0.05$).

Overall median survival on this study was 13 months with a median survival for responding patients of 30 months in contrast to one of 10 months for non-responders. This difference was highly significant when analyzed by means of the standard log-rank method, using response as an initial variable ($p < 0.001$). However, when the Mantel and Byar procedure and landmark analysis method were applied, this difference was no longer statistically significant. These data highlight the problems inherent in analysis therapeutic data in refractory MM in terms of response versus non-response. Demonstration of significant differences in survival between responders and non-responders to a given regimen may, using some statistical methods, introduce a lead-time bias in favor of responders (i.e. the time from the start of treatment necessary to detect the response). In addition, responders may survive longer than non-responders for reasons other than the treatment itself. In general, as with *de novo* patient treatment, there is a poor correlation between response and survival in refractory MM. The effectiveness of the VBAD regimen seems equivalent to that of VAD in the therapy of refractory MM.

On the basis of data suggesting a lower cardiotoxicity of mitoxantrone, many investigators have focused on mitoxantrone-containing variants of the VAD regimen.[10,28,92,97,98] The Nordic Myeloma Study Group (NMSG) have reported on a Phase II study in which fifty-eight patients with refractory MM were treated with an NOP combination chemotherapy regimen[92] consisting of mitoxantrone (16 mg/m^2 for the first 25 patients and 12 mg/m^2 for the subsequent 33), vincristine (2 mg), both as bolus injections on day 1 and prednisone (250 mg/day on days 1–4 and 17–20 of a 28-day cycle). An objective response (MTF criteria) was obtained in 26 per cent of patients and a minor response (clinical improvement but < 50 per cent reduction in serum paraprotein level) in 21 per cent. Median response duration was 6.7 months and median survival for all patients was 6.2 months. There were no differences in response rate or duration between patients receiving the higher or lower mitoxantrone dosages, but patients in the lower dose group had fewer serious infections. This variant of the NOP regimen seems inferior to VAD.

In a further Phase II study the NMSG treated ninety-two patients with a different NOP regimen:[10] mitoxantrone (bolus injection of 4 mg/m^2 on days 1–4), vincristine (continuous infusion of 0.4 mg/24 h on days 1–4) and prednisone (250 mg/day on days 1–4 and 17–20 of a 28 day cycle). Patients were treated after they were found refractory to treatment with MP (and occasionally vincristine) ($n = 50$) or more complex treatment regimens ($n = 42$) including anthracyclines ($n = 18$). Thirty-five patients stopped NOP treatment and left the protocol due to their own wish (8 per cent), toxicity of the treatment (25 per cent), or progressive disease (67 per cent). In the second cycle of NOP therapy only 60 per cent of patients received full-dose therapy with 30 per cent receiving half the intended mitoxantrone dose. In subsequent treatment cycles about 40 per cent of patients received the full intended dose. Objective

responses (MTF criteria) were obtained in 25 per cent of patients and minor responses in 24 per cent of patients. Median duration of response was 7.5 months. Response rates were not obviously affected by the type of previous treatment.

Among the fifty patients who had previously received MP ± V, 26 per cent obtained objective responses and a further 30 per cent minor responses. Of the twenty-four patients who had received prior intensive treatment regimens without anthracyclines, 25 per cent obtained OR and 17 per cent minor responses. Of those eighteen patients already treated with anthracyclines, objective responses were obtained in 20 per cent and minor responses in 17 per cent. Among the seven patients who had received VAD prior to NOP, only one obtained a minor response, one patient stable disease, and five had progressive disease or early death. The overall relapse-free survival (time from start of treatment until death or relapse) was 8.5 and 6.8 months for objective response and minor response patients, respectively. There was no significant difference in the survival of patients who previously had received MP ± V compared to those who had received other intensive treatment schedules either with or without anthracyclines. Overall median survival for the study cohort was 9.5 months and thus seems directly comparable to that seen with the VAD regimen.

Barlogie et al., when initially describing the VAD regimen, emphasized the potential importance of prolonged exposure to cytotoxic drugs, i.e. continuous intravenous infusions in the therapy of MM.[5] On our current CEVAD regimen (Table 12.1), etoposide, vincristine and doxorubicin will be given by continuous i.v. infusions in a manner similar to that of the EPOCH regimen developed by the NCI.[129] This latter regimen is very effective in the therapy of refractory lymphoma.[129,130] Cyclophosphamide, when used in combination with etoposide, doxorubicin, and high-dose betamethasone (EACB) as salvage therapy, has

Table 12.1 The CEVAD regimen

Therapy	Day	1	2	3	4	5	6	7→	21
Cyclophosphamide 1000 mg/m^2 i.v.							X		
L1 Etoposide 50 mg/m^2/day continuous i.v.		X	X	X	X				
L2 Vincristine 0.4 mg/day continuous i.v.		X	X	X	X				
L2 Doxorubicin 9 mg/m^2/day continuous i.v.		X	X	X	X				
Dexamethasone 40 mg p.o. daily		X	X	X	X				
rGM-CSF 250μ g/m^2 s.c.								X →*	

* Continue until ANC > 2 x10^9/l.
L1 – Line one of double lumen central venous catheter.
L2 – Line two of double lumen central venous catheter.
Patients should receive H2-blocker plus trimethoprim–sulfamethoxazole. Allopurinol should be prescribed for at least first cycle of chemotherapy. Patients must be vigorously hydrated during therapy.

given favorable results.[100] Cyclophosphamide combined with etoposide and RGM-CSF is effective therapy for VAD-resistant/refractory MM.[94] CEVAD will allow us to examine the effectiveness of bolus cyclophosphamide and infusional etoposide given in combination with the VAD regimen. Recombinant GM-CSF is used to allow on-schedule delivery of this intensive regimen. Interferon-α combined with dexamethasone, which will be used as a maintenance regimen on this protocol, is effective in refractory MM.[2,31] All responding patients receive an interferon-α2b/dexamethasone maintenance regimen. Data to date indicate a response rate at least equivalent to that of VAD alone.

Myeloablative cytotoxic therapy

High-dose melphalan, with or without total body irradiation (TBI), and similar myeloablative cytotoxic regimens are increasingly successful in minimally treated MM and are under intense study in patients with refractory MM.[21,22,110–117,131,132] Many patients with refractory MM have received myeloablative treatment supported by peripheral blood stem cells (PBSC) although, because of the high risk of serious complications, such treatments have usually been limited to patients younger than 60, with good performance and without serious medical complications. Most reports have included patients in diverse phases of disease, and only one has compared results with those of similar patients who did not receive intensive therapy.

Data on 135 patients with refractory MM who received one of three intensive regimens have clearly indicated the current optimal form of myeloablative therapy for this patient group.[113] The three regimens reported on were melphalan at doses of 90–100 mg/m^2 (M100) (47 patients) without autotransplant; TBI (850 cGy) with either melphalan 140 mg/m^2 or thiotepa 750 mg/m^2 and autologous bone marrow transplant (ABMT) (21 patients); melphalan 200 mg/m^2 (M200) supported by both peripheral blood stem cells (PBSC) and ABMT plus GM-CSF (67 patients). Mortality within 2 months of therapy was 20–25 per cent with the M100 and TBI regimens, but less than 1 per cent with the M200 regimen, mainly because the duration of severe neutropenia was shortened to less than 1 week due to infusion of PBSC and use of growth factor therapy.

Low ß$_2$-microglobulin (ß$_2$-M) levels (2.5 mg/l) and M200 therapy were identified as the two most important independent favorable variables associated with prolonged event-free survival (EFS) and overall sur-

vival (OS). On the basis of these parameters, three risk groups were defined: 29 patients with low ß$_2$-M receiving M200 had a median EFS duration of 37 months and projected OS of 43 months; 54 patients with either of the favorable parameters had EFS and OS durations of 16 and 36 months, respectively; and 52 patients with high ß$_2$-M who did not receive the M200 regimen had a very poor prognosis, with EFS of 3 months and OS of 5 months ($p < 0.0001$). Primary resistant disease, low serum LDH, less than 12 months of prior therapy, and younger age (less than 50 years) were additional favorable prognostic features. Further analysis that excluded treatment as a variable identified high ß$_2$-M and resistant relapse as the two major adverse prognostic factors, one of which was present in 80 per cent of the 135 patients. Among these 108 high-risk patients, prognosis was improved markedly with the M200 regimen because of both better supportive care (PBSC and hematopoietic growth factors) and more intensive therapy.

These data represent the largest series of intensive therapy for MM administered in the setting of refractory disease, using progressively more intensive therapy and improved supportive care. The early mortality of approximately 25 per cent with TBI has been dramatically reduced to 1 per cent, mainly as a result of rapid hematopoietic engraftment with PBSC added to ABMT and recombinant GM-CSF. In support of this conclusion was the observation of a similarly low mortality among the 42 second-transplant recipients in the M200 group, whether M200 was applied twice (no death among 24 patients) or TBI was added (one death among 18 patients). M200 also prolonged EFS and OS markedly, especially among patients with resistant relapse. In adjusting for the heterogeneity of the three treatment groups, multivariate regression analysis was applied and identified low ß$_2$-M, reflecting low tumor mass, and M200 (rather than TBI or M100) as the two most important independent favorable features associated with prolonged EFS and OS. A superior outcome after M200 was confirmed by a separate multivariate analysis that excluded treatment as a variable and identified low ß$_2$-M and primary resistant MM as principal favorable parameters. Among the 80 per cent of patients in the poor-risk category (less than two favorable features), M200 was superior to the remaining therapies both with lower mortality and marked extension of EFS and OS. Longer EFS and OS in patients with primary resistant disease as opposed to relapsed disease was partly explained by fewer months of prior treatment before intensive therapy in the former group (median of 10 months versus 33 months with resistant relapse, $p = 0.0001$). However, as shown by multivariate analysis of pretreatment variables, the type of resistance was the principal feature, whereas differences in prior therapy duration did not retain independent significance.

The major observation from this body of data is that dose intensification with M200 with tandem myeloablative/stem cell rescue procedures resulted in extended EFS and OS, independent of other well-recognized favorable prognostic variables. These results support the contention that resistance of therapy in MM patients can be overcome by increasing dose intensity. The use of cyclophosphamide-primed PBSC and recombinant GM-CSF assure rapid hematopoietic engraftment post myeloablative therapy, and thus virtually eliminating early procedure-related mortality.

A previous study by this group involved the use of TBI (850 cGy) with either melphalan 140 mg/m^2 or thiotepa 750 mg/m^2 and ABMT rescue in patients with refractory MM.[114] They reported that patients with resistant relapse had an especially poor response to this regimen, whereas those with primary resistance had an equivalent prognosis to those patients receiving the same regimen as response consolidation. These data again emphasize the need for appropriate classification of refractory MM patients and would suggest that all patients who exhibit primary resistance to an adequate trial of a recognized induction regimen should be considered for urgent myeloablative therapy. Current data on myeloablative therapy are a marked contrast to the failure of more modest dose increments of melphalan and other active agents in the therapy of advanced MM. Based on current data, the 'more is better' approach to therapy of refractory MM, utilizing myeloablative doses of active single agents, seems likely to dominate clinical studies in this area for the foreseeable future.

The optimal timing of myeloablative therapy, either as consolidation or as refractory therapy has been addressed recently.[117] The utility of myeloablative therapy supported by ABMT or PBSC was assessed in forty-nine patients with MM who had received at least 1 year of prior chemotherapy. Results were compared with those of similar patients who qualified for marrow transplantation in all respects but were denied treatment primarily for socioeconomic reasons. Because they continued to receive standard care, such patients appeared to provide a suitable comparison group for the patients who received intensive regimens. The clinical features, response, and survival time of the control patients were similar to those observed in previously reported studies of VAD in patients with refractory MM.

Among patients with disease in resistant relapse despite VAD treatment, a 61 per cent response rate was associated with a median remission time of 5 months. After primary resistance for more than 1 year, 6 of 15 patients responded and the overall survival was similar to that of control patients. For patients with melphalan-resistant disease that responded to VAD, the remission time was similar to that of control patients.

Intensive therapy in patients with resistant relapse induced responses in 61 per cent of 23 patients with disease that was not responding to VAD therapy. Treatment-related deaths occurred in 4 patients (17 per cent). No patient achieved a complete remission, and the median survival was 8 months. When all patients are considered, the median remission was 3 months (5 months for responding patients) and no patient responded for more than 15 months. In all responding patients, the remission time after transplant was shorter than the first remission. Despite the slightly less advanced disease among transplanted patients ($p = 0.18$), survival after prior VAD was similar to that of 33 control patients. Response rates and remission times were the same regardless of the degree of plasmacytosis in transplanted marrow (range, 0–25 per cent).

Among 15 patients with primary resistant disease for at least 1 year, there were 2 treatment-related deaths, 6 patients responded, and none achieved a complete remission. Among the 6 responding patients, the median remission was 17 months. Although 4 patients responded for more than 1 year, survival after VAD was similar for comparable patients who did or did not receive myeloablative treatment ($p = 0.47$). Transplanted bone marrow contained 11–20 per cent plasma cells in 5 patients, among whom one patient responded for 16 months; with fewer plasma cells or with PBSC transplantation, 5 of 10 patients responded ($p = 0.26$).

Myeloablative therapy seems to be most effective in patients treated during the first year after diagnosis and prospective controled studies dealing with this issue are now indicated. Further pilot studies of myeloablative therapy should be reserved for patients earlier in their disease course, either for primary resistant disease or during a remission that is likely to be short after initial chemotherapy.

Systemic radiation therapy

In 1971 Bergsagel suggested that systemic radiation might be of value in MM since the relatively prolonged doubling time of the tumor should permit a high logarithmic cell kill with a clinically attainable radiation dose.[133] The necessity for 'protection of the patient's marrow function' with such therapy was highlighted. Bergsagel proposed either division of TBI into two doses to allow repopulation of the irradiated marrow by circulating hematopoietic progenitor cells, or ABMT rescue from TBI. Currently systemic radiation therapy in MM is indeed used either as part of a conditioning regimen pre-allogeneic or ABMT or as a double hemi-body irradiation (DHBI) procedure. Bergsagel suggested that if a dose of approximately 725 cGy was administered to the upper half of the body and 1000 cGy to the lower half, a 3 log cell kill might be achieved. Tumor debulking of this magnitude was expected to induce significant palliation and hopefully extend survival.

On the basis of this postulate, systemic radiotherapy was included in clinical trials for patients with refractory MM.[85–89,102,103,134] The use of hemibody irradiation (HBI) allows for higher doses to be given than would be possible with TBI, since the marrow reserve in the unirradiated half of the body is sufficient for circulating blood elements to be maintained, and for the irradiated marrow to be reseeded. This technique has been shown to be of value in patients with widespread metastases from solid cancers.[135]

HBI has been shown by a number of groups to be an effective, well-tolerated modality of therapy in refractory MM while recent data confirm that DHBI gives a higher response rate in relapsed MM patients than a single HBI procedure.[86–88,103] An immediately apparent effect of systemic radiation is significant relief of bone pain in most patients. The most common adverse effects of DHBI are myelosuppression, nausea, diarrhea, vomiting, pneumonitis, and stomatitis.

We have previously reported on a Phase II DHBI study in which fifty-nine patients with relapsed MM were entered.[103] It was intended that all patients receive DHBI. Immediately prior to first HBI patients were prospectively randomized to receive or not to receive subsequent interferon-α2b maintenance therapy. Thirteen of the total of 59 (22 per cent) (8 of 31 (26 per cent) in the control arm, 5 of 28 (18 per cent) in the interferon arm) received a single HBI due to progressive disease and/or persistent cytopenias following the initial procedure. Mean time between upper and lower HBI was 69 days (range 35–294). Of 23 patients randomized to receive interferon and completing DHBI, 15 patients (65 per cent) achieved peripheral blood counts adequate to allow interferon administration as per study criteria, commencing at a mean of 116 days (range 61–241) from time of study entry. There was no significant difference in median survival durations (10 months) from time of initial radiotherapy between control and interferon patient populations.

It has been suggested that DHBI gives a more durable response in refractory MM patients than a single HBI procedure.[88,89] This claim is based on a comparison of response rates and median survival times of patients receiving either HBI or DHBI in single-arm retrospective studies. The degree to which the initially administered HBI had acted as a 'screening' procedure is therefore difficult to assess. Survival for sufficient time to attain a hematological, biochemical, and clinical state adequate to undergo the second HBI might well represent inadvertent selection of relatively good prognosis patients. Recent data support previous claims that DHBI may offer a significant advantage to these patients in terms of quality of life while a significant gain in survival duration in comparison to aggressive combination cytotoxic regimens remains to be established in a prospective randomized study.

Prolonged red cell and/or platelet transfusion support may be necessary for those patients with pre-DHBI cytopenias due to extensive marrow plasma cell infiltrate but, even in this patient cohort with very advanced disease, good response may be seen. These problems have also recently been highlighted in data from a SWOG study investigating the use of DHBI as consolidation therapy following combination cytotoxic induction therapy or as second-line therapy in partial or non-responding patients.[136] A clear distinction must be made between the use of DHBI as early consolidation where it has proven to be ineffective and its use in refractory MM patients where it is effective.

We have previously documented the significant lung toxicity of the conventional MP induction regimens.[137] This underlines the need to monitor lung function, particularly pre-UHBI, and the adjustment of UHBI dosimetry based on data from these tests. These data are also important in the context of post-DHBI therapy as maintenance or at relapse. Pulmonary and marrow toxicity could be expected to be the most significant limiting factors in this regard. Sequential lung function testing in those patients who have received DHBI, even where existing lung damage attributed to prior chemotherapy and recurrent sepsis exists, has shown no evidence of major deterioration in function. This finding is consistent with the very low rates of overt pneumonitis reported to date in patients receiving HBI therapy.

The need to irradiate the entire skeleton as part of the DHBI procedure has also been highlighted, as a rapid appearance and/or expansion of MM in some patients has been noted post HBI in areas, e.g. the skull, excluded from the initial radiation fields.[134] Many elderly patients suffering from MM are in poor physical condition or in pain and are unable or unwilling to tolerate intensive chemotherapeutic regimens. DHBI may be of particular value to these patients. Current data on systemic irradiation therapy is sufficiently positive to warrant prospective randomized studies of the DHBI technique versus cytotoxic chemotherapy regimens.

Cytotoxic drug resistance

The development of resistance to cytotoxic therapy in MM patients is a complex issue which is becoming of increasing clinical relevance.[105–109,138–145] Relevant factors in its development may be considered either at the systemic or patient level or at an individual MM cellular level. Factors conferring systemic resistance may involve the absorption, distribution, metabolism, or excretion of a cytotoxic agent. These factors will affect the level of a cytotoxic agent which actually reaches the malignant plasma cell. Poor gastrointestinal absorption of melphalan may contribute to lack of response in a given patient.[146] Tumor cells may reside in sites such as the central nervous system which are 'sanctuary sites' because of a relative inaccessibility to cytotoxic drugs. Rates of metabolic inactivation and excretion may be altered so that critical cytotoxic concentrations are not achieved or cytotoxic other agents may fail to be metabolized to an active form. Physiologic factors occurring at the generalized level may be manipulated to overcome drug resistance by changing the route of drug administration, drug dosage, or scheduling. The increasing use of intravenous melphalan represents one such attempt. As a general rule, attempts to manipulate systemic drug resistance have not been shown to represent major progress in MM therapy.

The more critical mechanisms of cytotoxic drug resistance in MM are considered to reside at the cellular level. Malignant cells may exhibit genetic instability allowing spontaneous generation of variant forms that may result in drug resistance.[147] Altered gene products may result in the development of resistance at the cellular level by causing reduced intracellular drug accumulation, altered drug distribution within the cell, modification of the drug target, or an enhanced ability to repair drug damage. Any single one or combination of these alterations may lead to clinically significant cellular drug resistance.

Cell lines which have been selected for drug resistance by incubating the cells to gradually increasing levels of single cytotoxic agents have greatly facilitated *in vitro* studies of cellular mechanisms of drug resistance. The cytotoxic agents eradicate drug-sensitive cells, and a drug-resistant clone will emerge. Drug-resistant clones are then expanded to provide drug-resistant cell lines which allow the development of probes identifying individual resistance mechanisms. These probes are tested in individual patient specimens to investigate whether a given mechanism is clinically relevant. Clinical investigations may then apply potential chemosensitizers to a standard regimen or test agents which may reduce further development of cellular drug resistance.

Melphalan exerts its cytotoxic effect by DNA adduct formation, resulting in inter- and/or intrastrand cross-links. A biochemical change noted in tumor cells resistant to alkylating agents involves drug detoxification. Glutathione (GSH) protects cells against toxic electrophiles by thioether formation and is catalyzed by the enzyme family of glutathione-S-transferases (GST).[148] An enhanced ability of resistant MM cells to detoxify melphalan via the GSH pathway may be an important mechanism of cytotoxic drug resistance. Thus, a possible means of combating melphalan resistance may be GSH depletion or GST inhibition. Buthionine sulfoximine (BSO) is an inhibitor of a key enzyme in the GSH synthesis pathway.[149] BSO is effective in GSH depletion in both drug-sensitive and drug-resistant 8226 MM cell lines.[150,151] The depletion of GSH by BSO has also been shown to significantly enhance melphalan toxicity in MM cell lines and data from clinical trials using BSO/melphalan are eagerly awaited.

Multidrug resistance (MDR) is the phenomenon of the development by a malignancy of resistance to a variety of agents that have little similarity in physical structure or mechanism of action.[138] The presence or absence of increased expression of p-glycoprotein (PGP), a 170,000 Da integral membrane transport glycoprotein, correlates with the presence of MDR.[109, 138–141] The natural substrate(s) and normal function of PGP are unknown. In addition to being expressed in cytotoxic-resistant malignant cells, PGP is expressed in some normal tissues including the adrenal cortex, proximal renal tubules, and hepatic bile canaliculi.[152,153] The patterns of expression of PGP in normal hematopoietic tissue are not fully defined. Initial data suggest that CD34-positive cells from normal bone marrow express PGP. New tools have become available to assess the intrinsic drug sensitivity of MM to chemotherapeutic agents including the immunohistochemical detection and quantification of PGP in individual MM cells.[141] In drug-resistant malignant cells, PGP lowers the intracellular concentration of cytotoxic drugs by binding the drug and actively pumping it out of the cell before it reaches a critical cytotoxic concentration.[109] The level of PGP on the surface of the MM cell correlates directly with the level of cytotoxic drug resistance.[154–156]

MDR represents resistance to a variety of biological substances, including anthracyclines, vincristine, etoposide, mitoxantrone, mitomycin-C, and actinomycin-D. Increased production of PGP correlates with clinical drug resistance. Less than 5 per cent of patients have detectable PGP on the MM cell surface at time of diagnosis.[144] When treated with single-agent alkylators, this remains essentially unchanged.

As therapy with anthracyclines and/or vincristine is introduced, the expression of PGP increases. By the time patients are overtly resistant to the VAD regimen, more than 75 per cent of MM cells express PGP.[144] Recently, Grogan et al. have documented that the frequency of PGP-positive MM cells correlates with the received cumulative dose of doxorubicin (> 340 mg) and vincristine (> 20 mg), respectively.[142] These data imply that patients who have been treated with the equivalent of 10 cycles of VAD have a 100 per cent probability of PGP expression. Thus the development of MDR in MM is an acquired phenomenon. Drug-resistance patterns are associated with a variety of phenotypic aberrations including expression of aberrant antigens, such as CALLA, myeloid, erythroid and megakaryocyte antigens, plus T-lymphocyte and NK lymphocyte markers. Recent data have highlighted the relatively low rate of induction of MDR by mitoxantrone, leading to an increasing incorporation of this agent into induction and relapse regimens.[157] Largely because of these data, our current induction regimen for MM patients pre-ABMT and/or PBSC is a combination of mitoxantrone and dexamethasone.

Efforts to overcome MDR include the use of high-dose chemotherapy, non-cross-resistant treatment regimens that include drugs not affected by MDR, collaterally sensitive drugs such as corticosteroids, and agents that are able to inhibit the function of PGP. In 1981, the ability of a calcium channel blocker, verapamil, to overcome MDR, both *in vitro* and *in vivo*, in P388 mouse leukemia cells was first reported.[158] The exact mechanism by which verapamil inhibits PGP function is unknown, but presumably it competitively inhibits the binding of cytotoxic drugs, thereby reducing drug efflux. PGP modulators are generally lipid-soluble at physiologic pH and possess a basic nitrogen atom and two planar aromatic rings. Several of these agents, including verapamil, cyclosporine, and quinine, have been determined by photoaffinity labeling studies to bind directly to PGP and thus may block cytotoxic drug binding and efflux by competitive inhibition.

Clinical studies using chemosensitizers plus chemotherapy are now being reported with initial reports focusing on the use of verapamil[108,145,159–162] or cyclosporine[105,156] plus chemotherapy in VAD-resistant patients.

Based on positive data from these pilot studies, prospective randomized controlled studies are now underway. Adverse effects of high-dose verapamil are of serious concern, being primarily cardiovascular including first-degree heart block, hypotension, sinus bradycardia, and junctional rhythms.[161] The most common non-cardiovascular adverse effects are weight gain, constipation, and peripheral edema.

Sonneveld et al. have recently reported on a study of MDR modulation *in vivo* and the effects of cyclosporine on resistant MM cells.[156] Data from this study data indicate that PGP overexpression is functional in refractory MM and that clinical modulation of MDR by cyclosporine is mediated through an inhibition of PGP-associated drug efflux. Eight patients with VAD-refractory MM were treated with VAD plus cyclosporine. PGP expression was determined by flow cytometry/immunocytochemistry before and after VAD/cyclosporine treatment. Functional PGP expression was determined by the effect of cyclosporine on the intracellular accumulation of doxorubicin and vincristine. Five of eight patients responded to VAD/cyclosporine. The percentage of PGP-positive MM cells was 30 per cent to 100 per cent (median, 90 per cent) prior to VAD/cyclosporine and 90 per cent (median, 40 per cent) post-therapy. When analyzing the entire group, clinical treatment with VAD/cyclosporine caused a relative reduction of PGP-positive MM cells in 7 of 8 patients. Although the relative increase of drug accumulation by cyclosporine is only 1.1- to 2.1-fold, such an increase is sufficient to restore the drug accumulation to levels usually observed in drug-sensitive cells.[145] In four responding patients, the PGP expression in residual MM cells after therapy was significantly lower than before treatment. It is not clear if the MDR-modulating effect of cyclosporine is largely mediated through an alteration of drug pharmacokinetics or by an inhibition of PGP function. These data indicate that clinical treatment with a MDR modifier may eliminate or greatly reduce PGP-positivity *in vivo*. When these patients ultimately relapse, the MM cell phenotype may be MDR-negative as was observed in two patients from this study.

PGP may be representative of a class of resistance mechanisms that reduce intracellular drug accumulation by enhanced drug efflux. Other strategies to overcome PGP-mediated resistance, such as antibodies to PGP and PGP antisense RNA,[163] may soon be subjected to clinical study. Combination of chemosensitizers, e.g. quinine and verapamil, may be more effective in blocking PGP.[164] Potential new chemosensitizers continue to be studied *in vitro* and *in vivo*.[165,166] Non-PGP-related mechanisms of resistance are involved in MM [167,168] and we can expect a new generation of therapeutic agents based on these newly discovered entities. The development of transgenic mouse models of MM and a strain in which the animals express the MDR phenotype in bone marrow cells and are resistant to leukopenia induced by anthracyclines is particularly promising.[169,170] The MDR mouse model should serve as a rapid reliable system to evaluate the bioactivity of potential chemosensitizers and thus accelerate the pace of clinical trials with these agents.

Conclusions

Approximately 30–50 per cent of MM patients do not respond to first-line therapy, and those who initially achieve a remission will eventually relapse. There is no standard approach to patients with refractory disease. All MM patients should be entered on protocol at all stages of therapy. The most promising approach utilizing conventional agents seems to be myeloablative therapy with PBSC/rGM-CSF therapy and increasing dose intensity seems to mark prolonged survival in refractory MM. For the many patients who cannot tolerate or afford such intensive therapy, chemosensitizers may enhance the effect of conventional dose cytotoxic regimens. These agents may also enhance the efficacy of myeloablative therapy. Systemic radiation therapy is very effective as a palliative procedure and warrants study at early stages of refractory MM. Promising experimental options including studies of biological response modifiers (e.g. anti-IL–6, anti-IL–2). The full potential of interferon-α to prolong remission in minimal tumor load MM has yet to be defined. Less toxic ways of achieving its effects, (e.g. interferon inducers, TNF blockade) may be of value. Newer recombinant molecules (e.g. IL–6, IL–11, PIXY321) may reduce dose-limiting thrombocytopenia associated with MM therapy. Fresh insights into the complex cytokine network in MM will probably provide the next significant advances in MM therapy. All current options for the treatment of refractory MM are sufficiently ineffective to render urgent the need for progress.

References

1. Alexanian R, Dimopolous M. The treatment of multiple myeloma. *New England Journal of Medicine* 1994; **330**: 484–9.
2. Salmon SE, Crowley J. Impact of glucocorticoids and interferon on outcome in multiple myeloma. *ASCO Proceedings* 1992; **11**: 316 (abstract).
3. Kyle RA, Greipp PR, Gertz MA. Treatment of refractory multiple myeloma and considerations for future therapy. *Seminars in Oncology* 1986; **13**: 326–33.
4. Buzaid AC, Durie BGM. Management of refractory myeloma: A review. *Journal of Clinical Oncology* 1988; **6**: 889–905.
5. Barlogie B, Smith L, Alexanian R. Effective treatment of advanced multiple myeloma refractory to alkylating agents. *New England Journal of Medicine* 1984; **310**: 1353–6.
6. Lokhorst HM, Meuwissen OJAT, Bast EJEG, Dekker AW. VAD chemotherapy for refractory multiple myeloma. *British Journal of Haematology* 1989; **71**: 25–30.
7. Forgeson GV, Selby P, Lakhani S et al. Infused vincristine and Adriamycin with high dose methylprednisolone (VAMP) in advanced previously treated multiple myeloma patients. *British Journal of Cancer* 1988; **58**: 469–73.
8. Petrucci MT, Avvisati G, Tribalto M et al. Intermediate dose (25 mg/m^2) intravenous melphalan for patients with multiple myeloma in relapse or refractory to standard treatment. *European Journal of Haematology* 1989; **42**: 233–7.
9. Palva IP, Ahrenberg P, Ala Harja K et al. Intensive chemotherapy with combinations containing anthracyclines for refractory and relapsing multiple myeloma. *European Journal of Haematology* 1990; **44**: 120–3.
10. Gimsing P, Bjerrum OW, Brandt E et al. Refractory myelomatosis treated with mitoxantrone in combination with vincristine and prednisone (NOP-regimen): a phase II study. *British Journal of Haematology* 1991; **77**: 73–9.
11. Lenhard RE, Oken MM, Barnes JM et al. High-dose cyclophosphamide. An effective treatment for advanced refractory myeloma. *Cancer* 1984; **53**: 1456–60.
12. Bonnet J, Alexanian R, Salmon S et al. Vincristine, BCNU, doxorubicin and prednisone (VBAP) combination in the treatment of relapsing or resistant multiple myeloma: a Southwest Oncology Group Study. *Cancer Treatment Reports* 1982; **66**: 1267–71.
13. Browman GP, Belch A, Skillings J et al. Modified Adriamycin–vincristine–dexamethasone (mVAD) in primary and relapsed multiple myeloma: an NCI (Canada) pilot study. *British Journal of Haematology* 1992; **82**: 555–9.
14. Bonnet JD, Alexanian R, Salmon SE et al. Addition of cisplatin and bleomycin to vincristine, carmustine, doxorubicin, prednisone (VBAP) combination in the treatment of relapsing or resistant multiple myeloma: A Southwest Oncology Group Study. *Cancer Treatment Reports* 1984; **68**: 481–5.
15. German A, Gomez MD, Tin Han MD et al. Salvage treatment for patients with multiple myeloma refractory to alkylating agents. *Medical Pediatric Oncology* 1985; **13**: 325–9.
16. Alberts DS, Stanley P, Balcerzak AK et al. Phase II trial of mitoxantrone in Multiple Myeloma: A Southwest Oncology Group Study. *Cancer Treatment Reports* 1985; **69**: 1321–3.
17. White D, Bergsagel D, Rapp EF et al. Failure of cyclophosphamide to produce response in melphalan resistant multiple myeloma. *Blood* 1981; **58**: 169 (abstract).
18. Blade J, Felin E, Rozman C et al. Cross resistance to alkylating agents in multiple myeloma. *Cancer* 1983; **51**: 1981–91.
19. Kyle RA, Gailani S, Seligman BR et al. Multiple myeloma resistant to melphalan: treatment with cyclophosfamide, prednisone and BCNU. *Cancer Treatment Reports* 1979; **63**: 1265–9.
20. Kyle RA, Pajak TF, Henderson ES et al. Multiple myeloma resistant to melphalan: treatment with dox-

orubicin, cyclophosphamide, carmustine and predni-sone. *Cancer Treatment Reports* 1982; **66**: 451–6.

21. Selby PJ, McElwain TJ. Multiple myeloma treated with high dose intravenous melphalan. *British Journal of Haematology* 1987; **66**: 55–62.

22. McElwain TJ, Powles RL. High dose intravenous melphalan for plasma cell leukemia and myeloma. *Lancet* 1983; **2**: 822–4.

23. Alexanian R, Barlogie B, Dixon D. High dose glucocorticoid treatment of resistant myeloma. *Annals of Internal Medicine* 1986; **105**: 8–11.

24. Barlogie B, Velasquez W, Alexanian R, Cabanillas F. Etoposide, dexamethasone, cytarabine and cispla-tin in vincristine, doxorubicin and dexamethasone-refractory myeloma. *Journal of Clinical Oncology* 1989; **7**: 1514–18.

25. Gockerman J, Bartolucci A, Nelson M. Phase II evaluation of etoposide in refractory multiple mye-loma. *Cancer Treatment Reports* 1986; **70**: 801–2.

26. Hall R, Barlogie B, Alexanian R. High dose alkylating agent for refractory myeloma. *ASCO Proceedings* 1985; **4**: 217 (abstract).

27. Leoni F, Ciolli S, Salti F et al. Teniposide, dexa-methasone and continuous infusion cyclophospha-mide in advanced refractory myeloma. *British Journal of Haematology* 1991; **77**: 180–4.

28. Musto P, Greco MM, Falcone A, Carotenuto M. Treatment of plasma cell leukaemia and resistant/relapsed multiple myeloma with vincristine, mitox-antrone and dexamethasone (VMD protocol). *British Journal of Haematology* 1991; **79**: 655–6.

29. Friedenberg WR, Kyle RA, Knospe WH et al. High dose dexamethasone for refractory or relapsing multiple myeloma. *American Journal of Haematology* 1991; **36**: 171–5.

30. Alexanian R, Barlogie B, Gutterman J. Alpha inter-feron combination therapy of resistant myeloma. *American Journal of Clinical Oncology* 1991; **14**: 188–92.

31. San Miguel JF, Moro M, Blade J et al. Interferon and dexamethasone in multiple myeloma patients refrac-tory to chemotherapy. *European Journal of Cancer* 1991; **27**: 48–9.

32. Wahlim A, Holm J. Rapid response to recombinant interferon alpha 2B and high dose prednisolone in multiple drug resistant multiple myeloma. *European Journal of Haematology* 1989; **43**: 352–4.

33. Blade J, San Miguel J, Sanz-Sanz MA et al. Treat-ment of melphalan-resistant multiple myeloma with vincristine, BCNU, doxorubicin, and high-dose dexamethasone (VBAD). *European Journal of Can-cer* 1992; **29**: 57–60.

34. Monconduit M, Le Loet X, Beernard JF, Michaux JL. Combination chemotherapy with vincristine, doxorubicin, dexamethasone for refractory or relap-sing multiple myeloma. *British Journal of Haema-tology* 1986; **63**: 599–601.

35. Steinke B, Busch FW, Becherer C et al. Melpha-lan-resistant multiple myeloma: results of treatment according to M2 protocol. *Cancer Chemotherapy and Pharmacology* 1985; **14**: 279–81.

36. Cavo M, Galieni P, Tassi C et al. M–2 protocol for melphalan resistant and relapsing multiple mye-loma. *European Journal of Haematology* 1988; **40**: 168–73.

37. Finnish Leukaemia Group. Aggressive combination chemotherapy in multiple myeloma. A multicentre trial. *Scandinavian Journal of Haematology* 1985; **35**: 205–9.

38. Buonanno O, Tortarolo M, Valente A et al. Drug resistant multiple myeloma. A trial with the M–2 cyclic alkylating agents polychemotherapy. *Haematologica* 1978; **63**: 45–55.

39. Blade J, Rozman C, Montserrat E et al. Treatment of alkylating agent resistant myeloma with vincristine, BCNU, Doxorubicin and prednisone (VBAP). *European Journal of Cancer and Clinical Oncology* 1986; **22**: 1193–7.

40. Alexanian R, Yap BS, Bodey GP. Prednisone pulse therapy for refractory myeloma. *Blood* 1983; **62**: 572–7.

41. Barlogie B, Jagannath S, Dixon DO et al. High-dose melphalan and granulocyte-macrophage colony-sti-mulating factor for refractory multiple myeloma. *Blood* 1990; **76**: 677–80.

42. Bergsagel D. Phase II trials of mitomycin C. AB–100, NSC–1026, L-sarcolysin, and meta-sarcolysin in the treatment of multiple myeloma. *Cancer Chemotherapy Reports* 1962; **16**: 261–6.

43. Moon JH, Edmonson JH. Procarbazine (NSC–77213) and multiple myeloma. *Cancer Chemotherapy Reports* 1970; **54**: 245–8.

44. Salmon SE. Nitrosoureas in multiple myeloma. *Cancer Treatment Reports* 1976; **60**: 789–94.

45. Jones SE, Tucker WG, Haut A et al. Phase II trial of piperazinedione in Hodgkin's disease, non-Hodg-kin's lymphoma, and multiple myeloma: A South-west Oncology Group study. *Cancer Treatment Reports* 1977; **61**: 1617–21.

46. Bennett JM, Silber R, Ezdinli E et al. Phase II study of Adriamycin and bleomycin in patients with multi-ple myeloma. *Cancer Treatment Reports* 1978; **62**: 1367–9.

47. Lake-Lewin D, Myers J, Lee BL et al. Phase II trial of pyrazofurin in patients with multiple myeloma refractory to standard cytotoxic therapy. *Cancer Treatment Reports* 1979; **63**: 1403–4.

48. Scarffe JH, Beaumont AR, Crowther D. Phase I-II evaluation of acronine in patients with multiple mye-loma. *Cancer Treatment Reports* 1983; **67**: 93–4.

49. Ahmann FR, Meyskens FL, Jones SE et al. Phase II evaluation of amsacrine (m-AMSA) in solid tumors, myeloma, and lymphoma: A University of Arizona and Southwest Oncology Group study. *Cancer Treatment Reports* 1983; **67**: 697–700.

50. Cornell CJ Jr, Pajak TF, McIntyre OR. Chlorozoto-cin. Phase II evaluation in patients with myeloma. *Cancer Treatment Reports* 1984; **68**: 685–6.

51. Forman WB, Cohen HJ, Bartolucci AA et al. Phase II evaluation of chlorozotocin in refractory multiple myeloma. *Cancer Treatment Reports* 1984; **68**: 1409–10.

52. Vinciguerra V, Anderson K, McIntyre OR.

Diaziquone for resistant multiple myeloma. *Cancer Treatment Reports* 1985; **69**: 331–2.

53. Alberts DS, Balcerzak SP, Bonnet ID et al. Phase II trial of mitoxantrone in multiple myeloma: A Southwest Oncology Group study. *Cancer Treatment Reports* 1985; **69**: 1321–3.

54. Esseesse I, Bartolucci AA, Gams RA et al. Weekly mitoxantrone therapy for refractory multiple myeloma: A Southeastern Cancer Study Group trial. *Cancer Treatment Reports* 1986; **70**: 669–70.

55. Tirelli U, Sorio R, Magri MD et al. Predmustine in elderly patients with multiple myeloma: A phase II study. *Cancer Treatment Reports* 1986; **70**: 537–8.

56. Alberts DS, Salmon E. Adriamycin in the treatment of alkylator-resistant multiple myeloma: a pilot study. *Cancer Chemotherapy Reports* 1975; **59**: 345–50.

57. Tirelli U, Zagonel V, Veronesi A et al. Phase II study of teniposide (VM–26) in multiple myeloma. *American Journal of Clinical Oncology* 1985; **8**: 329–31.

58. Durie BGM, Levy HB, Voakes I et al. Poly(I,C)-LC as an interferon inducer in refractory multiple myeloma. *Journal of Biologic Response Modifiers* 1985; **4**: 518–24.

59. Jackson DV, Case LD, Pope EK et al. Single agent vincristine by infusion in refractory multiple myeloma. *Journal of Clinical Oncology* 1985; **3**: 1508–12.

60. Belch AR, Henderson JF, Brox LW. Treatment of multiple myeloma with deoxycoformycin. *Cancer Chemotherapy and Pharmacology* 1985; **14**: 49–52.

61. Paccagnella A, Salvagno L, Chiarion-Sileni V et al. Peptichemio in pretreated patients with plasma cell neoplasms. *European Journal of Cancer and Clinical Oncology* 1986; **22**: 1053–8.

62. Case DC Jr, Oldham F, Ervin T et al. Phase I-II study of epirubicin in multiple myeloma. *ASCO Proceedings* 1987; **6**: 146 (abstract).

63. Salmon SE, Shadduck RK, Schilling A. Intermittent high-dose prednisone (NSC–10023) therapy for multiple myeloma. *Cancer Chemotherapy Reports* 1967; **51**: 179–87.

64. Perren TJ, Selby PJ, Mbidde EK et al. High dose chemotherapy of multiple myeloma (MM) with melphalan (HDM) and with methylprednisolone (HDMP). *ASCO Proceedings* 1986; **5**: 158 (abstract).

65. Bergsagel DE, Cowan DH, Hasselbach R. Plasma cell myeloma: Response of melphalan-resistant patients to high-dose intermittent cyclophosphamide. *Canadian Medical Association Journal* 1972; **107**: 851–5.

66. Kyle RA, Seligman BR, Wallace J et al. Multiple myeloma resistant to melphalan (NSC–8806) treated with cyclophosphamide (NSC–26271), prednisone (NSC–10023), and chloroquine (NSC–187208). *Cancer Chemotherapy Reports* 1975; **59**: 557–62.

67. Tornyos K, Silberman H, Solomon A. Phase II study of oral methyl-CCNU and prednisone in previously treated alkylating agent-resistant multiple myeloma. *Cancer Treatment Reports* 1977; **61**: 785-7.

68. Cohen HJ, Silberman HR, Larsen WE et al. Combination chemotherapy with intermittent 1–3-bis (2-chloroethyl) 1-nitrosourea (BCNU), cyclophosphamide, and prednisone for multiple myeloma. *Blood* 1979; **54**: 824–36.

69. Cohen HJ, Bartolucci AA. Hexamethylmelamine and prednisone in the treatment of refractory multiple myeloma. *American Journal of Clinical Oncology* 1982; **5**: 21–7.

70. Broun GO Jr, Petruska PJ, Hiramoto RN et al. Cisplatin. BCNU, cyclophosphamide, and prednisone in multiple myeloma. *Cancer Treatment Reports* 1982; **66**: 237–42.

71. Norberg B, Holm J, Winqvist E et al. Treatment of advanced bone marrow neoplasms with ifosfamide combinations. *Scandinavian Journal of Haematology* 1984; **32**: 95–100.

72. Houwen B, Marrink J, Nieweg HO. Vindesine therapy in melphalan-resistant multiple myeloma. *European Journal of Cancer* 1981; **17**: 227–32.

73. White D, Bergsagel DE, Rapp EF et al. Failure of cyclophosphamide to produce responses in melphalan-resistant multiple myeloma. *Blood* 1981; **58**: 169 (abstract).

74. Gomez GA, Han T, Ozer H et al. Salvage treatment for patients with multiple myeloma refractory to alkylating agents. *Medical Pediatric Oncology* 1985; **13**: 325–9.

75. Paccagnella A, Salvagno L, Sileni VC et al. Vincristine and Adriamycin 4-day continuous infusion with high-dose methylprednisolone (VAM regimen) in alkylator-resistant multiple myeloma. *ASCO Proceedings* 1987; **6**: 207 (abstract).

76. Wilson K, Shelley W, Belch A et al. Weekly cyclophosphamide and alternate-day prednisone: An effective secondary therapy in multiple myeloma. *Cancer Treatment Reports* 1987; **71**: 981–2.

77. Alberts DS, Durie BGM, Salmon SE. Doxorubicin/BCNU chemotherapy for multiple myeloma in relapse. *Lancet* 1987; **1**: 926–8.

78. Presant CA, Klahr C. Adriamycin, BCNU, cyclophosphamide plus prednisone (ABC-P) in melphalan-resistant multiple myeloma. *Cancer* 1978; **42**: 1222–7.

79. Karp JE, Humphrey RL, Burke PJ. Timed sequential chemotherapy of cytoxan-refractory multiple myeloma with Cytoxan and Adriamycin based on induced tumor proliferation. *Blood* 1981; **57**: 468–75.

80. Van Dobbenburgh OA, Houwen B, Piersma H et al. Cyclohosphamide, doxorubicin, prednisone and vindesine combination chemotherapy in melphalan-resistant multiple myeloma. *Netherlands Journal of Medicine* 1984; **27**: 25–30.

81. Sheehan T, Judge M, Parker AC. The efficacy and toxicity of VAD in the treatment of myeloma and related disorders. *Scandinavian Journal of Haematology* 1986; **37**: 425–8.

82. Monconduit M, Bauters F, Najman A. Evaluation of the association vincristine-Adriamycin plus high-dose dexamethasone (VAD) in severe previously treated myeloma. *Blood* 1986; **68**: 240 (abstract).

83. Costanzi JJ, Cooper R, Scarffe JH et al. Phase II study of recombinant alpha–2 interferon in resistant multiple myeloma. *Journal of Clinical Oncology* 1985; **3**: 654–9.

84. Oken MM, Kyle RA, Kay NE et al. A phase II trial of interferon alpha–2 in the treatment of resistant multiple myeloma. *ASCO Proceedings* 1985; **4**: 215 (abstract).

85. Jaffe JP, Bosch A, Raich PC. Sequential hemi-body radiotherapy in advanced multiple myeloma. *Cancer* 1979; **43**: 124–8.

86. Coleman M, Saletan S, Wolf D et al. Whole bone marrow irradiation for the treatment of multiple myeloma. *Cancer* 1982; **49**: 1328–33.

87. Thomas PJ, Daban A, Bontoux D. Double hemibody irradiation in chemotherapy-resistant multiple myeloma. *Cancer Treatment Reports* 1984; **68**: 1173–5.

88. Rostom AY, O'Cathail SM, Folkes A. Systemic irradiation in multiple myeloma. A report on nineteen cases. *British Journal of Haematology* 1984; **58**: 423–31.

89. Tobias JS, Richards JDM, Blackman GM et al. Hemibody irradiation in multiple myeloma. *Radiotherapy and Oncology* 1985; **3**: 11–16.

90. Kantarjian H, Dreicer R, Barlogie B et al. High-dose cytosine arabinoside in multiple myeloma. *European Journal of Cancer and Clinical Oncology* 1984; **20**: 227–31.

91. Lenhard RE, Tsiatis AA, Oken MM et al. Timed sequential high-dose cyclophosphamide and vincristine treatment of multiple myeloma. *ASCO Proceedings* 1985; **4**: 217 (abstract).

92. Wisloff F, Gimsing P, Hedenus M et al. Bolus therapy with mitoxantrone and vincristine in combination with high-dose prednisone (NOP-bolus) in resistant multiple myeloma. *European Journal of Haematology* 1992; **48**: 70–4.

93. Finnish Leukemia Group. Combination chemotherapy MOCCA in resistant and relapsing multiple myeloma. *European Journal of Haematology* 1992; **48**: 37–40.

94. Dimopoulous MA, Delasalle KB, Champlin R, Alexanian R. Cyclophosphamide and etoposide therapy with GM-CSF for VAD-resistant multiple myeloma. *British Journal of Haematology* 1993; **83**: 240–4.

95. Ganjoo RK, Johnson PWM, Evans ML et al. Recombinant interferon-alpha 2b and high dose methyl prednisolone in relapsed and resistant multiple myeloma. *Hematology and Oncology* 1993; **11**: 179–86.

96. Collin R, Greaves M, Preston FE. Potential value of vincristine–adriamycin–dexamethasone combination chemotherapy (VAD) in refractory and rapidly progressive myeloma. *European Journal of Haematology* 1987; **39**: 203–8.

97. Hippe E, Clausen NAT, Gimsing P, Haedersdal C. Resistant multiple myeloma treated with mitoxantrone in combination with vincristine and prednisolone (NOP-regime). *European Journal of Haematology* 1987; **39**: 88–9.

98. Phillips JK. Resistant multiple myeloma treated with mitoxantrone in combination with vincristine and dexamethasone (MOD). *European Journal of Haematology* 1989; **42**: 109–10.

99. Koskela K, Pelliniemi TT, Remes K. VAD regimen in the treatment of resistant multiple myeloma: slow or fast infusion? *Leukemia Lymphoma* 1993; **10**: 347–51.

100. Ohrling M, Bjorkholm M, Osterborg A et al. Etoposide, doxorubicin, cyclophosphamide and high-dose betamethasone (EACB) as outpatient salvage therapy for refractory multiple myeloma. *European Journal of Haematology* 1993; **51**: 45–9.

101. Cunningham D, Paz-Ares L, Milan S et al. High-dose melphalan for multiple myeloma: Long term follow-up data. *Journal of Clinical Oncology* 1994; **12**: 764–8.

102. Singer CRJ, Tobias JS, Giles FJ et al. Hemi-body irradiation – an effective second-line therapy in drug-resistant multiple myeloma. *Cancer* 1989; **63**: 2446–51.

103. Giles FJ, Richards JDM, Tobias JS et al. A prospective randomised study of the effect on survival in patients with relapsed multiple myeloma of double hemi-body irradiation with and without subsequent maintenance recombinant alpha 2b interferon therapy. *European Journal of Cancer* 1992; **28**: 1392–5.

104. Mansi J, Da Costa Fernando, Viner C et al. High-dose busulfan in patients with myeloma. *Journal of Clinical Oncology* 1992; **10**: 1569–73.

105. Sonneveld P, Durie BGM, Lokhorst HM et al. Modulation of multidrug-resistant multiple myeloma by cyclosporine. *Lancet* 1992; **340**: 255–9.

106. Cornelissen JJ, Sonneveld P, Schoester M et al. MDR–1 expression and response to vincristine, doxorubicin, and dexamethasone chemotherapy in multiple myeloma refractory to alkylating agents. *Journal of Clinical Oncology* 1994; **12**: 115–19.

107. Pilarski LM, Belch AR. Circulating monoclonal B cells expressing P glycoprotein may be a reservoir of multidrug-resistant disease in multiple myeloma. *Blood* 1994; **83**: 724–36.

108. Salmon SE, Dalton WS, Grogan TM et al. Multidrug-resistant myeloma : laboratory and clinical effects of verapamil as a chemosensitizer. *Blood* 1991; **78**: 44–50.

109. Dalton WS, Salmon SE. Drug resistance in myeloma: Mechanisms and approaches to treatment. *Hematology and Oncology Clinics of North America* 1992; **6**: 383–94.

110. Jagannath S, Vesole DH, Gleen L et al. Low-risk intensive therapy for multiple myeloma with combined autologous bone marrow transplantation and blood stem cell support. *Blood* 1992; **80**: 1666–72.

111. Alexanian R, Smallwood L, Cheson B et al. Prognostic factors with high dose melphalan for refractory multiple myeloma. *Blood* 1988; **72**: 2015–19.

112. Barlogie B, Jagannath S, Dixon D et al. High dose melphalan and granulocyte-macrophage colony-stimulating factor for refractory multiple myeloma. *Blood* 1990; **76**: 677–80.

113. Vesole DH, Barlogie B, Jagannath S et al. High-

dose therapy for refractory multiple myeloma: Improved prognosis with better supportive care and double transplants. *Blood* 1994; **84**: 950–6.

114. Jagannath S, Barlogie B, Dicke KA et al. Autologous bone marrow transplantation in multiple myeloma: Identification of prognostic factors. *Blood* 1990; **76**: 1860–5.

115. Harousseau JL, Milpied N, Laporte JP et al. Double intensive therapy in high risk multiple myeloma. *Blood* 1992; **79**: 2827–31.

116. Gianni AM, Tarella C, Bregni M et al. High-dose sequential chemoradiotherapy, a widely applicable regimen, confers survival benefit to patients with high-risk multiple myeloma. *Journal of Clinical Oncology* 1994; **12**: 503–9.

117. Alexanian R, Dimopoulos M, Smith T et al. Limited value of myeloablative therapy for late multiple myeloma. *Blood* 1994; **83**: 512–16.

118. Cunningham D, Powles R, Malpas JS et al. A randomized trial of maintenance therapy with Intron-A following high-dose Melphalan and ABMT in myeloma. *ASCO Proceedings* 1993; **12**: 364 (abstract).

119. Mandelli F, Arrisati G, Amadori S et al. Maintenance treatment with recombinant interferon alpha–2b in patients with multiple myeloma responding to conventional induction chemotherapy. *New England Journal of Medicine* 1990; **322**: 1430–4.

120. Browman GP, Rubin S, Walker I et al. Interferon alpha2b maintenance therapy prolongs progression-free and overall survival in plasma cell myeloma: results of a randomized trial. *ASCO Proceedings* 1994; **13**: 408 (abstract).

121. Ludwig H, Cohen AM, Huber H et al. Interferon with VMCP compared to VMCP for induction and interferon compared to control for remission maintenance in multiple myeloma. *ASCO Proceedings* 1994; **13**: 408 (abstract).

122. Westin J, Cortelezzi A, Hjorth M et al. Interferon therapy during the plateau phase of multiple myeloma: an update of the Swedish study. *European Journal of Cancer* 1991; **27** (suppl. 4): 45–8.

123. Chronic Leukemia–Myeloma Task Force, National Cancer Institute: Proposed guidelines for protocol studies. II. Plasma cell myeloma. *Cancer Chemotherapy Reports* 1973; **1**: 145–58.

124. Alexanian R, Bonnet J, Gehan E et al. Combination chemotherapy for multiple myeloma. *Cancer* 1972; **30**: 382–9.

125. Baldini L, Radaelli F, Chiorboli O et al. No correlation between response and survival in patients with multiple myeloma treated with vincristine, melphalan, cyclophosphamide and prednisone. *Cancer* 1991; **68**: 62–7.

126. Palmer M, Belch A, Hanson J, Brox L. Reassessment of the relationship between M-protein decrement and survival in multiple myeloma. *British Journal of Cancer* 1989; **59**: 110–12.

127. Boccadoro M, Marmont F, Tribalto M et al. Early responder myeloma : Kinetic studies identify a patient subgroup characterized by very poor prognosis. *Journal of Clinical Oncology* 1989; **7**: 119–25.

128. Hjorth M, Holmberg E, Rödjer S. Impact of active and passive exclusions on the results of a clinical trial in multiple myeloma. *British Journal of Haematology* 1991; **80**: 55–61.

129. Sparano JA, Wiernik P, Leaf A, Dutcher J. Infusional cyclophosphamide, doxorubicin, and etoposide in relapsed and resistant non-Hodgkin's lymphoma: Evidence for a schedule-dependent effect favouring infusional administration of chemotherapy. *Journal of Clinical Oncology* 1993; **11**: 1071–9.

130. Wilson WH, Bryant G, Bates S et al. Infusional etoposide, vincristine, and adriamycin, with cyclophosphamide, prednisone and R-verapamil in relapsed lymphoma. *ASCO Proceedings* 1991; **10**: 275 (abstract).

131. Dimopoulos M, Alexanian R, Przepiorka D et al. Thiotepa, busulfan and cyclophosphamide: A new preparative regimen for autologous marrow or blood stem cell transplantation in high-risk multiple myeloma. *Blood* 1993; **82**: 2324–8.

132. Fermand J, Chevret S, Levy Y et al. The role of autologous blood stem cells in support of high-dose therapy for multiple myeloma. *Hematology and Oncology Clinics of North America* 1992; **6**: 451–62.

133. Bergsagel DE. Total body irradiation in myelomatosis. *British Medical Journal* 1971; **1**: 325.

134. Giles FJ, De Lord C, Gaminara EJ et al. Systemic irradiation therapy of myelomatosis – the therapeutic implications of technique. *Leukemia and Lymphoma* 1990; **1**: 227–33.

135. Salazar OM, Rubin P, Keller B, Scarantino C. Systemic (half-body) radiation therapy; response and toxicity. *International Journal of Radiation Oncology, Biology and Physics* 1978; **4**: 937–50.

136. Salmon SE, Tesh D, Crowley J et al. Chemotherapy is superior to sequential Hemibody Irradiation for remission consolidation in multiple myeloma: a Southwest Oncology Group study. *Journal of Clinical Oncology* 1990; **8**: 1575–80.

137. Giles FJ, Singer CRJ, Goldstone AH et al. Lung toxicity of melphalan and steroid combination therapy in multiple myeloma. *British Journal of Cancer* 1988; **58**: 545 (abstract).

138. Kohno K, Sato S, Takano H et al. The direct activation of human multidrug resistance gene (mdr1) by anticancer agents. *Biochemistry and Biophysics Research Communications* 1989; **165**: 1415–21.

139. Chin C, Soffir R, Noonan KE et al. Structure and expression of the human MDR (P-glycoprotein) gene family. *Molecular and Cellular Biology* 1989; **9**: 3808–20.

140. Goldstein LJ, Galski H, Fojo AT et al. Expression of a multidrug resistance gene in human cancers. *Journal of the National Cancer Institute* 1989; **81**: 116–24.

141. Grogan TM, Dalton WS, Rybski JA et al. Optimization of immunocytochemical PGP assessment in

multidrug resistant plasma cell myeloma using 3 antibodies. *Laboratory Investigations* 1990; **63**: 815–23.

142. Grogan TM, Spier CM, Salmon SE et al. P-Glyco-protein expression in human plasma cell myeloma: Correlation with prior chemotherapy. *Blood* 1993; **81**: 490–5.

143. Epstein J, Xiao H, Oba BK. P-Glycoprotein expression in plasma-cell myeloma is associated with resistance to VAD. *Blood* 1989; **74**: 913–17.

144. Sonneveld P, Durie BGM, Lokhorst HM et al. Analysis of multidrug-resistance (MDR–1) glyco-protein and CD56 expression to separate monoclonal gammopathy from multiple myeloma. *British Journal of Haematology* 1993; **83**: 63–7.

145. Dalton WS, Grogan TM, Meltzer PS et al. Drug-resistance in multiple myeloma and non-Hodgkin's lymphoma: Detection of P-glycoprotein and potential circumvention by addition of verapamil to chemotherapy. *Journal of Clinical Oncology* 1989; **7**: 415–24.

146. Alberts DS, Chang FY, Chen HSG et al. Oral melphalan kinetics. *Clinical Pharmacology and Therapeutics* 1979; **6**: 737–45.

147. Goldie JH, Coldman AJ. Genetic instability in the development of drug resistance. *Seminars in Oncology* 1985; **12**: 222–30.

148. Habig WH, Pabst MN, Jakoby WB. Glutathione S-transferases: The first step in mercapturic acid formation. *Journal of Biological Chemistry* 1974; **249**: 7130–9.

149. Griffith OW. Mechanism of action, metabolism and toxicity of buthionine sulfoximine and its higher homologs, patent inhibitors of glutathione synthesis. *Journal of Biochemistry* 1988; **257**: 13704–12.

150. Bellamy WT, Dalton WS, Gleason MC et al. Development and characterization of a melphalan resistant human multiple myeloma cell line. *Cancer Research* 1991; **51**: 995–1002.

151. Bellamy WT, Dalton WS, Meltzer P, Dorr RT. The role of glutathione and its associated enzymes in multidrug resistant human myeloma cells. *Biochemistry and Pharmacology* 1989; **38**: 787–93.

152. Thiebaut F, Tsuruo T, Hamada H et al. Cellular localization of the multidrugresistance gene product P-glycoprotein in normal human tissues. *Proceedings of the National Academy of Sciences* 1987; **84**: 7735–8.

153. Cordon-Cardo C, O'Brien JP, Boccia J et al: Expression of the multidrug resistance gene product (P-glycoprotein) in human normal and tumor tissues. *Journal of Histochemistry and Cytochemistry* 1990; **38**: 1277–87.

154. Dalton WS, Durie BGM, Alberts DS et al. Characterization of a new drug resistant human myeloma cell line which expresses P-glycoprotein. *Cancer Research* 1986; **46**: 5125–30.

155. Dalton WS, Grogan TM, Rybski JA et al. Immunohistochemical detection and quantitation of P-glycoprotein in multiple drug-resistant human myeloma cells: Association with level of drug resistance and drug accumulation. *Blood* 1989; **73**: 747–52.

156. Sonneveld P, Schoester M, De Leeuw K. Clinical modulation of multidrug resistance in multiple myeloma: Effect of cyclosporine on resistant tumor cells. *Journal of Clinical Oncology* 1994; **72**: 7584–97.

157. Diez RA, Corrado C, Palacios MF et al. Treatment of relapsed/refractory multiple myeloma with mitoxantrone/dexamethasone: low induction of GP170 by western blot. *ASCO Proceedings* 1994; **13**: 411 (abstract).

158. Tsuruo T, Iida H, Tsukagoshi S et al. Overcoming of vincristine resistance in P388 leukemia *in vivo* and *in vitro* through enhanced cytotoxicity of vincristine and vinblastine by verapamil. *Cancer Research* 1981; **41**: 1967–72.

159. Durie BGM, Dalton WS. Reversal of drug-resistance in multiple myeloma with verapamil. *British Journal of Haematology* 1988; **68**: 203–5.

160. Gore ME, Selby PJ, Millar B et al. The use of verapamil to overcome drug in myeloma. *ASCO Proceedings* 1988; **7**: 228 (abstract).

161. Pennock GD, Dalton WS, Roeske WR et al. Systemic toxic effects associated with high-dose verapamil infusion and chemotherapy administration. *Journal of the National Cancer Institute* 1991; **83**: 105–10.

162. Trumper LH, Ho AD, Wulf G et al. Addition of verapamil to overcome drug resistance in multiple myeloma: Preliminary clinical observations in 10 patients. *Journal of Clinical Oncology* 1989; **7**: 1578–9.

163. Tong AW, Lee J, Wang R-M et al. Elimination of chemoresistant multiple myeloma clonogenic colony-forming cells by combined treatment with a plasma cell-reative monoclonal antibody and a P-glycoprotein-reactive monoclonal antibody. *Cancer Research* 1989; **49**: 4829–34.

164. Lehnert M, Dalton WS, Roe D et al. Synergistic inhibition by verapamil and quinine of P-glyco-protein-mediated multidrug resistance in a human myeloma cell line model. *Blood* 1991; **77**: 348–54.

165. Evans CH, Baker PD. Decreased P-glycoprotein expression in multidrug-sensitive and -resistant human myeloma cells induced by the cytokine leukoregulin. *Cancer Research* 1992; **52**: 5893–9.

166. Jonsson B, Nilsson K, Nygren P, Larsson R. SDZ PSC–833 – a novel potent *in vitro* chemosensitizer in multiple myeloma. *Anti-Cancer Drugs* 1992; **3**: 641–6.

167. Moalli PA, Pillay S, Weiner D et al. A mechanism of resistance to glucocorticoids in multiple myeloma: transient expression of a truncated glucocorticoid receptor mRNA. *Blood* 1992; **79**: 213–22.

168. Ishikawa H, Kawano MM, Okada K et al. Expressions of DNA topoisomerase I and II gene and the genes possibly related to drug resistance in human myeloma cells. *British Journal of Haematology* 1993; **83**: 68–74.

169. Tong AW, Huang YW, Zhang BQ et al. Heterotransplantation of human multiple myeloma cell lines in severe combined immunodeficiency (SCID) mice. *Anticancer Research* 1993; **13**: 593–7.

170. Bellamy WT, Odeleye A, Finley P et al. An *in vivo* model of human multidrug-resistant multiple myeloma in SCID mice. *American Journal of Pathology* 1993; **142**: 691–8.

CHAPTER 13

Use of interferon in the treatment of multiple myeloma

GIUSEPPE AVVISATI

Introduction	148	Therapeutic efficacy of IFNs in MM	150
History	148	IFN-β and IFN-γ in multiple myeloma	155
Biology, pharmacology, and pharmacokinetics of		Tolerability	155
IFNs	149	What next?	155
Mechanisms of action	149	References	156

Introduction

Despite progress in the biology and treatment of multiple myeloma (MM), the disease still remains incurable. Several new approaches including the use of high-dose chemotherapy followed by bone marrow transplantation procedures (allogeneic or autologous) or peripheral blood stem cells infusion, have been developed in the last few years with the aim of improving the prognosis of this disease. However, due to the high median age of MM patients these aggressive therapeutic procedures are applicable to less than 50 per cent of patients. Another investigational approach utilized in the last few years to improve the prognosis of MM has been the use of the interferons (IFNs), a family of biologic response modifiers with a wide range of activities that has also proved to be useful in the treatment of MM. Natural or recombinant IFN-α has been most widely used.

History

The term interferon (IFN) was originally used by Isaacs and Lindenmann in 1957 to identify a soluble factor produced by cells in response to viruses, which was able to interfere with viral infections.[1] The scientific community was initially very skeptical about the existence of a protein capable of interfering with viral infections, and due to the difficulties in chemical purification and characterization, the discovery was openly derided for some time. It was only in the late 1970s that the antitumor activity of IFN was discovered and in addition to antiviral activity, it became evident that IFN had both antigrowth and immunomodulatory properties.

The production by Kari Cantell of sufficient quantities of 'natural' human leukocyte IFN-α then made possible the use of this new protein in limited clinical trials in several types of malignancies. These showed promise, despite the fact that these early preparations of natural IFN contained less than 5 per cent of active protein.

Meanwhile, the rapid development of monoclonal antibodies, large-scale cell cultures and recombinant DNA technology, combined with improvements in protein and nucleic acid sequencing, provided new insights into the original concept of a single type of IFN with one biological function and also led to increasing clinical use of IFN in many hematological malignancies.

Biology, pharmacology and pharmacokinetics of IFNs

The human IFNs comprise a unique family of biologically active proteins. At present, three main subspecies of human IFN have been recognized, characterized by their physiochemical and antigenic properties: α, β and γ. These are produced by leukocytes, fibroblasts, and T lymphocytes, respectively.[2]

The IFN-α subtypes are acid-stable proteins of about 166 amino acids with a molecular weight of 15000–26000 naturally produced by B lymphocytes, monocytes, and macrophages. IFN-β is also an acid-stable protein of 166 amino acids with a molecular weight of 15 000–22 000 naturally produced by fibroblasts, epithelial cells, and macrophages. IFN-γ is an acid-unstable protein of 146 amino acids with a molecular weight of 17000 naturally produced by T lymphocytes and NK cells.

The genes coding for IFN-α and IFN-β are localized on chromosome 9 while the gene coding for IFN-γ is localized on chromosome 12. Up to twenty-three different IFN-α genes coding for fifteen functional proteins have been identified. All the IFN-α genes are closely related and clustered on chromosome 9 close to the IFN-β gene.[3] Because of the few chemical differences observed between the IFN-α subtypes and their similar clinical activity, it is still unclear why there are so many.

Natural leukocyte IFN-α is a mixture of IFN-α subtypes obtained as a supernatant from human white blood cells exposed to a virus. The molecule is glycosylated and has a higher molecular weight than recombinant IFN-α (37 kDa). In contrast, recombinant IFN-α is produced by genetic engineering techniques in *Escherichia coli*. This process involves the insertion of the human IFN-α gene into the bacterial genome with subsequent expression of a highly purified (> 99 per cent) single non-glycosylated subtype of IFN-α.

All IFNs have a short terminal plasma elimination half-life ranging from 0.2 (IFN-β) to 2 (IFN-α) h when administered intravenously. The half-life is prolonged, up to 6 h for IFN-α, when IFNs are administered intramuscularly and may reach 7 h when IFN-α is administered subcutaneously. Adverse effects do not differ between the three routes of administration.

Mechanisms of action

To exert their effect on cells, IFNs must bind to specific cell surface membrane receptors. Two distinct IFN receptors are present on the cell membrane: one receptor for IFN-α and IFN-β, the gene for which is located on chromosome 21, and another receptor for IFN-γ, with the coding gene located on chromosome 6.[4]

Following the binding of IFNs to their specific receptors, the IFN–receptor complex is internalized, IFN is partially degraded and the receptor is then returned to the surface. Inside the cells, IFN affects numerous biochemical processes by altering the expression of several genes, upregulating or downregulating their expression.

ANTIGROWTH EFFECT

The potent inhibitory effect of IFNs on cell growth has been clearly demonstrated in several laboratory studies.[5]

In hematopoietic cells, both natural and recombinant alpha-interferons have shown a greater control of cell growth and differentiation than IFN-β and IFN-γ.[6–8] This inhibitory effect is more evident in noncycling tumor cells (G_0–G_1) and in some cases an accumulation of cells in G_0 has been observed, accompanied by a decrease in transition to G_1 and the arrest of some cell types in G_1.[9,10] However the mechanisms responsible for this antigrowth effect of IFNs are unclear. Inhibition of c-*myc* and c-*fos* or other oncogenes may be involved and there is also induction of $2'$–$5'$ A synthetase and a protein kinase which inhibit protein translation.[11–16]

In particular, the $2'$–$5'$ A synthetase degrades the mRNA linked to double-stranded RNA by activating a latent endoribonuclease, thus inhibiting RNA transcription and translation.[13,15,16] Protein and DNA synthesis may also be inhibited as the result of activation of a specific protein kinase capable of altering other factors needed for protein synthesis.[13] Furthermore, the inhibition of the gene for ornithine decarboxylase may also explain the effect of IFN-α on overall cell cycle slowing and arrest of cell division during G_0.

IMMUNE REGULATION

IFNs have numerous immunoregulatory effects that may also play a role in tumor control, and in this area IFN-γ appears to be more active than IFN-α. These immunoregulatory effects include: (a) the activation of monocytes and macrophages; (b) the induction of antigen expression on cell surfaces, including tumor-associated antigen; (c) increased natural killer (NK) cell activity; and (d) enhanced cytotoxic T-lymphocyte activity. However, the effects of IFNs on T cells are complex,[17] and it must be remembered that, as members of the cytokines network, IFNs often induce the secretion of other cytokines. Furthermore, IFNs (mainly IFN-γ) induce or enhance the expression of cell surface antigens for Class I major histocompatibility complex (MHC) both in normal and malignant cells.[18] Class II MHC antigen expression is also increased on monocytes, T lymphocytes, fibroblasts, and endothelial cells following IFN-γ exposure. Finally, IFNs increase the expression of receptors for the Fc fragment of IgG on the surface of lymphocytes and macrophages, thus enhancing the tumoricidal activity of these cells.[19,20] However, the most important immune changes produced by IFNs appear to be the enhancement of the cytotoxic activity of NK cells and macrophages.[21,22]

EFFECTS OF IFNS ON MYELOMA CELLS

Because a number of *in vivo* as well as *in vitro* studies (using either myeloma cell lines or bone marrow plasma cells from myeloma patients) have demonstrated the effectiveness of IFN-α in this disease, the majority of clinical studies of IFN in the treatment of MM have used either natural or recombinant IFN-α. Most information about IFN in myeloma therefore concerns the effects of IFN-α.

In particular, it has been demonstrated that in patients with multiple myeloma IFN-α possesses the ability to increase the NK activity of peripheral lymphocytes[23] as well as raising the levels of β$_2$-microglobulin[24] and this should be taken into consideration when using this marker for the assessment of the response during treatment with IFN-α. *In vitro* studies have demonstrated that IFN-α decreases the production of monoclonal immunoglobulin by myelomatous plasma cells.[25,26] This effect appears to be independent of the cytotoxic effect of IFN-α on myeloma cells.[25] Furthermore, IFN-α decreases both *in vitro* colony formation and the labeling index of myeloma cells.[27–29] In addition, Swedish investigators have recently demonstrated that IFN-α inhibits the growth of both IL-6-dependent and IL-6-independent myeloma cell lines. IFN-γ, in contrast, inhibits only IL-6-dependent cell lines.[30] Klein et al.[31] have recently found that inhibition of myeloma cell proliferation by IFN-γ and high-dose (100 U/ml) IFN-α is probably mediated by antagonizing the response to IL-6.

Therapeutic efficacy of IFNs in MM

Since 1979, both natural and recombinant alpha-interferons have been widely used, either alone or in combination with conventional chemotherapy, for treating patients with multiple myeloma in different phases of the disease.

PREVIOUSLY UNTREATED PATIENTS

IFN-α as a single induction agent

The remarkable results obtained by Mellstedt et al.[32] when they first used natural IFN-α as a single agent for induction therapy in four previously untreated MM patients (all patients achieved a durable response lasting from 3 to 19 months), prompted several investigators to utilize this biological response modifier in the treatment of MM. However, these early clinical studies revealed a wide range (from 20 per cent to 100 per cent) of responses with an overall response rate of about 30 per cent.

Recently, the Myeloma Group of Central Sweden (MGCS) has reported the results of a randomized trial comparing the administration of human leukocyte IFN-α with the administration of oral melphalan and prednisone (MP). Responses were observed in 44 per cent of patients treated with MP but in only 14 per cent of the patients treated with IFN-α. This difference was mainly due to a low response rate in IFN-treated IgG myeloma, while in the IgA and Bence Jones myeloma subgroups the response rate did not differ significantly between the two treatment groups. Moreover, as the response rate to a second-line treatment was better in the group initially treated with IFN-α than in the MP group, the overall survival duration was similar in both groups.[33]

Based on these results and on a possible dose–response relationship, the MGCS designed a study for IgA and Bence Jones MM alone to verify this dose-dependent concept. Various treatment schedules were tested and fifty previously untreated IgA and Bence Jones MM patients entered the study. The majority of these patients received 10 MU/day of natural IFN-α intramuscularly for 7 consecutive days repeated every third week. A total of 36 per cent of patients responded (41 per cent of IgA and 23 per cent of Bence Jones). The median time to response was 1.5 months with a median response duration of 20 months.[34] The results of this study confirm the efficacy of IFN-α in MM and indicate that high-dose IFN-α might induce a slightly better response rate than low dose of IFN-α, when used alone during the induction treatment.

The observation that IFN-α might be particularly active in IgA and Bence Jones myeloma was also suggested in another randomized study in which forty-two patients were randomized to receive recombinant IFN-α2c or the combination chemotherapy regimen VMCP (for explanation of abbreviations, see Table 11.1, p. 109) as induction treatment. A response was achieved by 6 of 18 evaluable IFN-α-treated patients and by 17 of 19 evaluable patients treated with VMCP. In this study, patients unresponsive to IFN-α crossed over to VMCP, and despite the low response rate observed in the IFN-α-treated group, the survival duration of both groups was identical.[35]

However, this difference in response between various M-component isotypes has not been confirmed by other groups and may be dependent on the type of IFN-α used, being more evident in those studies using 'natural' IFN-α as single induction agent.

IFN-α associated with conventional induction therapy

The evidence of *in vitro* synergy between IFN-α and some chemotherapeutic agents, such as cyclophosphamide, melphalan, Adriamycin, and vinca alkaloids,[36–39] and the results obtained using IFN-α as a single agent for inducing a response in MM stimulated some investigators to evaluate the role of IFN-α combined with conventional chemotherapies in induction treatment for newly diagnosed MM patients.

The first Phase I–II trial using this modality of treatment was published in 1986.[40] The study was designed to explore the feasibility of combining recombinant IFN-α2b with melphalan and prednisone (MP) in previously untreated MM patients. An overall response rate of 75 per cent was obtained and it appeared that a dose of 2 MU/m^2 of IFN-α2b administered three times a week for 2 weeks of a 28-day cycle did not seriously reduce the amount of alkylating agent that could be administered. This very high response rate was confirmed by a subsequent pilot study carried out by the ECOG group. The induction treatment of this study alternated cycles of VBMCP with the administration of recombinant IFN-α2b (5 MU/m^2, three times weekly for 3 weeks). The high response rate obtained in this study (80 per cent) was accompanied by 30 per cent of complete remissions, defined as disappearance of the monoclonal immunoglobulin from the serum and/or from the urine and absence of plasma cells from the bone marrow.[41] Recently, Alexanian et al. have treated fifty-four newly diagnosed MM patients with IFN-

α2b and dexamethasone. The response rate was 57 per cent, which is significantly higher than the 25 per cent obtained in similar patients previously treated with IFN alone ($p = 0.02$).[42]

The results obtained in these studies have stimulated several cooperative groups to design randomized clinical trials aiming at better evaluation of the role of IFN-α combined with standard induction chemotherapies in untreated MM patients. The published results of these randomized trials (summarized in Table 13.1) are conflicting. Some studies do not indicate that the addition of IFN-α to standard induction chemotherapies improves either the response rate or the duration of response and/or survival.[43,44] Other studies, however, have confirmed a higher response rate and a longer response and/or survival duration in the arm containing IFN-α.[45–47] In particular, the CALGB study comparing MP + IFN-α2b versus MP as induction therapy of previously untreated 278 MM patients did not show any advantage to the concomitant delivery of MP + IFN-α2b as initial treatment for patients with multiple myeloma.[44] These results are in agreement with those observed by Corrado et al. in sixty-two patients randomized to receive MP or MP + recombinant IFN-α.[43] However, three other randomized studies indicate that by combining IFN-α to standard induction chemotherapy it is possible to obtain a better response rate as well as a longer survival duration.

In particular, Scheithauer et al. combining IFN-α2c + VMCP obtained a 67 per cent response rate compared with 35 per cent with VMCP alone.[45] This high response rate in the IFN-α2c + VMCP arm was also associated with a marginal survival benefit. Recently, Montuoro et al. have published the results of a study in which fifty previously untreated MM patients were randomized to receive the standard induction treatment MP (28 patients) or the combination induction treatment MP + IFN-α2a (22 patients). The rate of responders as well as the response and survival durations were significantly higher in the MP + IFN-α2a-treated group than in the MP group.[46] In particular, the median response duration was 30 weeks in the MP arm and not yet reached in the MP + IFN-α2a ($p < 0.025$), and the median survival duration was 80 weeks in the MP arm and not yet reached in the MP + IFN-α2a arm ($p < 0.025$).

Similarly, in a recent study, using natural IFN-α, the MGCS (Myeloma Group of Central Sweden) has demonstrated a response rate of 68 per cent in the MP + natural IFN-α treated group versus 42 per cent ($p = 0.0001$) in the MP-treated group.[47] This difference was more marked in patients with IgA or Bence Jones myeloma: 85 per cent of IgA myeloma and 71 per cent of Bence Jones myelomas responded to MP + natural IFN-α, compared with 48 per cent and 27 per cent, respectively, to MP treatment ($p = 0.001$).

Table 13.1 Resposne rates and survival durations in randomized studies comparing chemotherapy with chemotherapy + interferon-alpha

Reference	Response rate			Median survival duration (months)		
	MP	MP+IFN	*p*	MP	MP+IFN	*p*
Scheitauer et al.[45]	35	67	Not specified	22	> 27	NS
Montuoro et al.[46]	68	95	< 0.05	20	Not reached	< 0.025
Corrado et al.[43]	48	45	NS	38	38	NS
Cooper et al.[44]	44	37	NS	37	37	NS
Österborg et al.[47]	42	68	< 0.0001	27	29	NS*

* For IgA or Bence Jones myeloma, median survival duration in MP+IFN arm was 32 months as compared to 17 months in the MP arm ($p < 0.05$).

There was no difference in the overall survival between the two treatment groups. However, the survival of 72 patients with IgA or Bence Jones myeloma randomized to receive MP + natural IFN-α was significantly longer (median = 32 months) than that of 71 patients treated with MP (median = 17 months) ($p < 0.05$).[47]

Finally, data from the randomized study of the ECOG group comparing VBMCP polychemotherapy to VBMCP + IFN-α2b are not yet available.

The conflicting results noted here may be due either to the different schedules of IFN administration or to the different IFN dosages utilized by the investigators.

PATIENTS IN RELAPSE, REFRACTORY OR RESISTANT TO PREVIOUS TREATMENTS

The first observation of a transient response obtained in a patient with MM refractory to oral melphalan was reported in 1979 by Ideström et al. using human leukocyte IFN-α.[48]

Since then, several other studies in which IFN-α has been used as salvage treatment in MM have been published with an overall response rate of 20 per cent. This response rate observed in refractory or resistant MM is noteworthy and makes IFN-α one of the most effective second-line treatments for MM.

Moreover, data from these studies indicate that for some refractory patients it is possible to achieve responses lasting more than 3 years.[49] Furthermore, an improvement in the levels of normal immunoglobulins has been noted in some patients[49,50] and a few patients have demonstrated progressive healing and recalcification of lytic bone lesions.

Recently, three non-randomized studies dealing with the use of IFN-α to treat resistant MM have been published in the medical literature. In these studies IFN-α was given in combination with a VAD regimen,[51] with high-dose dexamethasone,[51,52] or alone.[53]

In the MD Anderson Hospital study, in which IFN-α was combined with either VAD or dexamethasone, the response rate as well as the survival times of the patients did not improve in comparison with similar salvage treatments without IFN-α. The authors conclude that adding IFN-α to salvage chemotherapy in patients with advanced disease has no benefit.[51]

However, a different conclusion was drawn by San Miguel et al., who used a combination schedule of IFN-α dexamethasone as salvage treatment in refractory MM patients. They observed that 15 (68 per cent) of evaluable patients responded to treatment and the response in 6 patients lasted more than 1 year. Moreover, in 4 patients there was a reduction of bone marrow plasma cells to less than 5 per cent. Response was observed in 5 of 11 patients previously refractory to a combination regimen containing high-dose dexamethasone.[52] The authors concluded that IFN-α dexamethasone combination is a promising therapeutic approach for patients with refractory MM.

A less enthusiastic conclusion about the efficacy of IFN-α alone in the salvage treatment of resistant MM was drawn by Rödjer et al., who observed two 'major responses' (reduction > 50 per cent of the monoclonal component) and three minor responses in fourteen evaluable patients. IFN-α was administered subcutaneously 3 times per week in escalating doses from 2 MU/m^2 up to 15 MU/m^2. Despite these results, the conclusion of the authors was that IFN-α treatment may be more effective alone or in combination to chemotherapy in patients with a low tumor cell burden.[53]

IFNα AS MAINTENANCE TREATMENT IN PATIENTS RESPONDING TO CONVENTIONAL INDUCTION THERAPY

The rationale for using IFN-α as maintenance treatment in MM patients responding to conventional induction chemotherapy is given by the *in vitro* evidence that at the end of the induction treatment the myeloma cells of responding patients are in a 'plateau phase' similar to the G_0-phase of the cell cycle.[54] Because IFN is capable of prolonging all phases of the cell cycle as well as the overall cell-generation time, and in some cases causes an accumulation of cells in the G_0-phase with marked reduction in the self-renewal capacity of myeloma-forming cells,[9,10,55] it appears particularly appropriate to utilize IFN-α as an agent for maintaining the response obtained with conventional induction treatments.

Among the numerous studies dealing with the use of IFN-α as maintenance treatment of MM patients, four have already been analyzed and have produced conflicting results in terms of response and survival duration (summarized in Table 13.2).

The Italian study[56]

This was the first published study in which IFN-α was employed as an agent for maintaining the response already obtained with conventional therapy, rather than as an agent for inducing a response.

During a period of 3 years (from April 1985 to May 1988), 101 MM patients responding to 12 monthly courses of traditional first-line induction chemotherapy (either classical oral melphalan plus prednisone or the alternating regimen VMCP/VBAP) were enrolled in this study. Patients were recruited from a group of 202 symptomatic MM patients evaluated at three participating institutions of the Universities of Rome, Bari, and Turin. Of these 202 patients, at the end of the induction treatment, 79 had refractory disease, 18 were lost to follow-up, and 4 refused to

participate in the study. Therefore, a total of 101 patients who had an objective response or a stabilization of the disease were randomized to receive ($n = 50$) or not ($n = 51$) recombinant IFN-α2b. Randomization was stratified by induction treatment.

The overall results of the study were as follows: the median duration of response (from the time of randomization to maintenance treatment) was 26 months in the patients given IFN and 14 months in the untreated patients ($p = 0.0002$), while the median duration of survival (from the time of randomization to maintenance therapy) was 52 months in the IFN group and 39 months in the control group ($p = 0.0526$). However, among the patients who had an objective response to induction chemotherapy the difference in survival was statistically significant ($p = 0.0352$).

The obtained results indicated that maintenance treatment with IFN-α2b prolonged response and survival in patients with MM who have objectively responded to conventional induction chemotherapy.

After 5 more years of follow-up from the inclusion of the last patient in this study, the updated results confirm that IFN-α2b is a useful therapeutic tool in controlling the plateau phase of responding multiple myeloma patients ($p = 0.0002$). However, despite the fact that the median survival duration for responding patients is 50.9 months in the IFN arm as compared to 36.6 months in the control arm, the difference between the two arms is not significant using the log-rank test, while using the Wilcoxon statistical test the difference between the two arms is marginally significant ($p = 0.05$) (personal data).

The German study[57]

In this study 71 of 140 MM patients who had obtained a stable disease after induction chemotherapy were randomized to receive or not IFN-α2b as maintenance treatment. At the time of analysis there was no difference in relapse rate which was 50 per cent in both groups after 7 months and it was too early for survival analysis.

Table 13.2 Median response and survival durations in randomized studies comparing IFN-α with no treatment as maintenance

Reference	Response duration (months)			Survival duration (months)		
	IFN	Control	p	IFN	Control	p
Italian study[56]	26	14	0.0002	52	39	0.0526
German study[57]	8	8	NS	Not specified	Not specified	NS
Swedish study[58]	13.9	5.7	< 0.0001	36	35	NS
SWOG study[59]	12	11	NS	32	38	NS
English study[61]	39	27	< 0.025	Not specified	Not specified	< 0.05

However, the absence of benefit in the IFN-α-maintained group in this study may be due to differences in the modalities of randomization with respect to the other studies such as: (1) the absence of a stratification for the induction treatment before the randomization to the maintenance arm; (2) the shorter duration of the 'plateau phase' required before randomization when compared to the longer duration of the 'plateau phase' of other studies.

The Swedish study[58]

This is the most recent published study, on the use of IFN-α as maintenance treatment for newly diagnosed MM patients responding to conventional induction chemotherapy. Following MP induction treatment 125 evaluable patients who achieved a 'plateau phase', defined as response (> 50 per cent reduction of MC) lasting 4 months were randomized to receive (n = 61 patients) or not (n = 64 patients) IFN-α2b at the dose of 5 MU 3 times per week subcutaneously until relapse. After a minimum observation time from randomization of 36 months, the analysis of the study revealed a highly significant difference in the duration of the 'plateau phase' between the two treatment arms. The median duration of plateau was 13.9 months in the IFN arm and 5.7 months in the no therapy arm ($p < 0.0001$). As for survival duration, the difference between the two arms was not significant.

This well-designed study confirms that the use of IFN-α as maintenance treatment in MM patients responding to initial induction treatment and achieving a 'true' 'plateau phase' can significantly prolong the response duration as indicated by the Italian study.

The SWOG study[59]

Patients achieving a response to VMCP/VBAP or VAD or VMCPP/VBAPP (defined as 75 per cent of myeloma mass regression) were randomized to receive (n = 97) or not (n = 96) IFN-α2b until relapse. No differences in response duration or survival duration for patients receiving IFN-α have been observed.

These data do not confirm the results of the Italian and Swedish studies. However, some differences among the SWOG and the Italian and Swedish studies may be responsible for this discrepancy.

First, in the SWOG study the responding patients are only those with a reduction of at least 75 per cent of myeloma mass, while in the Italian and Swedish studies the cut-off for response is a reduction of at least 50 per cent of myeloma mass. Second, the proportion of patients with Stage III myeloma is higher in the SWOG than in the Italian and Swedish studies. Consequently the population of high-risk myeloma is higher in the SWOG study. The proportion of Stage I myeloma is also lower in the SWOG study.

It is therefore possible that these differences in response definition and in patient population may partially explain the negative results of this study. However, encouraging results have been obtained by the same group using IFN-α + dexamethasone in those patients with a lower reduction (< 75 per cent) of myeloma mass after induction chemotherapy.

Other studies dealing with the use of IFN-α as maintenance treatment of MM patients are still ongoing in different parts of the world: however, definitive data from these trials are not yet available.

IFN-α AS MAINTENANCE TREATMENT FOLLOWING BONE MARROW TRANSPLANTATION PROCEDURES

In general, advanced age and/or the presence of major medical problems preclude allogeneic bone marrow transplantation for most myeloma patients. More myeloma patients are candidates for high-dose chemotherapy and autologous bone marrow transplantation (ABMT) or peripheral stem cell infusion (PSCI) rescue.

High-dose chemotherapy (HDC) was developed to improve survival of patients with aggressive myeloma. However, even though HDC is capable of giving a high response rate, it is associated with severe life-threatening toxicities that can be reduced by ABMT or PSCI. As a consequence, an increasing number of patients with MM are now treated with HDC followed by ABMT or PSCI. However, relapse remains a major problem, and so in an attempt to improve the response and survival durations of these patients, a number of investigators have begun to use IFN-α as maintenance treatment following HDC and ABMT or PSCI in myeloma.

Attal and co-workers were among the first investigators to adopt this type of therapy in previously untreated aggressive myeloma.[60] They have recently published the results of a trial initiated to determine the feasibility and efficacy of three-phase treatment consisting of induction chemotherapy followed by high-dose melphalan and total body irradiation supported by unpurged ABMT and IFN-α maintenance treatment. After a median follow-up of 24 months after diagnosis and 15.5 months after ABMT, among the 15 patients who had obtained a complete response (CR) only 1 has relapsed, while among the 15 patients who did not obtain a CR there were 7 relapses.

The results of Attal et al. have been recently confirmed in a randomized study by English investigators.[61] In this study 84 patients with myeloma were randomized to receive maintenance IFN-α2b, 3 MU/m^2 s.c. 3 times weekly or no treatment following induction therapy with C-VAMP (cyclophosphamide, vincristine, Adriamycin, methylprednisolone), consolidated with high-dose melphalan (HDM) 200 mg/m^2 and unpurged ABMT. At a median follow-up of 24 months, median progression-free survival from HDM was 27 months in the control group and 39 months in the IFN-α2b group ($p < 0.025$). For the 62 patients who achieved CR with HDM there was a significant prolongation of remission ($p < 0.01$) and 53 per cent of these patients who received IFN-α2b remain in remission 4 years after treatment. However, for PR and non-responders to HDM there was no significant prolongation of progression-free survival. Overall survival was also significantly longer for the IFN-α2b group, with 1 death versus 6 deaths in the control group ($p < 0.05$).

The results obtained in these two studies suggest that the small residual tumor mass of patients achieving CR after ABMT could be highly sensitive to IFN-α maintenance treatment.

IFN-β and IFN-γ in multiple myeloma

To date there are only a few negative published resports in which IFN-β and IFN-γ have been used as salvage treatment with an overall response rate of less than 5 per cent. Therefore, at present, it does not seem that these interferons have a role in MM.

Tolerability

The major toxic side-effects observed in MM patients treated with IFN-α have been a self-limiting flu-like syndrome, a generally mild hematological toxicity, weight loss, headaches, and mental depression. The dosage of 3–5 MU of IFN-α 3 times per week subcutaneously is generally well-tolerated and only in a minority of patients must the dosage be reduced or discontinued because of the presence of severe toxic side-effects.

However, the optimal dose, schedule, and administration of IFN-α for the management of myelomatosis remain to be established.

What next?

The optimal uses of IFN in the management of MM remain to be fully explored. Therefore, in the near future we need to determine which MM patients are the best candidates for IFN treatment. For the time being only the following general indications can be given.

PREVIOUSLY TREATED (RELAPSING AND RESISTANT) MM PATIENTS

The data so far available do not support the use of IFNs alone as single agent in these subgroups of patients. However, the combination IFN-α + dexamethasone has proved to be effective in some of these patients, although further studies are needed before recommending this as a useful regimen. We have to consider that the use of dexamethasone associated to IFN can help in reducing cytopenia which can otherwise limit IFN dosage.

NEWLY DIAGNOSED MM PATIENTS

IFN alone

Even though a small number of patients have proved to be responsive to induction treatment with IFN alone, its use as single induction agents cannot, at present, be recommended.

Combinations of IFN + conventional chemotherapies

The published studies using these approaches have shown conflicting results (Table 13.1). Therefore, controlled trials using the same inclusion, exclusion, as well as response criteria are needed to better identify those patients who may benefit from these combinations. Moreover, it is extremely important that these studies address the questions of which dose, schedule, and type of IFNs (natural or recombinant) are more useful in this setting.

Maintenance treatment with IFNs after conventional or high-dose chemotherapy

The studies addressing the question of IFN maintenance versus no maintenance have also shown conflicting results (Table 13.2). Two studies have shown no benefit of the maintenance treatment with IFN-α, while two have shown a significant prolongation in response duration. Moreover, in one of these last studies there was also a significant prolongation of survival. Furthermore, data from a randomized study[61] utilizing IFN-α as maintenance treatment after autologous bone marrow transplant indicate that the response and survival durations can be prolonged by IFN maintenance in those patients who have obtained a good response after the transplant.

Considering these studies, it will be interesting in the next years to have more data about the role of IFN when used as maintenance treatment after conventional chemotherapies, autologous or allogeneic bone marrow transplant, even though no data about allogeneic bone marrow transplant + IFN are so far available.

In summary, at present, the combined data suggest that IFN-α, when used as maintenance treatment, is more efficacious in those MM patients who have reached a good 'plateau phase' with low tumor burden at the end of the chemotherapeutic program, whether or not this includes high-dose therapy followed by a transplantation procedure.

References

1. Isaacs A, Lindemann J. Virus interference. *Proceedings of the Royal Society of London (Biology)* 1957; **147**: 249–67.
2. Stewart WE II, Blalock JE, Burke DC et al. Interferon nomenclature. *Journal of Immunology* 1980; **24**: 235 (letter).
3. Balkwill F. Interferons. *Lancet* 1989; **i**: 1060–3.
4. Aguet M, Morgensen KE. Interferon receptors. *Interferon* 1983; **3**: 1–22.
5. Gresser I. How does interferon inhibit tumor growth? *Interferon* 1985; **6**: 93–126.
6. Blalock J, Georgiades JE, Langford MP et al. Purified human immune interferon has more potent anticellular activity than fibroblast or leukocyte interferon. *Cellular Immunology* 1980; **49**: 390–4.
7. Borden EC, Hogan TF, Voelkel JG. Comparative antiproliferative activity in vitro of natural interferons alpha and beta for diploid and transformed human cells. *Cancer Research* 1982; **42**: 4948–53.
8. Chadha KC, Srivasta BI. Comparison of the antiproliferative effects of human fibroblast and leukocyte interferons on various leukemic cell lines. *Journal of Clinical Hematology and Oncology* 1981; **11**: 55–60.
9. Creasey AA, Bartholomew JC, Merigan TC. Role of G0–G1 arrest in the inhibition of tumor cell growth by interferon. *Proceedings of the National Academy of Sciences* 1980; **77**: 1471–5.
10. Horoszewicz JS, Leong SS, Carter WS. Non cycling tumor cells are sensitive targets for the antiproliferative activity of human interferon. *Science* 1979; **206**: 1091–3.
11. Bishoff JR, Samuel CE. Mechanism of interferon action. *Journal of Biological Chemistry* 1985; **260**: 8237–9.
12. Kimci A. Autocrine interferon and the suppression of the c-myc nuclear oncogene. *Interferon* 1987; **8**: 86–110.
13. Revel M, Kimci A, Shulman L. Role of interferon-induced enzymes in the antiviral and antimitogenic

effects of interferon. *Annals of the New York Academy of Sciences* 1980; **350**: 459–73.
14. Samid D, Chang EH, Friedman RM. Development of transformed phenotype induced by a human RAS oncogene is inhibited by interferon. *Biochemistry and Biophysics Research Communications* 1985; **126**: 509–16.
15. Senn CC. Biochemical pathways in interferon action. *Pharmacology and Therapeutics* 1984; **24**: 235–57.
16. William BRG. Biochemical actions of interferon. In: Sikora K ed. *Interferon and cancer.* New York: Plenum Press, 1983: 33–52.
17. Siegel JP. Effects of interferon-γ on the activation of human T lymphocytes. *Cellular Immunology* 1988; **11**: 461–72.
18. Rosa F, Fellous M. The effect of γ-interferon on MHC antigens. *Immunology Today* 1984; **5**: 261–2.
19. De Maeyer-Guinard J, De Maeyer E. Immunomodulation by interferons: recent developments. *Interferon* 1985; **6**: 69–86.
20. Fertsch D, Vogel SN. Recombinant interferons increase macrophage Fc receptor capacity. *Journal of Immunology* 1984; **132**: 2436–9.
21. Edwards BS, Hawkins MJ, Borden EC. Comparative *in vivo* and *in vitro* action of human NK cells by two recombinant α-interferons differing in antiviral activity. *Cancer Research* 1984; **44**: 3135–9.
22. Huddlestone JR, Merigan TC, Oldstone MBA. Induction and kinetics of natural killer cells in humans following interferon therapy. *Nature* 1979; **282**: 417–19.
23. Heinhorn S, Ahre A, Blomgren et al. Interferon and natural killer activity in MM. Lack of correlation between interferon-induced enhancement of natural killer activity and clinical response to human interferon-alpha. *International Journal of Cancer* 1982; **30**: 167–72.
24. Tienhaara A, Remes K, Pelliniemi TT. Alpha interferon raises serum beta-2-microglobulin in patients with multiple myeloma. *British Journal of Haematology* 1991; **77**: 335–8.
25. Grandér D, von Stedingk LV, von Stedingk M et al. Influence of interferon on antibody production and viability of malignant cells from patients with multiple myeloma. *European Journal of Haematology* 1991; **46**: 17–25.
26. Tanaka H, Tanabe O, Iwato K et al. Sensitive inhibitory effect of interferon-alpha on M-protein secretion of human myeloma cells. *Blood* 1989; **74**: 1718–22.
27. Brenning G, Ahre A, Nilsson K. Correlation between *in vitro* and *in vivo* sensitivity to human leukocyte interferon in patients with multiple myeloma. *Scandinavian Journal of Haematology* 1985; **35**: 543–9.
28. Brenning G. The *in vitro* effect of leukocyte alpha-interferon on human myeloma cells in a semisolid agar culture system. *Scandinavian Journal of Haematology* 1985; **35**: 178–85.
29. Salmon SE, Durie BG, Yung L et al. Effects of cloned leukocyte interferons in the human tumor

stem cell assay. *Journal of Clinical Oncology* 1983; **1**: 217–25.

30. Jernberg-Wiklund H, Pettersson M, Nilsson K. Recombinant interferon-gamma inhibits the growth of IL-6-dependent human multiple myeloma cell lines *in vitro*. *European Journal of Haematology* 1991; **46**: 231–6.

31. Klein B, Zhang XG, Jourdan M et al. Interleukin-6 is a major myeloma cell growth factor *in vitro* and *in vivo* especially in patients with terminal disease. *Current Topics in Microbiology and Immunology* 1990; **166**: 23–31.

32. Mellstedt H, Ahre A, Björkholm M et al. Interferon therapy in myelomatosis. *Lancet* 1979; **i**: 245–7.

33. Ahre A, Björkholm M, Mellstedt H et al. Human leukocyte interferon and intermittent high-dose melphalan-prednisone administration in the treatment of multiple myeloma: a randomized clinical trial from the Myeloma Group of Central Sweden. *Cancer Treatment Reports* 1984; **68**: 1331–8.

34. Ahre A, Björkholm M, Osterborg A et al. High dose of natural α-interferon (α-IFN) in the treatment of multiple myeloma. A pilot study from the Myeloma Group of Central Sweden (MGCS). *European Journal of Haematology* 1988; **41**: 123–30.

35. Ludwig H, Cortelezzi A, Scheithauer W et al. Recombinant interferon alpha-2c versus polychemotherapy (VMCP) for treatment of multiple myeloma: a prospective randomized trial. *European Journal of Cancer and Clinical Oncology* 1986; **22**: 1111–16.

36. Aapro Ms, Alberts DS, Salmon SE. Interactions of human leukocyte interferon with vinca alkaloids and other chemotherapeutic agents against human tumors in clonogenic assay. *Cancer Chemotherapy and Pharmacology* 1983; **10**: 161–6.

37. Balkwill FR, Monshowitz S, Seilman S. Positive interaction between interferon and chemotherapy due to direct tumor action rather than effects on host drug-metabolizing enzymes. *Cancer Research* 1984; **44**: 5249–55.

38. Balkwill FR, Moodey EM. Positive interactions between human interferon and cyclophosphamide or Adriamycin in a human tumor model system. *Cancer Research* 1985; **44**: 906–8.

39. Welander CE, Morgan TM, Homesley HD et al. Combined recombinant human interferon alpha 2 and cytotoxic agents in a clonogenic assay. *International Journal of Cancer* 1985; **35**: 721–9.

40. Cooper MR, Fefer A, Thompson J et al. Alpha-2 interferon/melphalan/prednisone in previously untreated patients with multiple myeloma: A phase I–II trial. *Cancer Treatment Reports* 1986; **70**: 473–6.

41. Oken MM, Kyle RA, Greipp PR et al. Complete remission (CR) induction with VBMCP+interferon (rIFNα2) in multiple myeloma: 3 year follow-up. *ASCO Proceedings* 1989; **8**: 272 (abstract 1062).

42. Alexanian R, Barlogie B, Gutterman J. Alpha interferon combination therapy for multiple myeloma. IFN therapy in B-cell malignancies. *V Hannover Interferon Workshop* 1990, Abstract book, p. 43.

43. Corrado C, Flores A, Pavlovsky S et al. Randomized trial of melphalan (L-PAM)-prednisone (PRED) with or without recombinant Alpha2 interferon (r α2 IFN) in multiple myeloma. *ASCO Proceedings* 1991; **10**: 304 (abstract 1064).

44. Cooper RB, Dear K, McIntyre RO et al. A randomized trial comparing melphalan/prednisone with or without interferon alfa-2b in newly diagnosed patients with multiple myeloma: a cancer and leukemia group B study. *Journal of Clinical Oncology* 1993; **11**: 155–60.

45. Scheithauer W, Cortelezzi A, Fritz E et al. Combined α-2c-interferon/VMCP polychemotherapy versus VMCP polychemotherapy as induction therapy in multiple myeloma: A prospective randomized trial. *Journal of Biol Response Mod* 1989; **8**: 109–15.

46. Montuoro A, De Rosa L, De Blasio A et al. Alpha-2a interferon/melphalan/prednisone versus melphalan/prednisone in previously untreated patients with multiple myeloma. *British Journal of Haematology* 1990; **76**: 365–8.

47. Österborg A, Björkholm M, Björeman M et al. Natural interferon-α in combination with melphalan/prednisone versus melphalan/prednisone in the treatment of multiple myeloma stage II and III: a randomized study from the myeloma group of central Sweden. *Blood* 1993; **81**: 1428–34.

48. Ideström K, Cantell K, Killander D et al. Interferon therapy in multiple myeloma. *Acta Medica Scandinavica* 1979; **205**: 149–54.

49. Costanzi JJ, Cooper R, Scarffe JH et al. Phase II study of recombinant alfa-2-interferon in resistant multiple myeloma. *Journal of Clinical Oncology* 1985; **3**: 654–9.

50. Quesada JR, Alexanian R, Hawkins M et al. Treatment of multiple myeloma with recombinant alpha-interferon. *Blood* 1986; **67**: 275–8.

51. Alexanian R, Barlogie B, Gutterman J. Alpha-interferon combination therapy of resistant myeloma. *American Journal of Clinical Oncology* 1991; **14**: 188–92.

52. San Miguel JF, Moro M, Blade J et al. Combination of interferon and dexamethasone in refractory multiple myeloma. *Hematology and Oncology* 1990; **8**: 185–9.

53. Rödjer S, Vikrot O, Wahlin A et al. Effect of interferon alpha-2b in advanced multiple myeloma. *Journal of Internal Medicine* 1990; **227**: 45–8.

54. Durie BGM, Russell DH, Salmon SE. Reappraisal of plateau phase in myeloma. *Lancet* 1980; **ii**: 65–7.

55. Bergsagel DE, Haas RH, Messner HA. Interferon alpha-2b in the treatment of chronic granulocytic leukemia. *Seminars in Oncology* 1986; **13** (suppl. 2): 29–34.

56. Mandelli F, Avvisati G, Amadori S et al. Maintenance treatment with recombinant interferon alfa-2b in patients with multiple myeloma responding to conventional induction chemotherapy. *New England Journal of Medicine* 1990; **322**: 1430–4.

57. Peest D, Deicher H, Coldewey R et al. Melphalan and prednisone (MP) versus vincristine, BCNU, adriamycin, melphalan and dexamethasone

(VBAMDex) induction chemotherapy and interferon maintenance treatment in multiple myeloma. Current results of a multicenter trial. *Onkologie* 1990; **13**: 458–60.

58. Westin J, Rödjer S, Turesson I et al. Interferon alpha versus no maintenance therapy during the plateau phase in multiple myeloma: a randomized study. *British Journal of Haematology* 1995; **89**: 561–8.

59. Salmon SE, Crowley JJ, Grogan TM et al. Combination chemotherapy, glucocorticoids and interferon alpha in the treatment of multiple myeloma: a South-West Oncology Group Study. *Journal of Clinical Onclology* 1994; **12**: 2405.

60. Attal M, Huguet F, Schlaifer D et al. Intensive combined therapy for previously untreated aggressive myeloma. *Blood* 1992; **79**: 1130–6.

61. Cunningham D, Viner C, Montes A et al. A randomized trial of maintenance therapy with Intron-A following high-dose melphalan and ABMT in myeloma. *ASCO Proceedings* 1993; **12**: 364 (abstract 1232).

CHAPTER 14

Cytokines in the treatment of multiple myeloma

HEINZ LUDWIG, ELKE FRITZ

Introduction	159	Myeloid growth factors	171
Interferons	159	Erythropoietin	172
Interleukin-2	167	Conclusions	174
Interleukin-4	169	References	175
Inhibition of IL-6	169		

Introduction

The introduction of active chemotherapy for multiple myeloma in the 1960s[1] led to a significant prolongation of patients' survival up to a period of 30–35 months. Attempts to improve these results with combination chemotherapy have so far remained largely unsuccessful. Only recently a new therapeutic concept, namely highly aggressive treatment regimens followed by allogeneic or autologous bone marrow transplantation, has triggered considerable hope for improved treatment results. This is the background to the high expectations raised by another distinct therapeutic approach, the clinical application of interferon and other cytokines subsequently introduced into the treatment of multiple myeloma.

Cytokines are important mediators of a wide variety of cellular functions and have, in fact, already been shown to possess significant therapeutic potential in myeloma. In the following review, particular emphasis will be placed on the efficacy of interferon in the treatment of myeloma and additional information will be given on interleukin-2 (IL-2), IL-4, myeloid growth factors and the attempts to inhibit the biologic sequels of IL-6. Finally, the role of recombinant human erythropoietin in the treatment of chronic myeloma-associated anemia will be discussed. The chief characteristics of these cytokines are listed in Table 14.1.

Interferons

THE BIOLOGY OF INTERFERONS

Three main subspecies of interferon are presently recognized: IFN-α, IFN-β, and IFN-γ. Physiologically, the production of IFN-α mainly by leukocytes and of IFN-β by fibroblasts is induced through viral stimulation; the production of IFN-γ by T lymphocytes constitutes an immune reaction.[2] All of the 15 or so subtypes of IFN-α so far identified are presumably contained in the preparations of natural IFN-α available for clinical use. This is in contrast to the genetically engineered types of IFN-α which consist of only one subtype, namely either α2a, α2b, or α2c. These subtypes differ only in one or two positions of their amino acid sequence,[3] and are nowadays preferentially used for clinical practice

Table 14.1 Cytokines and their inhibitors with proven or potential therapeutic activity in myeloma

Cytokine or inhibitor	Produced by	Chromosome	Amino acids	Potentially therapeutic functions
Interferon-α (leukocyte, alpha-2a, -2b, -2c)	Monocytes Macrophages Virus-infected cells	9p22	156–166[*]	Growth inhibition of myeloma cells Stimulation of immune system Downregulation of IL-6 and oncogenes Synergism with drugs Enhanced expression of cell surface antigens
Interferon-β	Fibroblasts Virus-infected cells	9p22	166	Same as above
Interferon-γ	T cells NK cells	12	143	Immunomodulation
Interleukin-2	T cells	4q26–q27	133	Stimulation and activation of NK and LAK cells Increase in LAK-mediated lysis of myeloma cells
Interleukin-4	T cells	5q23–q32	129	Downregulation of IL-6 production
IL-6 inhibitors	Variable			Inhibition: blockade of IL-6 and IL-6 receptors
G-CSF	Monocytes Fibroblasts	17q11–q22	207	Stimulation and differentiation of neutrophilic-committed progenitors
GM-CSF	T cells Fibroblasts Endothelial cells Monocytes	5q21–q32	127	Stimulation and differentiation of myeloid and monocytoid progenitors
Erythropoietin	Renal interstitial cells	7q11	165	Stimulation of erythropoiesis

[*] 23 variants in natural interferon-α.

and in clinical trials. For IFN-β two subtypes have been identified.[4] Both IFN-α and IFN-β bind to the same receptor type on the cell membrane while IFN-γ binds to another interferon receptor type.[5]

After the binding of interferon to its receptor, the intracellular signal transduction activates a rather complex transcriptional pathway[6] which causes differential modulation of the expression of several genes. As a result of these gene modulations, the synthesis of several distinct proteins will either be enhanced or suppressed. The ensuing biologic sequels are manifold and mainly depend on the affected cell type.

For the treatment of patients with multiple myeloma, several biologic activities of interferon may be relevant. Among these, the direct, dose-dependent[7] growth inhibitory effect of IFN-α on myeloma cell lines[8] and on myeloma colony formation *in vitro*[9] seems to be particularly important. Also, a direct cytotoxic effect of interferon on myeloma cell cultures has been shown in some

systems.[10] While IFN-α is able to inhibit the growth of both IL-6-dependent and IL-6-independent myeloma cells *in vitro*, IFN-γ inhibits only the growth of IL-6-dependent cell lines.[11] When a small number of myeloma patients were treated with IFN-γ, practically none showed a significant response.[12] IFN-β, a glycoprotein,[13] is not freely available as a natural interferon preparation. Thus, the vast majority of clinical trials have used recombinant IFN-α and, from this point on, whenever our review states 'interferon', we are referring to that interferon subtype.

It has been shown that interferon downregulates monoclonal protein production of cultured[14] as well as *in vivo* myeloma cells during long-term treatment,[15] an effect which sometimes may be superimposed on the direct inhibition of cell growth. This possible decrease in the amount of monoclonal protein produced per myeloma cell should be considered when the clinical effects of interferon are evaluated in individual patients by monitoring the level of their

M-component. However, in patients responding to interferon, the clinical data usually indicate a concordant decrease in the number of myeloma cells as well as in the level of myeloma protein.

A limited number of *in vitro* experiments point to a synergistic activity when interferon is combined with melphalan.[16] This synergy seems to be even further enhanced with the use of a triple combination consisting of interferon, melphalan, and prednisone.[17] In the clinical setting, the combination of interferon with melphalan results in a decrease in the area under the plasma concentration–time curve of melphalan.[18] This observation may be interpreted in terms of increased cellular uptake of the cytostatic drug which should yield increased cytotoxic activity. However, an alternative explanation for this finding could be an increased catabolism due to the frequent interferon-induced elevation of body temperature and/or other metabolic processes.

Interferon-α was also found to downregulate the expression of the IL-6 receptor and to reduce its sensitivity to its ligand.[19] In a few selected cell lines, however, low concentrations of IFN-α induced an autocrine production of IL-6 and thereby enhanced the proliferation of the myeloma cell clone.[20] In some myeloma patients the oncogenes c-*myc* and/or N-*ras* are overexpressed. Since in other cell systems interferon has been shown to downregulate these oncogenes,[21,22] that mechanism may also play a role in multiple myeloma.

Alpha-interferons are potent stimulators of several immune functions. Among these activities the stimulation of the cytolytic activity of natural killer (NK)-cells[23] and of macrophages[24] may be particularly important in myeloma, since NK cell function is markedly reduced in patients with a high tumor load.[25] *In vitro* studies indicated the presence of cytolytic T cells which react to autologous myeloma cells and possess the potential to lyse them,[26] a mechanism which may also be enhanced by interferon. In addition, interferon increases the expression of the major histocompatibility antigens[6] and of tumor-associated antigens[26] which may facilitate the recognition of tumor cells by the immune system.

Other biological effects of IFN-α regard its role as a differentiation agent with the potential both to inhibit[27] and to induce[28] differentiation of several cell lines. At present, no data are available to substantiate the role of these interferon effects in myeloma.

THERAPEUTIC EFFICACY: INTERFERON MONOTHERAPY

The first report on the activity of interferon in myeloma was published in 1979 by Mellstedt and colleagues.[29] Those authors treated four chemotherapy-resistant patients with daily intramuscular injections of 3×10^6 U leukocyte interferon and observed objective responses in all of them; two patients achieved complete and two partial remissions. Due to these encouraging results, they subsequently designed a prospective randomized study comparing interferon monotherapy, at a dose of 3×10^6 U leukocyte interferon daily, with melphalan/prednisone (MP).[30] The overall response rate was only 14 per cent in the interferon arm as compared to 44 per cent in the MP arm. Interestingly, patients with IgA or light-chain myeloma achieved a comparable response rate in both treatment arms. This prompted the authors to apply high-dose leukocyte interferon only to patients with IgA or light-chain myeloma.[31] Their treatment regimen called for seven daily injections of interferon up to doses as high as 30×10^6 units per day which should be repeated every 3 weeks. The maximal dosage tolerated by a larger group of patients was 10×10^6 units. The overall response rate was 36 per cent with the highest number of responses in patients who tolerated the highest doses. From these observations two conclusions can be drawn. First, there seems to be a dose–response effect for interferon, at least when used as a single agent. Second, higher doses frequently lead to intolerable side-effects in the usually elderly population of myeloma patients.

Our own prospective, randomized trial compared interferon monotherapy with VMCP (for explanation of abbreviations, see Table 11.1 on p.109) chemotherapy.[32] The responses obtained were similar to those of the Swedish trial with a response rate of 14 per cent in the interferon and 57 per cent in the VMCP arm. When these results became evident in an interim analysis of forty-two patients, we terminated the trial and continued our investigations with the combined treatment protocols.

Other authors who had employed interferon monotherapy reported response rates between 10 per cent[33] and 37 per cent,[34] with one study on a small number of patients claiming 50 per cent responses.[35] However, this figure could not be confirmed in subsequent trials. In general, several clues emerged from these investigations. Recombinant human IFN-αs render response rates comparable to those of natural leukocyte interferon, although the latter may be particularly active in patients with IgA or light-chain myeloma. Patients with newly diagnosed disease possess a higher sensitivity to interferon as compared to those who had been heavily pretreated with cytostatic drugs. Finally, interferon seems to be more active in patients with a lower tumor burden.

Collectively, these studies convincingly demonstrated an antitumor activity of interferon in myeloma. The overall response rate seems to be in the vicinity of 20 per cent[36] for IFN-α, but it is significantly lower (6 per cent) for natural IFN-β.[37] Thus, the activity of IFN-α is comparable to that of some cytostatic drugs when used as single agents; only monotherapy with melphalan and possibly with dexamethasone yields better results.

INTERFERON WITH CHEMOTHERAPY FOR INDUCTION TREATMENT

The positive outcome of the interferon monotherapy trials as well as *in vitro* studies showing a synergistic activity of interferon when combined with certain cytostatic agents[16,38] led to the design of combined interferon chemotherapy protocols. The first trial utilizing the concept of biomodulation during induction chemotherapy was published in 1986 by Cooper and colleagues,[39] who reported a remarkable 75 per cent response rate in MP+IFN-treated patients. Three years later, an important trial was first published by the ECOG group.[36] Those authors obtained with a

regimen alternating VBMCP with IFN-α2b a response rate of 80 per cent and a remarkable rate of 30 per cent complete responses with the complete disappearance of the serum M-component and reductions of bone marrow plasma cells to less than 5 per cent. Follow-up data, however, failed to meet the high expectations raised by the initially observed complete responses. The median survival time was 42 months with 42 per cent of the patients surviving after 5 years (personal communication).

The positive results of these Phase I–II trials combining IFN with chemotherapy led to several large-scale prospective, randomized clinical trials comparing induction chemotherapy with or without the addition of interferon. Three larger Phase III trials have so far been concluded, two have already been published, our own study has as yet only been reported as interim results.[40] In addition, there are several completed minor studies on smaller patient populations. All but our own trial, which used VMCP, employed melphalan/prednisone for chemotherapy. The main features of the larger studies are shown in Table 14.2.

Table 14.2 Randomized trials on interferon for induction therapy in patients with multiple myeloma

Authors/regimens	*n*	Interferon	Response rate (%)	Duration of response	Survival
Österborg et al.[41]					
MP	160	–	42	No info.	29
Leukocyte IFN + MP	157	7 MU/m^2 days 1–5, 23–26	68 $p < 0.0001$	No info.	27
Cooper et al.[46]					
MP	134	–	44	22	37
IFN-α2b + MP	134	2 MU/m^2 days 1,3,5,8,10,12	37	18	36
Ludwig et al.[40]					
VMCP	131	–	60	13	30
IFN-α2b + VMCP	125	2 MU/m^2 days 1–5 every week, continuously	67	19 $p < 0.02*$ $p < 0.06**$	39
Montuoro et al.[43]					
MP	44	–	68	No info.	No info.
IFN-α2b + MP	51	6 MU/m^2 days 1,3,5,8,10,12	86 $p < 0.05$	No info.	No info.
Corrado et al.[47]					
MP	29	–	48	39	38
IFN-α2b + MP	33	5 MU/m^2 days 1,3,5,8,10,12	45	84	39
Garcia-Larana et al.[48]					
MP	28	–	54	45	No info.
Leukocyte IFN + MP	26	7 MU/m^2 days 1–5, 23–26	62	74	No info.

* Breslow, ** Mantel-Cox.

Of the representative studies, two revealed results in favor of the interferon chemotherapy combination. Österborg et al.[41] combined 7×10^6 U/m² leukocyte interferon, given for 5 days every 3 weeks, with standard melphalan/prednisone at 6-week intervals. In case of response, the interferon dose was reduced to 3×10^6 U/m² 3 times per week continuously, MP every 6 weeks was continued until progression or relapse. In the control group, the equivalent melphalan/prednisone treatment was given. The authors found a significant increase of the response rate in the combined cytokine chemotherapy group (65 per cent) as compared to the MP treatment arm (45 per cent). As previously reported,[42] that observed difference was mainly due to enhanced response rates in patients with Stage II, whereas in Stage III only a tendency towards improved response to MP+IFN was seen. The success in terms of higher response rates, however, could not be translated into increased survival times. Similar results were reported by Montuoro et al.[43] from a trial involving 50 patients. These authors claim that the twenty-two patients randomized to receive MP+IFN showed not only a significantly higher response rate (95 per cent as compared to 68 per cent under MP only) and prolonged progression-free survival, but also significantly longer overall survival. Aitchison et al.,[44] who treated 34 patients with cyclophosphamide (600 mg/m² every 21 days) and 17 patients with additional IFN-α, found a significantly higher response rate (25 per cent in the former, 53 per cent in the latter treatment arm), but no prolongation of response duration or survival times.

In our trial,[40] we compared a flat dose of 2×10^6 U IFN-α 5 times per week + VMCP with VMCP alone. Those patients who achieved remission or disease stabilization were later randomized to a maintenance protocol. The response rates to the interferon chemotherapy protocol (67 per cent) and to the VMCP regimen (62 per cent) were comparable. However, in the interferon+VMCP group the proportion of patients with progressive disease (11 per cent) was significantly lower in comparison to that in the group that received VMCP chemotherapy only (22 per cent). This observed effect was mainly due to a significant decrease in the number of patients with progressive disease in Stages I and II, but not in Stage III. The duration of response was significantly prolonged in the combined treatment arm (19 months versus 13 months), but median survival showed only a tendency towards prolongation (40 months versus 30 months for VMCP+IFN and VMCP, respectively). Similar results have been reported recently from a trial comparing maintenance with or without interferon while consolidation treatment with MP was continued until at least 4 months of stable response phase had been reached.[45] The relative risks of progression and death were significantly reduced in the interferon arm, with progression-free survival of 16 and 12 months for IFN and controls, respectively, and the equivalent overall survival values of 39 and 34 months.

In contrast to the encouraging results reported above, the outcome of the third large Phase III trial was completely negative.[46] In that study, interferon was applied at an unusually low dosage (12×10^6 U/ m² per month) which most probably is insufficient to induce a significant inhibitory effect on the growth of the myeloma cell clone. Other trials on much smaller samples of about 50 patients in total[47,48] yielded no significant differences between MP with interferon or without it. Considering the relatively small margin of the expected improvement, significant results can hardly be expected from such small patient populations.

We tentatively conclude that there may be some advantage in combined interferon chemotherapy protocols, particularly for patients with a low tumor burden. With regard to unselected myeloma patients of all stages, however, the advantage of the addition of interferon to standard chemotherapy seems to be moderate at best.

INTERFERON FOR REMISSION MAINTENANCE

As mentioned above, some data suggest a preferential activity of interferon in patients with low numbers of myeloma cells. Such a state of disease is usually achieved in patients in remission after successful induction chemotherapy. The relevance of this consideration was first shown by Mandelli and his associates.[49] These authors randomized 101 patients, who had achieved PR, CR or stable disease after either MP or VMCP/VBAP, to either 10×10^6 U/m² interferon 3 times a week or to an untreated control group. The uncommonly high interferon dosage was only poorly tolerated by the first twelve patients of this trial, so the authors reduced the dose to 3×10^6 U/m² 3 times a week for subsequent individuals. The outcome of the trial was striking. Remission duration was substantially prolonged from 14 months in the control group to 26 months in the interferon arm. The survival data of the entire group showed a tendency towards prolonged survival in patients maintained with interferon (median: 52 months) as compared to unmaintained patients (median: 39 months). Subgroup analysis revealed, however, that significantly prolonged survival was only seen in patients who had responded to induction treatment, but not in those who had merely achieved disease stabilization. Thus, only patients with a low tumor load at

Table 14.3 Trials on interferon-α2b for maintenance therapy in patients with responsive or stabilized myeloma

Authors Induction regimen	Reduction of M-component	Interferon dose (3 times a week)	n Controls n IFN arm	Progression- free survival	Overall survival
Mandelli et al.[49] MP or VMCP/VBAP	> 50% or stabilization	3 MU/m²	51 50	Controls 14 IFN 26 $p<0.0002$	39 50 $p<0.056$
Peest et al.[52] MP for stage II MP or VBAMDex	> 75% tumor cells	5 MU/m²	55 46	Controls 9 IFN 11 n.s.	No info. No info.
Westin et al.[50] MP	> 50%	5 MU/m²	61 59	Controls 6 IFN 17 $p<0.05$	38 39 n.s.
Salmon et al.[53] VMCP/VBAP or VMCPS/VBAPP or VAD	> 75%	3 MU/m²	96 96	Controls 12 IFN 11 n.s.	34 37 n.s.
Cunningham et al.[51] hd-M	100% or > 50%	3 MU/m²	31 30	Controls 27 IFN 39 $p<0.1$	93%[a] 75%[a] $p<0.001$
Browman et al.[45] MP		2 MU/m²	92 85	Controls 12 IFN 16 $p<0.01$	34 39 n.s.
Ludwig et al.[40] VMCP or IFN + VMCP	> 75% or > 50% or stabilization	2 MU/m²	54 46	Controls 8 IFN 21 $p<0.02$[*] $p<0.08$[**]	45 51 $p<0.07$[*] $p=0.17$[**]

[a] Percentage surviving at last analysis. [*] Breslow, [**] Mantel–Cox.

randomization benefited from interferon in terms of prolonged survival. The main features of this trial and of other studies, which compared the IFN maintenance arm to an untreated control group, are shown in Table 14.3.

Similar findings emerged from the Swedish trial.[50] In that study, only patients who had achieved objective response and a plateau phase after melphalan/prednisone induction treatment were randomized to either interferon maintenance therapy (5×10^6 U flat dose 3 times a week) or to 'wait and see'. The median duration of the plateau phase was 17 months in patients on interferon maintenance and only 6 months in the controls. Again, the most pronounced effect (median plateau phase: 24 months) was observed in patients with an optimal response to previous chemotherapy (< 5 per cent plasma cells and disappearance of M-component as determined by ordinary electrophoresis). Median overall survival, however, was practically identical in the two groups, at 42 months in interferon-maintained and 44 months in unmaintained patients.

In our own trial,[40] we randomized patients who had achieved CR, PR, or stabilized disease under either IFN+VMCP or VMCP induction treatment, to interferon (2×10^6 units flat dose 3 times a week) maintenance therapy or to the control group of patients without maintenance treatment. Both arms were closely monitored for one year. The median progression-free survival (Fig. 14.1) was significantly longer in interferon-maintained patients (21 months) compared to the control arm (8 months), and overall survival showed a tendency towards prolongation in interferon-maintained patients (50 months versus 34 months in controls). Our study design allowed the comparison of four different treatment protocols: (1) interferon during both induction and maintenance therapy; (2) sole VMCP induction therapy followed by interferon in the maintenance phase; (3) VMCP+IFN for induction and no maintenance therapy; and (4) induction treatment and maintenance phase without interferon. Progression-free survival as determined from the start of the maintenance phase was significantly longer in patients who had received interferon both for induction and maintenance treatment. Progression-free survival of patients who had not received any interferon at all was significantly poorer; protocols 2 and 3 yielded

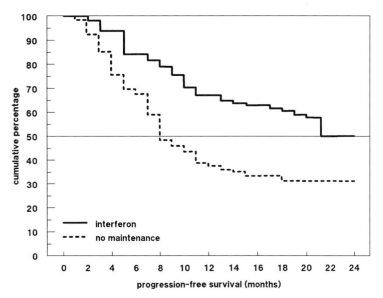

Fig. 14.1 Cumulative progression-free survival in myeloma patients treated with interferon or without therapy during maintenance phase. Interferon maintenance therapy resulted in significantly prolonged progression-free survival (median: 21 months) as compared to controls (median: 8 months).

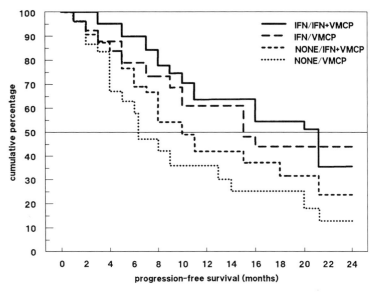

Fig. 14.2 Cumulative progression-free survival in myeloma patients treated with or without interferon during induction and maintenance phase. When the results of interferon maintenance therapy (IFN) versus controls (NONE) are stratified according to induction treatment with interferon (IFN+VMCP) or without it (VMCP), continuous interferon therapy effects the best results, but even unmaintained patients seem to benefit from the interferon treatment received during the induction phase.

intermediate results (Fig. 14.2). A similar tendency was observed for overall survival. Patients who had from the start of induction treatment been continuously under interferon therapy showed prolonged survival, the results of the other treatment protocols have, within the relatively short observation period, not yet become distinguishable.

Additional and intriguing arguments for interferon maintenance treatment come from a trial conducted in patients after high-dose melphalan treatment.[51] Of the patients randomized to either interferon treatment or to the control group, those who had achieved a complete response (complete disappearance of the M-component and < 5 per cent myeloma cells in the bone marrow) showed a significantly longer median response duration (46 months) under interferon maintenance therapy than the controls (27 months). The survival rates after a median follow-up time of 35 months in the IFN arm and 41 in controls were also significantly higher (93 per cent and 75 per cent, respectively). When the patients who had only achieved partial responses after high-dose chemotherapy were also included in the analysis, the outcome was still in favor of interferon maintenance but it failed to reach statistical significance.

Two larger studies contradict the above favorable results for interferon maintenance treatment in myeloma. The study by the German Myeloma Group[52] showed no benefit whatsoever in 101 patients of whom 46 had been randomized to 5×10^6 U interferon 3 times per week until relapse. Of the 101 cases, all 41 Stage II patients had been induced with MP, 29 of the patients with Stage III myeloma had received MP and 31 of them VBAM-Dex as initial treatment. Median remission duration was 11 months on interferon maintenance and 9 months without maintenance therapy. The overall survival was comparable in unmaintained patients and in patients on interferon remission treatment. However, that study can hardly be compared to the others, because the authors defined response according to a certain decrease in the number of tumor cells. This method, based on estimations, is less precise and much less reliable than the widely used simple and relatively accurate quantification of the changes in the serum paraprotein level. Because of the imbalance in pretreatment during the induction phase, it remains unclear whether the category of patients who in other studies benefited most from interferon maintenance therapy, namely patients with the best response to induction chemotherapy, achieved better results under interferon maintenance or not.

Definitely negative results were derived in a recent SWOG study.[53] A sample of 192 patients who had achieved complete remissions (> 75 per cent reduction of paraprotein) were randomized to interferon (3×10^6 U flat dose 3 times a week) until progression or to an untreated control group. Remission duration was 12 months in the former and 11 months in the latter arm; median survival was 34 months in the interferon-maintained and 37 months in the unmaintained population. Patients with a reduction of paraprotein between 25 per cent and 75 per cent after one of three different induction regimens (VMCP/VBAP, VMCPP/VBAPP or VAD) were subsequently treated with interferon–dexamethasone. Almost every 5th patient (19 per cent) showed a further decrease in M-component to <25 per cent of the pretreatment level.

An additional trial compared interferon maintenance after double hemi-body irradiation in relapsed patients, but accrued only 15 patients in the interferon and 13 in the control arm.[54] There was no difference in median survival times (10 months in both arms), but the number of investigated patients is, of course, too small to allow unequivocal conclusions.

INTERFERON COMBINED WITH CHEMOTHERAPY FOR REMISSION MAINTENANCE TREATMENT

Two major trials applied interferon in combination with chemotherapy during the maintenance phase. One of them analyzed whether the results of maintenance with interferon (3×10^6 U flat dose 3 times a week) could be further improved by the combination of interferon with alternating cycles of chemotherapy regimens.[55] In eighty-five evaluable patients, no difference in maintenance duration (9 months with sole interferon and 6 months with interferon plus chemotherapy) could be detected.[56] Similar results were obtained by Österborg et al.,[41] who continued the regimen used for induction therapy (either MP or MP+IFN) in 146 responding patients. Under that mode of maintenance treatment, only a tendency towards statistical significance was achieved for the differences in remission duration (29 and 20 months for MP+IFN and MP, respectively) and survival starting from the day of response (30 and 23 months, respectively).

In conclusion, the majority of the data presently available indicate that interferon is an active drug for remission maintenance. It also seems to have an impact on survival, but probably only in those patients who present with the most favorable prognostic factor, namely maximal tumor reduction after induction chemotherapy. Further information is required to put the negative results obtained in two trials into perspective with the other positive data reported. As interferon maintenance treatment with conventional doses (about 3×10^6 U IFN 3 times a week) causes only moderate toxicity, many physicians will presently decide to exploit the benefits of interferon in a patient with an optimal response to the induction chemotherapy.

INTERFERON COMBINED WITH DEXAMETHASONE

Corticosteroids have been shown to reduce the toxicity of interferon without ameliorating its treatment efficacy.[57] In addition, dexamethasone exerts significant activity as a single agent in myeloma and ranks, only surpassed by melphalan, in second place of the short list of active drugs for that disorder. These facts and a possible synergism with interferon make both substances an interesting choice for combination treatment in myeloma. The results so far obtained in untreated patients are in line with these expectations. Alexanian et al.[58] achieved with IFN-α2b+dexamethasone responses in 57 per cent of newly diagnosed patients, a result that is significantly higher than the usual outcome of interferon monotherapy. As the combination of interferon and high-dose dexamethasone may cause considerable toxicity, a direct comparison of this treatment mode with the standard MP protocol is needed, in order to facilitate decision-making of whether interferon–dexamethasone surpasses the gold standard of MP for the treatment of newly diagnosed myeloma patients.

San Miguel et al.[59] were able to achieve with interferon–dexamethasone a response rate of 68 per cent in patients already resistant to chemotherapy. However, Alexanian et al.[60] could not confirm those results. They treated sixty-eight patients with resistant myeloma with IFN-α and either VAD for patients in relapse or high-dose dexamethasone for those who had never responded and found no improvement in the response rate or in survival times by the addition of interferon in comparison to similar treatments without it. A recent Phase III trial,[61] in which refractory or relapsed myeloma patients received either VAD or VAD+IFN, rendered similar results, namely an objective response rate of 25 per cent in the VAD arm and 30 per cent in the VMCP arm, and no differences in progression-free and overall survival.

When Salmon et al.[62] applied interferon–dexamethasone to patients who had only reached partial remission during induction treatment with conventional chemotherapy, they could induce complete remission (> 75 per cent reduction of the initial M-component concentration) in 42 per cent of them. Similar observations have been reported by Palumbo and colleagues.[63] After induction with conventional chemotherapy, they maintained thirteen patients on interferon (3×10^6 units 3 times a week) plus prednisone, if the patient was older than 70 years, or high-dose dexamethasone (40 mg/day for 4 days every 4 weeks) in younger patients. This protocol resulted in a further reduction of the M-component by > 50 per cent in seven (54 per cent) patients.

INTERFERON IN MGUS AND IN ASYMPTOMATIC MYELOMA STAGE I

The potential of interferon to inhibit the progression of MGUS or asymptomatic myeloma to overt symptomatic myeloma has been tested in a few minor trials. Unfortunately, the number of investigated patients was too small and follow-up times were too short to render conclusive results. It may tentatively be summarized that up to now no convincing activity of interferon in preventing disease progression of early stage myeloma has been observed.

TOXICITY OF INTERFERON

The initial treatment phase of interferon therapy is often accompanied by flu-like symptoms, including fever, chills, and fatigue, which characteristically subside after 1–2 weeks of treatment. Hematotoxicity of interferon may necessitate dose reduction or even discontinuation of interferon treatment in a number of patients. Further side-effects are fatigue, muscle pain, and, less frequently, nausea, neurologic or psychiatric problems, loss of appetite, and, after long-term treatment, mild alopecia. Interferon-induced toxicity is dosage-dependent,[64] with doses of the magnitude of 30×10^6 U daily causing unacceptable adverse reactions,[31] while low to moderate doses (e.g. 10×10^6 U/week) generally induce only minor, tolerable side-effects. The toxicity of interferon may also be curbed by the concomitant application of corticosteroids.[57]

Interleukin-2

Interleukin-2 is an important growth factor for T cells, NK cells, and for certain B cell subpopulations.[65] It induces the antigen-selected expansion of T-cell clones and, in conjunction with other cytokines such as IL-4, also of B-cell clones.[66] IL-2 stimulates the proliferation and activation of NK cells and of certain cytotoxic lymphocytes which are known as LAK cells.[67] All those functions render IL-2 an active stimulant of the immune system and, thus, a possible agent for cancer immunotherapy. Besides interferon, IL-2 is presently the only other cytokine which regularly induces tumor regression in about 10–20 per cent of patients with metastatic melanoma or renal cancer[68] and in some patients with non-Hodgkin's lymphomas.[69]

IL-2 is also an interesting candidate for immunotherapy of myeloma. In that disease, several findings point to the involvement of the immune system in regulating proliferation and differentiation of the myeloma cell clone.[70] Myeloma patients show a relative and absolute decrease of CD4+ T cells and an increase of CD8+ suppressor cells, but both cell types express the activation marker CD9 and show the emergence of HLA-DR+ and an expansion of CD11b+ and CD56+ cells, reflecting a state of activation.[71] There is also evidence of the presence of autoreactive T cells capable of suppressing growth and immunoglobulin secretion of human myeloma cells[72,73] and of T cells with cytotoxic activity against autologous myeloma cells.[74] The latter effect can be enhanced in *ex vivo* cultures by the addition of IL-2.[75] An investigation of NK activity in myeloma patients yielded normal findings of their peripheral blood, but increased NK activity in their bone marrow.[76] A similar distribution was observed regarding the generation of LAK activity. As compared to controls, LAK activity was not enhanced in the peripheral blood of myeloma patients, but higher levels of LAK activity could be induced in their mononuclear bone marrow cells.[70] The magnitude of LAK enhancement, however, depends on the course of the disease, the treatment regimen, and response to treatment of the individual patient. It is important that autologous myeloma cells as well as plasma cell lines are sensitive to lysis by IL-2-induced LAK cells generated from mononuclear cells of the peripheral blood.[77] The addition of IL-2 to the patient's peripheral blood mononuclear cells resulted also in the increased release of tumor necrosis factor and IFN-γ, both cytokines which shorten the survival time of malignant plasma cells and reduce their thymidine uptake in culture.[77,78]

Serum levels of IL-2 are of prognostic relevance in myeloma. In a study by Cimino et al.,[79] higher levels of IL-2 (> = 10 U/ml) were associated with a 5-year survival rate of 87 per cent, whereas only 13 per cent of patients with low levels (< 10 U/ml) were still alive after 5 years. The serum levels of the circulating receptor, sIL-2R, on the other hand, display the reverse pattern with the highest levels in patients with progressive disease and the lowest during the plateau phase.[80] Clinical data suggest that T-cell and NK-cell activation may be important in controlling myeloma and influencing the patient's survival.[79]

In spite of these arguments in favor of the clinical use of interleukin-2 in myeloma, two facts should be kept in mind. First, IL-2 therapy induces the production of IL-6,[81] which is a potent stimulator of myeloma precursor cells. Second, IL-2 receptors have been detected on myeloma cells in 16 per cent of the investigated patients.[80] Both conditions carry, at least theoretically, the possibility of an unintentional stimulation of the myeloma clone by IL-2 treatment.

So far only few and rather small clinical studies have been conducted on IL-2 in myeloma patients. In one trial, Togawa et al.[82] applied 3.3×10^5 U/day intravenously or subcutaneously to three patients with myeloma. They observed an increase of Leu7+ and CD16+ cells in the peripheral blood and enhanced NK- and LAK-cell activities, but could not find any reduction of the M-component levels. Subsequently, the authors treated three further patients intravenously with a combination of LAK cells and IL-2. All three patients showed an increase in lymphocyte counts and NK activity. Their M-component decreased to 17 per cent, 76 per cent, and 40 per cent, but those objective improvements were only transient with a return to the baseline values after 35, 55, and 200 days, respectively.[82] So far, the largest series has been reported by Peest et al.[83] who treated eighteen patients with chemoresistant advanced disease. Subcutaneous injections of rIL-2 were given twice daily at a dosage of 9×10^6 U/m² on days 1 and 2 and of 0.9×10^6 U/m², 5 times a week, from day 3 to day 56. This treatment was repeated every 12 weeks until disease progression. Two of the 17 evaluable patients experienced an objective response (72 per cent and 35 per cent reduction in tumor cell mass) and 4 patients showed a stabilization of their disease. The median survival of these 6 responding patients had not been reached after an observation period of 36 months, the median survival of the remaining 11 cases was 12 months.

In our own trial,[84] we entered ten heavily pretreated patients on a protocol of 100 μg IFN-γ subcutaneously on day 1 and 1.8×10^6 U/m² IL-2 subcutaneously from day 1 to 5. This treatment was repeated every week for a total treatment period of 8 weeks. All patients showed a significant increase in their number of white blood cells, lymphocytes, NK cells (CD3−/CD16+/CD56+ and CD16+/CD38+), IL-2 receptor-positive (CD25+) cells and activated B and plasma cells (CD19+/CD38+) in their peripheral blood. In the bone marrow, a significant increase of activated T helper cells (CD4+/CD69+) and HLA-DR+ cells as well as a trend towards more NK cells (CD3−/CD16+/CD56+) and fewer plasma cells (CD19+/CD38+) were seen. This profound influence on immune regulation did not translate into objective tumor response, although in the total number of patients a tendency towards a decrease in LDH levels and serum M-component concentrations was noted. The side-effects were considerable, particularly on day 1, when we tried to enhance IL-2 receptors with the additional administration of IFN-γ. The toxicity included WHO Grade III–IV fever, myalgia, flu-like symptoms, pruritus, local indurations, pain, activation of viral infections, nausea, vomiting, diarrhea, and moderate deteriorations of renal function.

From the limited clinical experience available at the present, one may tentatively conclude that rIL-2 treatment shows only a poor to moderate efficacy in patients with chemoresistant progressive myeloma. It may, however, prove to be more efficient in patients with low tumor mass and stable disease. Such conditions are generally achieved after successful conventional or high-dose induction treatment has affected a complete remission or a partial response with a plateau phase. An intriguing additional field of application is the setting of allogeneic bone marrow transplantation in which IL-2 may exert dual activity. It may simultaneously suppress the graft-versus-host disease and increase the graft-versus-myeloma effect.[85] As a consequence of the relatively high toxicity of the present regimens, one may speculate whether ultra-low doses of IL-2 might turn out to be superior to the treatment regimens with conventional doses. Tolerance would probably be improved while, hopefully, IL-2 would still be active enough to induce a significant stimulation of the patient's immune system. It remains to be seen whether such an approach or other novel strategies will survive the critical clinical testing and help to improve the treatment results in multiple myeloma.

Interleukin-4

Interleukin-4 is a cytokine with pleotropic activity. Originally, it was identified as a stimulatory factor of B cells,[86] but subsequently it has been shown to have characteristic effects on various functions of B cells, T cells, and macrophages.[87–89]

IL-4 is mainly produced by CD4+ T cells[90] and, to a lesser extent, by mast cells[91] and tumor-associated natural killer cells.[92] Its main functions are the activation of resting B cells,[93] the enhancement of IgE production,[94] and increased expression of low-affinity receptors for IgE (CD23).[95] In addition, it stimulates antigen processing, the expression of HLA Class II antigens, the production of CSF by monocytes, and the tumoricidal activity of macrophages.[89,96] IL-4 promotes T-cell proliferation and enhances the generation of antigen-specific cytotoxic T cells.[97] It also causes a preferential switch to IgG1 and IgE production[98,99] and it antagonizes the IL-2-induced proliferation of B cells.[100]

Of particular interest for the therapeutic use of IL-4 in myeloma is its capacity to downregulate the clonogenic growth of myeloma cells in culture[101] and its inhibitory effect on the *in vitro* proliferation of myeloma cells by suppressing IL-6 expression.[102] These antitumor activities make IL-4 a candidate for clinical trials in myeloma.

In fact, IL-4 has already been tested in an American–European multicenter Phase II trial in more than forty myeloma patients, but official data have not yet been published. Therefore, it should only be men-

tioned that no objective response was observed in these patients who were all heavily pretreated with chemotherapy. The best responses were disease stabilizations in about 20 per cent of the cases. It is of interest that some patients developed hypercalcemia after the start of IL-4 treatment. This side-effect could be related to IL-4 acting as a B-cell growth factor. In previous Phase I trials, the maximally tolerated dose of IL-4 was 5 µg/day.[102] In another trial, toxicity was found to be moderate at doses of 0.25–1.0 µg/kg/day.[103]

The most common side-effects of IL-4 treatment were headaches, fever, nausea, vomiting, rigor, fatigue, myalgia, pain, edema, rhinitis, and other nasal symptoms. The incidence of these symptoms was dose-dependent.[104] From the only preliminary data derived from myeloma patients, it does not seem very likely that IL-4 will play a major role in the conventional treatment of that disease. The gene for IL-4 may, however, be an interesting candidate for transfection in gene therapy studies. In experimental models, the IL-4 gene has been transfected into tumor cells which subsequently were transplanted to nu/nu mice, either alone or mixed with a variety of non-transfected tumor cells. The preliminary results of that study indicate an objective antitumor effect which was primarily mediated by an inflammatory infiltrate of eosinophils and macrophages. The antitumor activity correlated strongly with the level of IL-4 production.[105]

Inhibition of IL-6

The molecular mechanisms leading to the malignant transformation in myeloma confer a definite growth advantage to the malignant clone, but several humoral factors seem to modify myeloma proliferation.[106] Among them, IL-6 is presently considered the most important growth factor.[107] This cytokine is predominantly produced by the bone marrow, i.e. by myeloid, monocytoid,[107] and stromal cells,[106] and functions mainly as a paracrine growth factor.[108] Only rarely is it expressed by myeloma cell clones. Receptors for IL-6, on the other hand, are abundant on bone marrow and circulating B cells.[109] They are found in only a minority of myeloma bone marrow cells, but some myeloma cell lines express them in a large percentage of cells.[110] When IL-6 binds to its cellular receptor, the latter combines with a 130 kDa transmembrane glycoprotein (gp130) which seems to be essential for the intracellular signal transduction.[110] The biologic role of the soluble IL-6 receptor (sIL-6R) is less well-understood, but evidence has emerged that after interaction with its

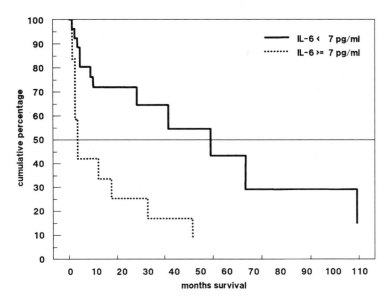

Fig. 14.3 Baseline levels of interleukin-6 as a prognostic factor in patients with multiple myeloma. Serum IL-6 levels of 7 pg/ml or higher are associated with significantly shorter survival times than values below 7 pg/ml.

ligand, it also binds to gp130, thus inducing the intracellular signal.[111]

When IL-6 is added in combination with IL-3 to cultures of peripheral blood mononuclear cells of myeloma patients, a considerable proportion of them differentiate to plasma cells which carry the immune phenotype of the bone marrow myeloma cell clone.[112,134] High serum levels of IL-6 have predominantly been found in patients with aggressive disease[113] and they correlated inversely with prognosis (Fig. 14.3).[114] Thus, there is considerable evidence for a pathogenic role of IL-6 in myeloma and also for potential therapeutic benefits from interventions which inhibit the biologic sequels of that important myeloma growth factor.

Several strategies for the accomplishment of that goal have already been successfully applied *in vitro* and one treatment concept has already proved to be efficient *in vivo*. Klein et al.[115] reported the first successful application of a mouse polyclonal antibody against IL-6 in a chemotherapy-resistant myeloma patient with progressive disease. The anti-IL-6 treatment resulted in a temporary inhibition of myeloma cell proliferation with the reduction of tumor cells in S-phase from 4.5 per cent to 0.0 per cent and a decrease in serum calcium, serum M-component, and C-reactive protein. Similar encouraging results were obtained by the same study group in a total of 3 of 9 patients.[116] However, the clinical improvement was transient and all patients rapidly succumbed to their disease.

In vitro experiments have exploited additional approaches to inhibit IL-6. Monoclonal antibodies against sIL-6R were shown to significantly reduce the proliferation of myeloma cells *in vitro*[117] and *in vivo*.[118] One group reported the production of humanized monoclonal antibodies against sIL-6R which also exerted significant growth inhibition *in vitro*.[119] Saito et al. produced monoclonal antibodies against the IL-6 signal transducer gp130 and showed that they can inhibit IL-6-mediated functions (120). The attempts to inhibit IL-6 production at the subcellular level with antisense oligonucleotides yielded a significant downregulation of IL-6 and the inhibition of *in vitro* growth of IL-6-dependent cell lines.[121] Another interesting strategy is the generation of toxic IL-6 fusion proteins. When *Pseudomonas* exotoxin[122] or diphtheria toxin[123] is coupled to IL-6, the resulting hybrid protein binds selectively to IL-6 receptor-positive cells. After binding, the fusion proteins are internalized into the lysosomal vacuoles where they exert most of their cytotoxic activity. This principle has also been shown to be highly active in bone marrow preparations of myeloma patients and thus has been recommended for *ex vivo* purging.[122] Table 14.4 summarizes the most common options currently available for inhibition of IL-6-stimulated myeloma growth and for IL-6-guided specific toxicity.

In spite of these promising first results, a word of caution is needed against overoptimistic expectations. Ballester et al.[124] suggest on the basis of their *in vitro* and *in vivo* findings that the importance of IL-6 as a myeloma growth factor presently is grossly overemphasized and they propose instead that the primary role of IL-6 is as a myeloma differentiation factor. Even if that assumption may not be confirmed, it is possible that in the course of IL-6-blockade treatment

Table 14.4 Options for inhibition of IL-6-stimulated myeloma growth and for IL-6-guided specific toxicity

Principle	Technical approach	Efficacy reported
Inhibition of extracellular IL-6	Polyclonal antibody (Ab) against IL-6	*in vivo*; Klein et al., 1991[115]
Blockade of the IL-6 receptor	mo-Ab against IL-6R	Huang and Vitetta, 1993[117]
	mo-Ab against IL-6R	Suziki et al., 1992[118]
	Reshaped the hu-Ab against IL-6R	Saito et al., 1993[120]
Inhibition of IL-6-induced intracellular signal transduction	mo-Ab against signal transducer protein gp130	Saito et al., 1993[120]
Inhibition of intracellular IL-6	Antisense oligonucleotides	Levy et al., 1991[121]
IL-6-targeted cytotoxicity	Fusion of *Pseudomonas* exotoxin with IL-6	Kreitman et al., 1992[122]
	Fusion of diphtheria toxin with IL-6	Chadwick et al., 1993[123]

a myeloma subclone will evolve which is independent of IL-6 and/or resistant to further attempts of IL-6 blockade. Such a subclone would probably sooner or later overgrow the IL-6-sensitive cell fraction and would then dominate the further clinical course of the individual patient. Nevertheless, the results achieved so far with the above-mentioned treatment concepts are promising enough to foster further intensive attempts to analyze and exploit the clinical potential of inhibiting the biologic activity of IL-6.

Myeloid growth factors

The myeloid growth factors GM-CSF, G-CSF, M-CSF, and IL-3 are a family of glycoprotein hormones which regulate self-renewal, proliferation, differentiation, and survival of myeloid progenitor cells as well as the functional activities of more mature myeloid cells.[125–127] GM-CSF shows a broad spectrum of activities and induces in semi-solid marrow cultures bilineage colonies, containing both neutrophils and mononuclear phagocytes, as well as pure neutrophil and eosinophil colonies.[128] In addition, it primes neutrophils for enhanced function and response to triggering agents[129] and increases the cytotoxicity of macrophages against tumor cells. G-CSF induces *in vitro* the growth of pure neutrophil colonies and stimulates the functional activity of polymorphonuclear cells *in vitro*[130] and *in vivo*, including phagocytosis, intracellular killing, and antibody-dependent cytotoxicity. M-CSF enhances the function of monocytes and macrophages, particularly their intracellular killing ability.[131] Finally, IL-3 is a pluripotent growth factor which significantly affects myelopoiesis and, to a lesser degree, also erythro- and megakaryopoiesis.[132]

There is evidence that some of these growth factors, when applied *in vitro*, synergize with IL-6 in supporting the proliferation of myeloma cells. In one study, GM-CSF was found to enhance the endogenous IL-6-mediated myeloma cell proliferation observed in 5-day-old myeloma cell cultures.[133] In another *in vitro* investigation of myeloma patients, IL-3 and IL-6 were found to enhance the differentiation of peripheral blood mononuclear cells into plasma cells.[134] However, no stimulation of freshly obtained myeloma cells was seen when either GM-CSF or G-CSF was directly added to the cultures. Only IL-3 induced the growth of myeloma cells in 2 of the 10 patients studied.[135] The latter results are in accordance with clinical experience which did not reveal any evidence of stimulated growth in myeloma during *in vivo* treatment with GM-CSF or G-CSF. On the contrary, these cytokines have been proven to be valuable tools for several treatment situations. First, GM-CSF and G-CSF have greatly helped to shorten and overcome the critical phase of bone marrow aplasia after high-dose chemotherapy with[136] or without bone marrow transplantation.[137,138] The administration of GM-CSF or G-CSF usually starts at a daily dose of 3–4 μg/kg, one day after the infusion of peripheral stem cells or bone marrow, and continues as long as myeloid aplasia prevails. As soon as the leukocyte production recovers and sufficient values have been reached, the treatment is discontinued. With this approach, the mean duration of significant granulocytopenia (< 500/μl) can be shortened by 5–6 days. This accelerated granulocyte recovery leads to a reduction in the number of infectious episodes, fewer requirements of antibiotic treatment, and shorter durations of hospitalization. Definite proof of a significant impact on survival, however, is still pending.

Another important indication for the use of myeloid growth factors stems from their potential to mobilize stem cells.[139,140] Again, both growth factors significantly enhance the shift of stem cells from the bone marrow to the peripheral circulation and thereby increase the yield of peripheral stem cells. This procedure of myeloid growth factor-primed stem cell collection has already become routine in most transplantation centers and, thus, is also widely used in myeloma patients.

One can assume that myeloid growth factors are also commonly used in patients with leukopenic phases after conventional chemotherapy, in order to reduce their risk of infections. However, no data have been published so far, which might prove the superiority of this approach to the situation prior to the era of myeloid growth factors.

In conclusion, myeloid growth factors now constitute an important therapeutic principle in the management of patients with multiple myeloma. They have been proven to significantly shorten the period of bone marrow aplasia and have also been shown to be the most active drugs available for stem cell mobilization.

Erythropoietin

Erythropoietin (EPO), a 30 kDa glycoprotein, is the major growth factor of erythropoiesis.[141] It stimulates proliferation and differentiation of erythroid progenitor cells[142,143] and also amplifies the production of red blood cells by inhibiting the premature death (apoptosis) of their precursor cells.[144] The main sites of EPO production are the peritubular cells of the kidney;[145] liver macrophages contribute approximately 10–15 per cent of the total production.[146] Erythropoietin synthesis is mainly stimulated by the low arterial and, in particular, low tissue oxygen tension[147] characteristic of anemia.

Soon after the application of recombinant DNA production technologies rendered sufficient supplies of EPO for therapeutic interventions, substitution of this hormone became the standard treatment for the anemia of chronic renal failure,[148,149] a typical manifestation of erythropoietin deficiency. Subsequently, erythropoietin treatment has been applied to a variety of anemias of cancer,[150,151] including chronic anemia associated with multiple myeloma.

In multiple myeloma about 20–50 per cent of patients present at diagnosis with anemia, which usually subsides upon the achievement of complete remission but persists, and may even worsen, if response to treatment is inferior. As a consequence of the limited success of chemotherapy, the proportion of anemic myeloma patients increases with increasing duration of the disease.[152] As myeloma

commonly is a disease of the elderly, who often suffer from reduced cardiopulmonary function, myeloma-associated anemia frequently becomes symptomatic even if the reduction of hemoglobin is only minor.

The pathogenesis of myeloma-associated anemia is presently only poorly understood. In many anemic myeloma patients the number of erythroid precursors (BFU-E) is significantly reduced in their bone marrow[153] and peripheral blood.[154] This deficiency probably contributes to or accounts for the reduced responsiveness of the erythron to erythropoietin[154–156] which is more frequently, but not exclusively, seen in patients with a large tumor mass, e.g. in 50 per cent of patients with myeloma Stage III,[155] and in those with a high percentage of plasma cells.[157] Endogenous EPO production is inadequate in approximately 25 per cent of unselected myeloma patients. However, that proportion increases to 50 per cent in patients with myeloma Stage III[155] and it is even higher in patients who suffer from severe renal impairment.[155,158] This blunted erythropoietin response with regard to the patient's degree of anemia has been suggested to be related to the increased plasma viscosity frequently found in myeloma patients.[159] In the remaining majority of patients, biogenesis of erythropoietin is normal in relation to the degree of anemia.[154,158] Thus, chronic myeloma-associated anemia is due to either a decrease in the erythroid precurser compartment or the reduced sensitivity of the erythron to erythropoietin stimulation or a blunted erythropoietin response to the anemic condition or it may result from a combination of these factors.[155]

Erythropoietin treatment of patients with myeloma-associated anemia frequently corrects their anemia, apparently irrespective of the underlying pathomechanisms described above. In our initial series of 13 and then 18 anemic myeloma patients, we were able to achieve a response rate of 85 per cent and 78 per cent, respectively, with 150 U/kg erythropoietin twice a week.[153,160] These results still hold true with our present population of 34 investigated patients of whom 26 (76.5 per cent) responded to treatment with recombinant erythropoietin. Barlogie's group achieved comparable results by applying the same treatment regimen.[161] Additional confirmations of these results come from a placebo-controlled trial by Abels[162] and from other studies which investigated only relatively small numbers of anemic myeloma patients but yielded comparable results.[151,163–165] In our study, the median time to response was 4.1 weeks with increases in hemoglobin of at least 2 g/dl seen as early as 2 weeks, but also as late as 12 or even 20 weeks after the start of erythropoietin treatment (Fig. 14.4). The likelihood of response was higher in patients with stable disease, low baseline serum levels of endogenous erythropoietin, and no intercur-

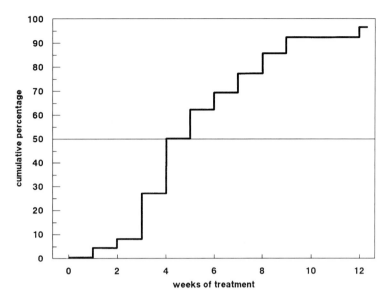

Fig. 14.4 Time to response to recombinant human erythropoietin therapy in anemic myeloma patients. Fifty per cent of responsive patients have achieved their response (a gain of at least 2 g/dl hemoglobin) by the fifth week of therapy, but response may occur as early as during the second week or as late as after 3 months.

rent infections or other concomitant illnesses. In one series,[161] patients with prolonged alkylator treatment (>1 year) were found less likely to respond than those with a shorter treatment duration. Possibly, long-term treatment with alkylating agents leads to an impairment of stem cell functions. Patients who develop progressive disease or undergo surgery usually lose their sensitivity to erythropoietin substitution, even if the dosage is increased. After recovery from surgery and the successful management of complications, responsiveness to erythropoietin substitution therapy may be restored.

Transfusion-dependent patients seem to respond almost as well as those without prior transfusion requirements. Although the bone marrow iron stores tend to be overloaded initially,[166] the pronounced stimulation of erythropoieses in responding patients rapidly increases the demand for endogenous iron, resulting in repleted iron stores[167] which may require iron supplementation during EPO treatment. It is interesting that some of the non-responding patients present with relatively high serum levels of endogenous erythropoietin and that during the course of therapy the serum EPO levels tend to rise in nonresponders, while they hardly change in responding patients (Fig. 14.5).

Our data show that correction of anemia in myeloma patients can be maintained as long as the patient's basic disease remains stable. In several responding patients the dose of erythropoietin could be tapered off to as low as 100 U/kg twice a week, in order to avoid overshooting of the hemoglobin values and to reduce the expense of that treatment mode.

Our group investigated whether responding patients subjectively benefit from EPO treatment by experiencing an improved quality of life or whether only a correction of laboratory parameters occurs. In accordance with previous reports on patients with chronic renal failure,[168] the outcome of EPO treatment in myeloma patients revealed a substantial improvement in the patients' physical capability, their subjective sense of well-being, their scope of social activities, and in other parameters indicating quality of life.[169]

Besides these conventional indications, EPO may also be applied successfully for the amelioration of anemia and/or the reduction of required red blood cell transfusions in patients after high-dose chemotherapy with or without bone marrow transplantation. The effectiveness of this principle, particularly in the setting of allogeneic bone marrow transplantation, has already been shown in patients with various malignant diseases,[170] but it awaits further confirmation in myeloma. EPO may also be given for the collection of autologous red blood cells and their precursors to be used as substitution after aggressive chemotherapy.[171] In one pilot study, this concept has already been extended to the use of erythropoietin in both donors and recipients of allogeneic bone marrow in malignant diseases.[172] During a 5-week period a median of 6 units of blood could be obtained from the donors under EPO support. This blood was available to the bone marrow recipient or to the donor, as

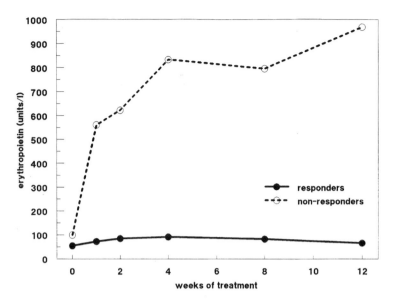

Fig. 14.5 Serum levels of erythropoietin before and during treatment with recombinant human erythropoietin. The median level of responders is below 100 units/l at baseline and remains at that magnitude during the course of therapy. Non-responders, on the other hand, have a slightly higher median value at baseline and show pronounced increases during the substitution therapy with exogenous erythropoietin.

needed. In addition, the transplant recipients were treated with EPO post-BMT, in order to hasten erythropoiesis. With this regimen, homologous red blood cell transfusions could be avoided in half of the transplanted patient population.

Theoretically, exposure of myeloma cells to EPO or, in fact, to any of the cytokines discussed above, could induce an unwanted growth stimulation of the myeloma clone. This may happen when the myeloma cell expresses aberrant receptors which are specific for the cytokine. Such a constellation has already been observed in a myeloma cell line which expressed EPO-binding receptors and which reacted with increased proliferation to the addition of EPO to the culture medium.[173] A similar phenomenon has been observed in one patient in whom significant disease progression had occurred in connection with EPO treatment.[174] Whether there was really a causal relationship or merely a coincidence of time remains unclear. It is important that none of the other studies so far published have ever reported that phenomenon, but it still calls for clinical vigilance.

In the clinical situation, therapy with recombinant human erythropoietin has convincingly been demonstrated to rank among the most efficient treatment modalities available for the improvement of myeloma-associated complications. Thus, erythropoietin therapy is likely to become widely applied to patients with symptomatic chronic myeloma-associated anemia, particularly when the present commercial constraints can be overcome. In our experience, those myeloma patients who respond to this novel treatment mode are unwilling to discontinue the treatment as long as they experience a substantial improvement of their physical condition and of their general sense of well-being.

Conclusions

An enormous amount of information has been accumulated on the biology of cytokines and their actual or possible clinical usefulness in myeloma since the first tentative application of natural leukocyte interferon in 1979. In fact, the interferons, particularly IFN-α, still hold the promise that they may permanently contribute to the therapeutic armament against multiple myeloma. Interferon-α seems to significantly prolong remission duration and possibly also survival, at least in patients with excellent response to their initial chemotherapy. In addition, interferon in combination with dexamethasone seems to induce excellent responses in chemoresistant patients and in those who responded only poorly to their previous treatment. But even as more evidence of the clinical usefulness of interferons in myeloma is being generated, there remain important questions to be answered. How should we use interferons in terms of patient selection, optimal dosage and schedule and in terms of combinations with other drugs? These issues need to be clarified before the role of IFN-α in myeloma can be definitely settled.

Several theoretical arguments favor the use of interleukin-2 in myeloma, e.g. its ability to stimulate NK activity and induce LAK cells which may lyse autologous myeloma cells. The existing clinical data, however, are scarce and show at best moderate activity paired with considerable toxicity. For the successful clinical exploitation of this cytokine, stringent patient selection is necessary. The most promising criteria seem to be a low tumor load and stable disease, as encountered after successful conventional induction treatment and, in particular, after high-dose chemotherapy with or without bone marrow transplantation. IL-2 may also play a role in allogeneic bone marrow transplantations by increasing the graft-versus-myeloma effect while curbing graft-versus-host disease.

The clinical usefulness of IL-4 has as yet not been confirmed in multiple myeloma. However, the concept of inhibiting the biologic sequels of IL-6 is intriguing and has been supported by *in vitro* and *in vivo* studies. The limited clinical data available indicate that the inhibition of IL-6 may transiently inhibit the proliferation of the myeloma clone. *In vitro* experiments on inhibiting the IL-6 receptor by monoclonal antibodies or antisense molecule technology yielded the desired inhibition of myeloma proliferation. It is questionable, however, whether this success can be achieved permanantly *in vivo*. Considering the hypermutability of myeloma cells, one of the main characteristics of tumor clones, one has to expect the *in vivo* evolution of IL-6-independent subclones which would yield only transient therapeutic effects.

The myeloid growth factors, in particular G-CSF and GM-CSF, have contributed to the presently widely established use of aggressive chemotherapy with or without transplantation in myeloma patients. These growth factors clearly reduce the length of aplasia, the number of infectious episodes, the consumption of antibiotics, and the length of hospitalization. In other words, they make high-dose therapy safer and reduce the need for in-patient treatment. Since myeloid growth factors induce the proliferation of precursors and shift stem cells from the bone marrow into the peripheral circulation, they are able to enhance the yield of stem cells from peripheral blood. With the aid of these growth factors, an increasing number of the stem cell preparations needed for transplantation are now obtained from peripheral blood.

Erythropoietin is gradually becoming more accepted in the clinical management of patients with advanced myeloma because of its capability to normalize myeloma-associated anemia. Thus, it can improve the patient's performance status, their subjective sense of well-being, and their entire quality of life.

Acknowledgements: This work was supported by the Wilhelminen-Cancer Research Institute for the Austrian Forum Against Cancer.

Note added in proof: We have updated the published data on interferon treatment in myeloma. The meta-analysis was based on 1518 patients randomized to compare combined interferon–chemotherapy with chemotherapy alone and on 924 patients randomized to either interferon maintenance or control. Significant improvements were found in both combined interferon–chemotherapy induction treatment (+ 10 per cent remission rate, 4 months gain relapse-free, and 6 months gain in overall survival) and interferon maintenance therapy (7 months gain relapse-free and 3 months gain in overall survival).

Ludwig, H, Cohen AM, Polliack A et al. Interferon-alpha for induction and maintenance in multiple myeloma: Results of two multicenter randomized trials and summary of other studies. *Ann Oncol* 1995; **6**: 467–76.

References

1. Alexanian R, Haut A, Khan A. Treatment for multiple myeloma: combination chemotherapy with different melphalan dose regimens. *Journal of the American Medical Association* 1969; **208**: 1680–5.

2. Stewart WE II, Blalock JE, Burke DC et al. Interferon nomenclature. *Journal of Immunology* 1980; **125**: 2353 (letter).

3. Von Gabain A, Lundgren E, Ohlsson M et al. Three human interferon-alpha2 subvariants disclose structural and functional differences. *European Journal of Biochemistry* 1990; **190**: 257–61.

4. Sehgal PB, May LT. Human interferon-beta2. *Journal of Interferon Research* 1987; **7**: 521–7.

5. Aguet M, Morgensen KE. Interferon receptors. *Interferon* 1983; **5**: 1–22.

6. Harada H, Fujita T, Miyamoto M et al. Structurally similar but functionally distinct factors, IRF-1 and IRF-2, bind to the same regulatory elements of IFN and IFN-inducible genes. *Cell* 1989; **58**: 729–39.

7. Einhorn S, Strander H. Interferon therapy for neoplastic diseases in man: *in vitro* and *in vivo* studies. *Advances in Experimental Medicine and Biology* 1978; **110**: 159–74.

8. Brenning G, Jernberg H, Gidlund M et al. The effect of alpha and gamma-interferon on proliferation and production of IgE and beta 2–micro-globulin in the human myeloma cell line U-266 and in an alpha-interferon resistant U-266 subline. *Scandinavian Journal of Haematology* 1986;37: 280–8.

9. Ludwig H, Swetly P. In vitro inhibitory effect of interferon on multiple myeloma stem cells. *Cancer Immunology and Immunotherapy* 1980; **9**: 139–43.

10. Senn CC. Biochemical pathways in interferon

action. *Pharmacology and Therapeutics* 1984; **24**: 235–40.

11. Jernberg-Wiklund H, Pettersson M, Nilsson K. Recombinant interferon-gamma inhibits the growth of IL-6-dependent human multiple myeloma cell lines *in vitro*. *European Journal of Haematology* 1991; **46**: 231–9.

12. Quesada JR, Alexanian R, Kurzrock R et al. Recombinant interferon gamma in hairy cell leukemia, multiple myeloma and Waldenstrom's macroglobulinemia. *American Journal of Hematology* 1988; **29**: 1–4.

13. Knight E Jr, Fahey D. Human interferon-beta: effects of deglycosylation. *Journal of Interferon Research* 1982; **2**: 421–9.

14. Grander D, von Stedingk LV, von Stedingk M et al. Influence of interferon on antibody production and viability of malignant cells from patients with multiple myeloma. *European Journal of Haematology* 1991; **46**: 17–25.

15. Teichmann JV, Sieber G, Ludwig WD, Ruehl H. Immunosuppressive effects of recombinant interferon-alpha during long-term treatment of cancer patients. *Cancer* 1989; **63**: 1990–3.

16. Aapro MS, Alberts DS, Salmon SE. Interactions of human leukocyte interferon with vinca alkaloids and other chemotherapeutic agents against human tumors in clonogenic assay. *Cancer Chemotherapy and Pharmacology* 1983; **10**: 161–6.

17. Cooper MR, Welander CE. Interferon in the treatment of multiple myeloma. *Seminars in Oncology* 1986; **13**: 334–40.

18. Ehrsson H, Eksborg S, Wallin I et al. Oral melphalan pharmacokinetics: influence of interferon-induced fever. *Clinical Pharmacology and Therapeutics* 1990; **47**: 86–90.

19. Princler GL, Brini AT, Kung H. IFN-alpha down-regulates the IL-6 receptor (IL-6R) by induction of endogenous IL-6. *Lymphokine Research* 1990; **9**: 570.

20. Jourdan M, Zhang XG, Portier M et al. IFN-alpha induces autocrine production of IL-6 in myeloma cell lines. *Journal of Immunology* 1991; **147**: 4402–7.

21. Kimchi A. Autocrine interferon and the suppression of the c-myc nuclear oncogene. *Interferon* 1987; **8**: 85–110.

22. Samid D, Chang EH, Friedman RM. Development of transformed phenotype induced by a human ras oncogene is inhibited by interferon. *Biochemistry and Biophysics Research Communications* 1985; **126**: 509–16.

23. Einhorn S, Ahre A, Blomgren H et al. Interferon and natural killer activity in multiple myeloma. Lack of correlation between interferon-induced enhancement of natural killer activity and clinical response to human interferon-alpha. *International Journal of Cancer* 1982; **30**: 167–72.

24. Jett R, Mantovani A, Herberman RB. Augmentation of human monocyte-mediated cytolysis by interferon. *Cellular Immunology* 1980; **54**: 425–32.

25. Österborg A, Nilsson B, Björkholm M et al. Natural killer cell activity in monoclonal gammopathies: relation to disease activity. *European Journal of Haematology* 1990; **45**: 153–7.

26. De Maeyer-Guignard J, De Maeyer E. Immuno-modulation by interferons: recent developments. *Interferon* 1985; **6**: 69–91.

27. Stadler R, Muller R, Orfanos CE. Effect of recombinant alpha A-interferon on DNA synthesis and differentiation of human keratinocytes *in vitro*. *British Journal of Dermatology* 1986; **114**: 273–7.

28. Michalevicz R, Revel M. Interferons regulate the *in vitro* differentiation of multilineage lympho-myeloid stem cells in hairy cell leukemia. *Proceedings of the National Academy of Sciences* 1987; **84**: 2307–11.

29. Mellstedt H, Ahre A, Björkholm M et al. Interferon therapy in myelomatosis. *Lancet* 1979; **i**: 245–8.

30. Ahre A, Björkholm M, Mellstedt H et al. Human leukocyte interferon and intermittent high-dose melphalan/prednisone administration in the treatment of multiple myeloma: a randomized clinical trial. *Cancer Treatment Reports* 1984; **68**: 1331–8.

31. Ahre A, Björkholm M, Österborg A et al. High doses of natural alpha-interferon (alpha-IFN) in the treatment of multiple myeloma: a pilot study from the Myeloma Group of Central Sweden (MGCS). *European Journal of Haematology* 1988; **41**: 123–32.

32. Ludwig H, Cortelezzi A, Scheithauer W et al. Recombinant interferon alpha-2c versus polychemotherapy (VMCP) for treatment of multiple myeloma: a prospective randomized trial. *European Journal of Cancer and Clinical Oncology* 1986; **22**: 1111–16.

33. Oken MM, Kyle RA, Kay NE et al. Interferon in the treatment of refractory multiple myeloma: an Eastern Cooperative Oncology Group study. *Leukemia and Lymphoma* 1990; **1**: 95–100.

34. Costanzi JJ, Cooper MR, Scarffe JH et al. Phase II study of recombinant alpha-2 interferon in resistant multiple myeloma. *Journal of Clinical Oncology* 1985; **3**: 654–9.

35. Quesada JR, Alexanian R, Hawkins M et al. Treatment of multiple myeloma with recombinant IFN-α. *Blood* 1986; **67**: 275–8.

36. Oken MM, Kyle RA. Strategies for combining interferon with chemotherapy for the treatment of multiple myeloma. *Seminars in Oncology* 1991; **18**(suppl. 7): 30–2.

37. Cox EB, Laszlo J, Krown S et al. Phase II study of human lymphoblastoid interferon in patients with multiple myeloma. *Journal of Biological Response Modifiers* 1988; **7**: 318–25.

38. Balkwill FR, Moodey EM. Positive interactions between human interferon and cyclophosphamide or adriamycin in a human tumor model system. *Cancer Research* 1985; **44**: 906–8.

39. Cooper MR, Fefer A, Thompson J et al. Alpha-2 interferon/melphalan/prednisone in previously untreated patients with multiple myeloma: a phase I–II trial. *Cancer Treatment Reports* 1986; **70**: 473–6.

40. Ludwig H, Cohen AM, Huber H et al. Interferon alpha-2b with VMCP compared to VMCP alone for induction and interferon alpha-2b compared to controls for remission maintenance in multiple myeloma: Interim results. *European Journal of Cancer* 1991; **27**: S40–5.

41. Österborg A, Björkholm M, Björeman M et al. Natural interferon-alpha in combination with melphalan/prednisone versus melphalan/prednisone in the treatment of multiple myeloma stages II and III: a randomized study from the Myeloma Group of Central Sweden. *Blood* 1993; **81**: 1428–34.

42. Mellstedt H, Österborg A, Björkholm M et al. Treatment of multiple myeloma with interferon alpha; the Scandinavian experience. *British Journal of Haematology* 1991; **79** (suppl. 1): 21–5.

43. Montuoro A, De Rosa L, De Blasio A et al. Alpha-2a-interferon/melphalan/prednisone versus melphalan/prednisone in previously untreated patients with multiple myeloma. *British Journal of Haematology* 1990; **76**: 365–8.

44. Aitchison R, Williams A, Schey S, Newland AC. A randomised trial of cyclophosphamide with and without low dose alpha-interferon in the treatment of newly diagnosed myeloma. *Leukemia and Lymphoma* 1993; **9**: 243–6.

45. Browman GP, Rubin S, Walker I et al. Interferon-alpha-2b (IFN) maintenance therapy prolongs progression-free and overall survival in plasma cell myeloma. *ASCO Proceedings* 1994; **13**: 408 (abstract).

46. Cooper MR, Dear K, McIntyre OR et al. A randomized clinical trial comparing melphalan/prednisone with or without interferon α-2b in newly diagnosed patients with multiple myeloma: a Cancer and Leukemia Group B study. *Journal of Clinical Oncology* 1993; **11**: 155–60.

47. Corrado C, Pavlovsky S, Saslasky J et al. Randomized trial comparing melphalan-prednisone with or without recombinant alpha-2 interferon (r-alpha2IFN) in multiple myeloma. *ASCO Proceedings* 1989; **8**: 258 (abstract).

48. Garcia-Larana J, Steegmann JL, Perez Oteyza J et al. Treatment of multiple myeloma with melphalan/prednisone (MP) versus melphalan/prednisone and alpha-2b-interferon (MP-IFN). Results of a Cooperative Spanish Group. *Congress of the International Society for Haematology ISH* 1992; **24**: 301 (abstract).

49. Mandelli F, Avvisati G, Amadori S et al. Maintenance treatment with recombinant interferon α-2b in patients with multiple myeloma responding to conventional induction chemotherapy. *New England Journal of Medicine* 1990; **322**: 1430–4.

50. Westin J, Cortelezzi A, Hjorth M et al. Interferon therapy during the plateau phase of multiple myeloma: an update of the Swedish study. *European Journal of Cancer* 1991; **27**(suppl. 4): S45–8.

51. Cunningham D, Powles R, Viner C et al. High dose chemotherapy and autologous bone marrow transplantation in multiple myeloma. Paper presented at the IV International Workshop on Multiple Myeloma, Rochester, MN, USA, 1993; p.102 (abstract).

52. Peest D, Deicher H, Coldewey R et al. Melphalan and prednisone (MP) versus vincristine, BCNU, adriamycin, melphalan and dexamethasone (VBAM Dex) induction chemotherapy and interferon maintenance treatment in multiple myeloma. Current results of a multicenter trial. The German Myeloma Treatment Group. *Onkologie* 1990; **13**: 458–60.

53. Salmon SE, Crowley J. Evaluation of interferon (IFN) in maintenance therapy for myeloma. Paper presented at the IV International Workshop on Multiple Myeloma, Rochester, MN, USA, 1993; p.91 (abstract).

54. Giles FJ, McSweeney EN, Richards JD et al. Prospective randomised study of double hemi-body irradiation with and without subsequent maintenance recombinant alpha 2b interferon on survival in patients with relapsed multiple myeloma. *European Journal of Cancer* 1992; **28A**: 1392–5.

55. Maniatis A. Maintenance treatment of multiple myeloma with alpha interferon versus an alternating schedule of alpha interferon and chemotherapy. *European Journal of Cancer* 1991; **27**(suppl. 4): S49–50.

56. Maniatis A. Maintenance therapy in multiple myeloma: interferon alpha-2b (IFN) vs an alternating IFN/chemotherapy schedule. Paper presented at the International Conference on Malignant Lymphoma, Lugano, Switzerland, 1993; p.105 (abstract).

57. Fossa SD, Gunderson R, Moe B. Recombinant interferon alpha combined with prednisone in metastatic renal cell carcinoma: Reduced toxicity without reduction of the response rate. – A phase II study. *Cancer* 1990; **65**: 2451–4.

58. Alexanian R, Barlogie B, Gutterman J. Alpha-interferon combination therapy of resistant myeloma. *American Journal of Clinical Oncology* 1991; **14**: 188–92.

59. San Miguel JF, Moro M, Bladé J et al. Combination of interferon and dexamethasone in refractory multiple myeloma. *Hematology and Oncology* 1990; **8**: 185–9.

60. Alexanian R, Barlogie B, Gutterman J. Alpha-interferon combination therapy of resistant myeloma. *American Journal of Clinical Oncology* 1991; **14**: 188–92.

61. Gertz MA, Kalish LA, Kyle RA. A phase III study comparing VAD chemotherapy with VAD plus recombinant alpha-2 interferon in refractory or relapsed multiple myeloma. *ASCO Proceedings* 1994; **13**: 375 (abstract).

62. Salmon SE, Beckord J, Pugh RP et al. Alpha interferon for remission maintenance: preliminary report on the Southwest Oncology Group study. *Seminars in Oncology* 1991; **18**(suppl. 7): 33–6.

63. Palumbo A, Boccadoro M, Garino LA et al. Multiple myeloma: intensified maintenance therapy with recombinant interferon-alpha-2b plus glucocorticoids. *European Journal of Haematology* 1992; **49**: 93–7.

64. Ohno R. Interferons in the treatment of multiple myeloma. *International Journal of Cancer* 1987;(suppl. 1) 14–20.

65. Smith KA. Interleukin-2: inception, impact, and implications. *Science* 1988; **240**: 1169–76.

66. Smith KA. Lowest dose interleukin-2 immunotherapy. *Blood* 1993; **81**: 1414–23.

67. Gillis S, Smith KA. Long term culture of tumourspecific cytotoxic T cells. *Nature* 1977; **268**: 154–6 (letter).

68. Rosenberg SA, Lotze MT, Muul LM et al. A progress report on the treatment of 157 patients with advanced cancer using lymphokine-activated killer cells and interleukin-2 or high-dose interleukin-2 alone. *New England Journal of Medicine* 1987; **316**: 889–97.

69. Gisselbrecht C, Maraninchi D, Pico JL et al. Interleukin-2 treatment in lymphoma: a phase II multicenter study. *Blood* 1994; **83**: 2081–5.

70. Hoover RG, Kornbluth J. Immunoregulation of murine and human myeloma. *Hematology and Oncology Clinics of North America* 1992; **6**: 407–24.

71. Massaia M, Dianzani U, Pioppo P et al. Emergence of activated lymphocytes in CD4 and CD8 subpopulations of multiple myeloma: correlation with the expansion of suppressor T-cells (CD8+ OKM1+) and ecto-5'nucleotidase deficiency. *Journal of Clinical and Laboratory Immunology* 1988; **26**: 89–95.

72. Page R, Kornbluth J, Epstein J. CD3+ CD16+ lymphocytes in patients with IgG myeloma: Regulation of growth and differentiation of human IgG myeloma tumor cells by soluble CD16. *Blood* 1991; **78**: 125a (abstract).

73. Peest D, Gasch S, Thiele C et al. Regulation of the in vitro monoclonal immunoglobulin production in cultures of peripheral blood mononuclear cells and bone marrow cells from myeloma patients mediated by T cell dependent mitogens. *Clinical and Experimental Immunology* 1986; **65**: 120–7.

74. Paglieroni T, MacKenzie MR. *In vitro* cytotoxic response to human myeloma plasma cells by peripheral blood leukocytes from patients with multiple myeloma and benign monoclonal gammopathy. *Blood* 1979; **54**: 226–37.

75. Peest D, de Vries I, Hölscher R et al. Effect of interleukin-2 on the *ex vivo* growth of human myeloma cells. *Cancer Immunology and Immunotherapy* 1989; **30**: 227–32.

76. Uchida A, Yagita M, Sugiyama H et al. Strong natural killer (NK) cell activity in bone marrow of myeloma patients: accelerated maturation of bone marrow NK cells and their interaction with other bone marrow cells. *International Journal of Cancer* 1984; **34**: 375–81.

77. Gottlieb DJ, Prentice HG, Mehta AB et al. Malignant plasma cells are sensitive to LAK cell lysis: pre-clinical and clinical studies of interleukin 2 in the treatment of multiple myeloma. *British Journal of Haematology* 1990; **75**: 499–505.

78. Heslop HE, Gottlieb DJ, Bianchi AC et al. *In vivo* induction of gamma interferon and tumor necrosis factor by interleukin-2 infusion following intensive chemotherapy or autologous marrow transplantation. *Blood* 1989; **74**: 1374–80.

79. Cimino G, Avvisati G, Amadori S et al. High serum IL-2 levels are predictive of prolonged survival in multiple myeloma. *British Journal of Hematology* 1990; **75**: 373–7.

80. Vacca A, Di Stefano R, Frassanito A et al. A disturbance of the IL-2/IL-2 receptor system parallels the activity of multiple myeloma. *Clinical and Experimental Immunology* 1991; **84**: 429–34.

81. Weidmann E, Bergmann L, Stock J et al. Rapid cytokine release in cancer patients treated with interleukin-2. *Journal of Immunotherapy* 1992; **12**: 123–31.

82. Togawa A, Sawada S, Amano M et al. Treatment of multiple myeloma with LAK cells plus interleukin 2 or interleukin 2 alone. *Rinsho-Ketsueki* 1989; **30**: 650–8.

83. Peest D, Leo R, Hein R et al. Low-dose recombinant interleukin-2 treatment of advanced multiple myeloma patients. Paper presented at the International Workshop on Multiple Myeloma, Rochester, MN, USA, 1993; p.37 (abstract).

84. Pecherstorfer M, Jilch R, Leitgeb C et al. Combined interleukin-2–interferon-gamma treatment for chemotherapy resistant multiple myeloma. *Proceedings of the American Association for Cancer Research* 1994; **35**: 197 (abstract).

85. Sykes M, Romick ML, Hoyles KA, Sachs DH. *In vivo* administration of interleukin 2 plus T cell-depleted syngeneic marrow prevents graft-versushost disease mortality and permits alloengraftment. *Journal of Experimental Medicine* 1990; **171**: 645–58.

86. Howard M, Farrar J, Hilfiker M et al. Identification of a T cell-derived B cell growth factor distinct from interleukin 2. *Journal of Experimental Medicine* 1982; **155**: 914–23.

87. Hu-Li J, Shevach EM, Mizuguchi J et al. B cell stimulatory factor 1 (interleukin 4) is a potent costimulant for normal resting T lymphocytes. *Journal of Experimental Medicine* 1987; **165**: 157–72.

88. Widmer MB, Grabstein KH. Regulation of cytolytic T-lymphocyte generation by B-cell stimulatory factor. *Nature* 1987; **326**: 795–8.

89. Crawford RM, Finbloom DS, Ohara J et al. B cell stimulatory factor-1 (interleukin 4) activates macrophages for increased tumoricidal activity and expression of Ia antigens. *Journal of Immunology* 1987; **139**: 135–41.

90. Ben-Sasson SZ, Le Gros G, Conrad DH et al. IL-4 production by T cells from naive donors. Il-2 is required for Il-4 production. *Journal of Immunology* 1990; **145**: 1127–36.

91. Brown MA, Pierce JH, Watson CJ et al. B cell stimulatory factor-1/interleukin-4 mRNA is expressed by normal and transformed mast cells. *Cell* 1987; **50**: 809–18.

92. Lorenzen J, Lewis CE, McCracken D et al. Human tumor-associated NK cells secrete increased

amounts of interferon-gamma and interleukin-4. *British Journal of Cancer* 1991; **64**: 457–62.

93. Rabin EM, Ohara J, Paul WE. B-cell stimulatory factor 1 activates resting B-cells. *Proceedings of the National Academy of Sciences* 1985; **82**: 2935–9.

94. Coffman RL, Ohara J, Bond MW et al. B cell stimulatory factor-1 enhances the IgE response of lipopolysaccharide-activated B cells. *Journal of Immunology* 1986; **136**: 4538–41.

95. Kikutani H, Inui S, Sato R et al. Molecular structure of human lymphocyte receptor for immunoglobulin E. *Cell* 1986; **47**: 657–65.

96. Wieser M, Bonifer R, Oster W et al. Interleukin-4 induces secretion of CSF for granulocytes and CSF for macrophages by peripheral blood monocytes. *Blood* 1989; **73**: 1105–8.

97. Spits H, Yssel H, Paliard X et al. IL-4 inhibits IL-2–mediated induction of human lymphokine-activated killer cells, but not the generation of antigen-specific cytotoxic T lymphocytes in mixed leukocyte cultures. *Journal of Immunology* 1988; **141**: 29–36.

98. Moon HB, Severinson E, Heusser C et al. Regulation of IgG1 and IgE synthesis by interleukin 4 in mouse B cells. *Scandinavian Journal of Immunology* 1989; 355–61.

99. Siebenkotten G, Esser C, Wabl M, Radbruch A. The murine IgG1/IgE class switch program. *European Journal of Immunology* 1992; **22**: 1827–34.

100. Karray S, DeFrance T, Merle-Beral H et al. Interleukin 4 counteracts the interleukin 2-induced proliferation of monoclonal B cells. *Journal of Experimental Medicine* 1988; **168**: 85–94.

101. Taylor CW, Grogan TM, Salmon SE. Effects of interleukin-4 on the *in vitro* growth of human lymphoid and plasma cell neoplasms. *Blood* 1990; **75**: 1114–18.

102. Herrmann F, Andreeff M, Gruss HJ et al. Interleukin-4 inhibits growth of multiple myelomas by suppressing interleukin-6 expression. *Blood* 1991; **78**: 2070–4.

103. Atkins MB, Vachino G, Tilg HJ et al. Phase I evaluation of thrice-daily intravenous bolus interleukin-4 in patients with refractory malignancy. *Journal of Clinical Oncology* 1992; **10**: 1802–9.

104. Gilleece MH, Scarffe JH, Ghosh A et al. Recombinant human interleukin 4 (IL-4) given as daily subcutaneous injections – a phase I dose toxicity trial. *British Journal of Cancer* 1992; **66**: 204–10.

105. Tepper RI, Pattengale PK, Leder P. Murine interleukin-4 displays potent anti-tumor activity *in vivo*. *Cell* 1989; **57**: 503–12.

106. Merico F, Bergui L, Gregoretti MG et al. Cytokines involved in the progression of multiple myeloma. *Clinical and Experimental Immunology* 1993; **92**: 27–31.

107. Klein B, Zhang XG, Jourdan M et al. Interleukin-6 is the central tumor growth factor *in vitro* and *in vivo* in multiple myeloma. *European Cytokine Network* 1990; **1**: 193–201.

108. Klein B, Zhang XG, Jourdan M et al. Paracrine rather than autocrine regulation of myeloma-cell growth and differentiation by interleukin-6. *Blood* 1989; **73**: 517–26.

109. Brown RD, Gorenc B, Gibson J, Joshua D. Interleukin-6 receptor expression and saturation on the bone marrow cells of patients with multiple myeloma. *Leukemia* 1993; **7**: 221–5.

110. Bazan JF. A novel family of growth factor receptors: a common binding domain in the growth hormone, prolactin, erythropoietin, and IL-6 receptors, and the p75 IL-2 receptor beta-chain. *Biochemistry and Biophysics Research Communications* 1989; **164**: 788–95.

111. Tamura T, Udagawa N, Takahashi N et al. Soluble interleukin-6 receptor triggers osteoclast formation by interleukin 6. *Proceedings of the National Academy of Sciences* 1993; **90**: 1924–8.

112. Goto H, Shimazaki C, Ashihara E et al. Effects of interleukin-3 and interleukin-6 on peripheral blood cells from multiple myeloma patients and their clinical significance. *Acta Haematologica* 1992; **88**: 129–35.

113. Bataille R, Jourdan M, Zhang XG, Klein B. Serum levels of interleukin-6, a potent myeloma cell growth factor, as a reflection of disease severity in plasma cell dyscrasias. *Journal of Clinical Investigation* 1989; **84**: 2008–11.

114. Ludwig H, Nachbaur DM, Fritz E et al. Interleukin-6 is a prognostic factor in multiple myeloma. *Blood* 1991; **77**: 2794–5 (letter).

115. Klein B, Wijdenes J, Zhang XG et al. Murine anti-interleukin-6 monoclonal antibody therapy for a patient with plasma cell leukemia. *Blood* 1991; **78**: 1198–204.

116. Klein B, Lu ZY, Bataille R. Clinical applications of IL-6 inhibitors. *Research in Immunology* 1992; **143**: 774–6.

117. Huang YW, Vitetta ES. A monoclonal anti-human IL-6 receptor antibody inhibits the proliferation of human myeloma cells. *Hybridoma* 1993; **12**: 621–30.

118. Suzuki H, Yasukawa K, Saito T et al. Anti-human interleukin-6 receptor antibody inhibits human myeloma growth *in vivo*. *European Journal of Immunology* 1992; **22**: 1989–93.

119. Sato K, Tsuchiya M, Saldanha J et al. Reshaping a human antibody to inhibit the interleukin 6-dependent tumor cell growth. *Cancer Research* 1993; **53**: 851–6.

120. Saito T, Taga T, Miki D et al. Preparation of monoclonal antibodies against the IL-6 signal transducer, gp130, that can inhibit IL-6-mediated functions. *Journal of Immunology Methods* 1993; **163**: 217–23.

121. Levy Y, Tsapis A, Brouet JC. Interleukin-6 antisense oligonucleotides inhibit the growth of human myeloma cell lines. *Journal of Clinical Investigation* 1991; **88**: 696–9.

122. Kreitman RJ, Siegall CB, Fitzgerald DJ et al. Interleukin-6 fused to a mutant form of pseudomonas exotoxin kills malignant cells from patients with multiple myeloma. *Blood* 1992; **79**: 1775–80.

123. Chadwick DE, Jean LF, Jamal N et al. Differential

sensitivity of human myeloma cell lines and normal bone marrow colony forming cells to a recombinant diphtheria toxin-interleukin 6 fusion protein. *British Journal of Haematology* 1993; **85**: 25–36.

124. Ballester OF, Moscinski LC, Lyman GH et al. High levels of interleukin-6 are associated with low tumor burden and low growth fraction in multiple myeloma. *Blood* 1994; **83**: 1903–8.

125. Metcalf D. The granulocyte-macrophage colony-stimulating factors. *Science* 1985; **229**: 16–22.

126. Sachs L. The molecular control of blood cell development. *Science* 1987; **238**: 1374–9.

127. Sieff CA. Hematopoietic growth factors. *Journal of Clinical Investigation* 1987; **79**: 1549–57.

128. Glaspy JA, Golde DW. Clinical applications of the myeloid growth factors. *Seminars in Hematology* 1989; **26**(suppl. 2): 14–7.

129. Weisbart RH, Golde DW, Clark SC et al. Human granulocyte–macrophage colony-stimulating factor is a neutrophil activator. *Nature* 1985; **314**: 361–3.

130. Platzer E, Welte K, Gabrilove JL et al. Biological activities of a human pluripotent hemopoietic colony stimulating factor on normal and leukemic cells. *Journal of Experimental Medicine* 1985; **162**: 1788–801.

131. Ralph P, Warren MK, Nakoinz I et al. Biological properties and molecular biology of the human macrophage growth factor, CSF-1. *Immunobiology* 1986; **172**: 194–204.

132. Ganser A, Lindemann A, Seipelt G et al. Effects of recombinant human interleukin-3 in patients with normal hematopoiesis and in patients with bone marrow failure. *Blood* 1990; **76**: 666–76.

133. Zhang XG, Bataille R, Jourdan M et al. Granulocyte–macrophage colony-stimulating factor synergizes with interleukin-6 in supporting the proliferation of human myeloma cells. *Blood* 1990; **76**: 2599–605.

134. Caligaris-Cappio F, Bergui L, Gaidano GL, et al. Circulating malignant precursors in monoclonal gammopathies. *European Journal of Haematology* 1989; **51** (suppl.): 27–9.

135. Anderson KC, Jones RM, Morimoto C et al. Response patterns of purified myeloma cells to hematopoietic growth factors. *Blood* 1989; **73**: 1915–24.

136. Jagannath S, Barlogie B. Autologous bone marrow transplantation for multiple myeloma. *Hematology and Oncology Clinics of North America* 1992; **6**: 437–49.

137. Dimopoulos MA, Delasalle KB, Champlin R, Alexanian R. Cyclophosphamide and etoposide therapy with GM-CSF for VAD-resistant multiple myeloma. *British Journal of Haematology* 1993; **83**: 240–4.

138. Barlogie B, Jagannath S, Dixon DO et al. High dose melphalan and granulocyte-macrophage colony-stimulating factor for refractory multiple myeloma. *Blood* 1990; **76**: 677–80.

139. Ossenkoppele GJ, Jonkhoff AR, Huijgens PC et al. Peripheral blood progenitors mobilised by G-CSF (filgrastim) and reinfused as unprocessed autologous whole blood shorten the pancytopenic period following high-dose melphalan in multiple myeloma. *Bone Marrow Transplantation* 1994; **13**: 37–41.

140. Ohler L, Scholten C, Reiter E et al. Mobilization of circulating hematopoietic stem cells by granulocyte colony stimulating factor after chemotherapy in multiple myeloma. *Wiener Klinische Wochenschrift* 1993: **105**: 580–4.

141. Goldwasser E. Erythropoietin and its mode of action. *Blood Cells* 1984; **4**: 89–103.

142. Erslev A. Erythropoietin coming of age. *New England Journal of Medicine* 1987; **316**: 101–3.

143. Dessypris EN, Graber SE, Krantz SB, Stone WJ. Effects of recombinant human erythropoietin on the concentration and cycling status of human marrow hematopoietic progenitor cells *in vivo*. *Blood* 1988; **72**: 2060–2.

144. Koury MJ, Bondurant MC. Maintenance by erythropoietin of viability and maturation of murine erythroid precursor cells. *Journal of Cell Physiology* 1988; **137**: 65–74.

145. Bauer C, Kurtz A. Oxygen sensing in the kidney and its relation to erythropoietin production. *Annual Reviews in Physiology* 1989; **51**: 845–56.

146. Erslev AJ. Erythropoietin. *New England Journal of Medicine* 1991; **324**: 1339–44.

147. Schuster SJ, Badiavas EV, Costa-Giomi P et al. Stimulation of erythropoietin gene transcription during hypoxia and cobalt exposure. *Blood* 1989; **73**: 13–16.

148. Winearls CG, Oliver DO, Pippard MJ et al. Effect of human erythropoietin derived from recombinant DNA on the anaemia of patients maintained by chronic haemodialysis. *Lancet* 1986; **ii**: 1175–7.

149. Eschbach JW, Egrie JC, Downing MR et al. Correction of the anemia of end-stage renal disease with recombinant human eythropoietin. Results of a combined phase I and II clinical trial. *New England Journal of Medicine* 1987; **316**: 73–8.

150. Spivak JL. Application of recombinant human erythropoietin in oncology. *Cancer Investigation* 1990; **8**: 301–2.

151. Oster W, Herrmann F, Gamm H et al. Erythropoietin for the treatment of anemia of malignancy associated with neoplastic bone marrow infiltration. *Journal of Clinical Oncology* 1990; **8**: 956–62.

152. Kyle RA. Diagnosis and management of multiple myeloma and related disorders. *Progress in Hematology* 1986; **14**: 257–82.

153. Ludwig H, Fritz E, Kotzmann H et al. Erythropoietin treatment of anemia associated with multiple myeloma. *New England Journal of Medicine* 1990; **322**: 1693–9.

154. Majumdar G, Westwood NB, Bell-Witter C et al. Serum erythropoietin and circulating BFU-E in patients with multiple myeloma and anaemia but without renal failure. *Leukemia and Lymphoma* 1993; **9**: 173–6.

155. Beguin Y, Yerna M, Loo M et al. Erythropoiesis in multiple myeloma: defective red cell production

due to inappropriate erythropoietin production. *British Journal of Haematology* 1992; **82**: 648–53.

156. Ariad S, Clifford D, Penfold G et al. Erythropoietin response in anaemic patients with multiple myeloma and other lymphoid malignancies infiltrating the bone marrow. *European Journal of Haematology* 1992; **49**: 59–62.

157. Aoki I, Nishijima K, Homori M et al. Responsiveness of bone marrow erythroid progenitors (CFU-E and BFU-E) to recombinant human erythropoietin (rh-Epo) in vitro in multiple myeloma. *British Journal of Haematology* 1992; **81**: 463–9.

158. Nielsen OJ, Brandt M, Drivsholm A. The secretory erythropoietin response in patients with multiple myeloma and Waldenstrom's macroglobulinaemia. *Scandinavian Journal of Clinical and Laboratory Investigation* 1990; **50**: 697–703.

159. Singh A, Eckhardt KU, Zimmermann A et al. Increased plasma viscosity as a reason for inappropriate erythropoietin formation. *Journal of Clinical Investigation* 1993; **91**: 251–6.

160. Ludwig H, Fritz E, Leitgeb C et al. Erythropoietin treatment for chronic anemia of selected hematological malignancies and solid tumors. *Annals of Oncology* 1993; **4**: 161–7.

161. Barlogie B, Beck T. Recombinant human erythropoietin and the anemia of multiple myeloma. *Stem Cells* 1993; **11**: 88–94.

162. Abels RI. Recombinant human erythropoietin in the treatment of the anaemia of cancer. *Acta Haematologica* 1992; **87**(suppl. 1): 4–11.

163. Ruedin P, Pechere-Bertschi A, Chapuis B et al. Safety and efficacy of recombinant human erythropoietin treatment of anaemia associated with multiple myeloma in haemodialysed patients. *Nephrology, Dialysis, Transplantation* 1993; **8**: 315–8.

164. Taylor J, Mactier RA, Stewart WK, Henderson IS. Effect of erythropoietin on anaemia in patients with myeloma receiving haemodialysis. *British Journal of Medicine* 1990; **301**: 476–7.

165. Cazzola M, Ponchio L, Beguin Y et al. Subcutaneous erythropoietin for treatment of refractory anemia in hematologic disorders. Results of a phase I/II clinical trial. *Blood* 1992; **79**: 29–37.

166. Cartwright GE. The anemia of chronic disorders. *Seminars in Hematology* 1966; **3**: 351–75.

167. Vreugdenhil G, Manger B, Nieuwenhuizen C et al. Iron stores and serum transferrin receptor levels during recombinant human erythropoietin treatment of anemia in rheumatoid arthritis. *Annals of Hematology* 1992; **65**: 265–8.

168. Delano BG. Improvements in quality of life following treatment with r-HuEPO in anemic hemodialysis patients. *American Journal of Kidney Disease* 1989; **14**(suppl. 1): 14–18.

169. Leitgeb C, Pecherstorfer M, Fritz E, Ludwig H. Quality of life in chronic anemia of cancer during treatment with recombinant human erythropoietin. *Cancer* 1994; **73**: 2535–42.

170. Demuynck H, Boogaerts MA. Recombinant human erythropoietin in allogeneic and autologous bone marrow transplantation. *Erythropoiesis* 1993; **4**: 73–9.

171. Brugger W, Mocklin W, Heimfeld S et al. Ex vivo expansion of enriched peripheral blood CD34+ progenitor cells by stem cell factor, interleukin-1 beta (IL-1 beta), IL-6, IL-3, interferon-gamma, and erythropoietin. *Blood* 1993; **81**: 2579–84.

172. Mitus AJ, Antin JH, Rutherford CJ et al. Use of recombinant human erythropoietin in allogeneic bone marrow transplant donor/recipient pairs. *Blood* 1994; **83**: 1952–7.

173. Okuno Y, Takahashi T, Suzuki A et al. Expression of the erythropoietin receptor on a human myeloma cell line. *Biochemistry and Biophysics Research Communications* 1990; **170**: 1128–34.

174. Rogers S, Russell NH, Morgan AG. Effect of erythropoietin in patients with myeloma. *British Medical Journal* 1990; **301**: 667 (letter).

CHAPTER 15

Autologous transplantation in multiple myeloma

MICHEL ATTAL, JEAN LUC HAROUSSEAU

Introduction	182	Clinical results of autologous transplantation	185
Dose–response effects and development of high-dose regimens	182	Prognostic factors with high-dose therapy	189
The source of autologous hematopoietic stem cells	184	The choice of treatment in the management of multiple myeloma	189
The role of autologous graft purging	185	Conclusions	191
		References	192

Introduction

For the past 25 years, intermittent melphalan/prednisone has been the standard treatment for multiple myeloma (MM). Extensive clinical trials with other drug combinations have been used without major improvements, and MM still remains a universally fatal malignancy with a median survival that does not usually exceed 2–3 years with conventional chemotherapy.

During the last few years, high-dose therapy has been introduced to improve survival of aggressive myeloma. In 1983, McElwain and Powles reported that high-dose melphalan (HDM) could induce a high response rate even in patients refractory to conventional dose of melphalan.[1] However, this result was achieved at the expense of severe myelosuppression with a high rate of toxic death.[2] In 1986, Barlogie et al. demonstrated that the hematologic toxicity of HDM was reduced in patients receiving an autologous bone marrow support.[3] They also reported that a myeloblative therapy including HDM and total body irradiation, could be administered relatively safely to elderly patients when supported

with autologous bone marrow transplantation.[4] As a result of these initial observations, several other investigators began to explore high-dose therapy supported by autologous hematopoietic stem cells in MM. This article will focus on the important aspects addressed by these ongoing research programs, including the preparative regimens, the source of stem cells, the role of autograft purging, the clinical results, and the prognostic factors with high-dose therapy. Finally, we will conclude this chapter with the preliminary results of a French prospective randomized protocol, designed to compare the results of antologous bone marrow transplantation verus conventional chemotherapy in aggressive MM.

Dose–response effects and development of high-dose regimens

In 1983, McElwain and Powles were the first to evaluate the efficacy of a single high-dose melphalan

Table 15.1 Response rate of different alkylant agents in refractory patients

Regimen	Graft	n	% ED	% Response	% CR	Reference
Melphalan 70 mg/m^2	–	23	26	34	4	6
Melphalan 90–100 mg/m^2	–	45	20	38	7	6
Melphalan 140 mg/m^2	–	15	13	53	13	2
	–	44	7	66	20	36
Melphalan 200 mg/m^2	+	24	0	50	4	6
Cyclophosphamide 6g/m^2	–	75	9	23	0	6
Busulfan 16 mg/kg	+	15	20	33	20	14

% ED, percentage early death; % CR, percentage complete response.

(140 mg/m^2), without hematopoietic stem cell support.[1] They treated 9 patients and obtained 3 complete remissions (CR). This preliminary experience was amplified, and Selby et al. reported updated results on fifty-eight patients: 26 per cent achieved CR and 50 per cent partial response. However, this unprecedented rate of CR was obtained at the expense of a severe myelosuppression with a high rate of toxic death (17 per cent).[2] Furthermore, the median duration of remission was only 19 months in previously untreated patients. In order to decrease this hematologic toxicity, Barlogie et al. explored the role of an unpurged autologous bone marrow graft administered after HDM. They demonstrated that the duration of neutropenia was significantly shortened.[4] They also demonstrated that with a better selection of patients (good performance status and renal status) the mortality due to HDM was virtually eliminated.[3] These results prompted investigators to perform additional studies with various doses of melphalan. Table 15.1 lists the larger trials of HDM in patients refractory to conventional chemotherapy. These trials, although not randomized, suggest that a dose of 140 mg/m^2 is more effective than a dose $\leqslant$ 100 mg/m^2. However, further dose escalation to 200 mg/m^2 does not appear to improve overall or complete response rates compared with 140 mg/m^2.

Cyclophosphamide is an alkylating agent reported to be as effective as melphalan when administered at a conventional dosage. In 1984, Lenhard et al. demonstrated a dose effect with this agent.[5] They observed a 44 per cent response rate in sixteen patients refractory to conventional dose of cyclophosphamide, when high-dose cyclophosphamide (2.4 g/m^2) was used. Recently, doses up to 6–7 g/m^2 (with or without GM-CSF) have been evaluated in order to collect circulating hematopoietic progenitor cells. Although an effective mobilization of progenitor cells was observed, the hematologic toxicity was severe (median duration of severe neutropenia ranging between 2 and 3 weeks), and the response rate appeared disappointing (about 20 per cent).[6] Thus, cyclophosphamide appears to be inferior to melphalan at the maximum tolerated dose.

Etoposide at a dose of 250 mg/m^2 has been ineffective in patients with advanced myeloma who were resistant to standard chemotherapy.[7] Recently, Gianni et al. reported that high-dose etoposide (2–2.4 g/m^2) induced a severe but transient bone marrow hypoplasia followed by a marked mobilization of the progenitor pool.[8] Thus, high-dose etoposide appears to be a promising agent for adequate collection of circulating progenitors. Furthermore, these authors reported that among 7 MM patients with assessable disease, bone marrow plasma cell population became indetectable in 4 patients and was reduced by 60–90 per cent in the remaining 3 patients, within 3 weeks of etoposide administration. The M. D. Anderson group found the combination of high-dose etoposide (900 mg/m^2) and cyclophosphamide (3 g/m^2) to be highly effective in refractory myeloma. Among fifty-two VAD-refractory patients treated with this regimen, 42 per cent achieved a partial response and adequate stem cell collection (bone marrow or blood) was possible.[8] These results suggest that etoposide and cyclophosphamide are more effective together than either drug alone. Thus, high-dose etoposide combined with cyclophosphamide may be of interest as initial part of high-dose therapy programs, not only to decrease tumor mass but also to collect hematopoietic stem cells.

Hemi-body or total-body irradiation (TBI), without stem cell support, has been reported to be effective in previously treated MM. Barlogie et al. demonstrated that, in refractory myeloma, TBI with melphalan (140 mg/m^2) was superior to melphalan alone (140 mg/m^2), with higher CR rates and longer duration of relapse-free and overall survival.[10] However, the early mortality was high (28 per cent), despite the use of an autologous bone marrow graft.[6] We reported that when this TBI–melphalan regimen was administered to patients in first response to conventional chemotherapy, the early mortality was abrogated and a promising CR rate was obtained (50 per cent).[11] McElwain and colleagues proposed a higher

Table 15.2   The role of total body irradiation in preparative regimens

Status of disease	Regimen	*n*	% ED	% Response	%CR	Reference
Refractory	Melphalan 200	24	0	50	4	6
Refractory	Melphalan 140+TBI	18	28	72	11	6
Sensitive	Melphalan 200	11	0	55	0	6
Sensitive	Melphalan 140+TBI	19	5	58	37	6
First response	Melphalan 140+TBI	20	0	85	50	11

dose version of HDM: 200 mg/m^2.[12] Preliminary comparisions between TBI plus melphalan, 140 mg/m^2 and melphalan alone at a dose of 200 mg/m^2 still indicate a higher incidence of complete response in TBI-containing regimens (Table 15.2).

Busulfan is an alkylating agent that has been used successfully as a conditioning agent before allogeneic bone marrow transplantation in the treatment of leukemia. Busulfan (12–14 mg/kg) was initially evaluated in combination with cyclophosphamide (120–174 mg/m^2), in MM patients undergoing allogeneic BMT by the Seattle group.[13] Although a high CR rate was observed, this regimen was noted to produce an undue rate of toxic death (8 of 19). In the Royal Marsden experience, busulfan (16 mg/kg) with autologous bone marrow transplantation was evaluated in 15 MM patients (8 of whom relapsed after HDM).[14] The response rate was satisfactory (46 per cent), but severe hepatic toxicity was observed in 6 patients and 3 toxic deaths occurred. In the Vancouver experience, 14 VAD-responsive patients received a purged bone marrow graft prepared with high-dose busulfan (16 mg/kg), cyclophosphamide (120 mg/m^2), and melphalan (90 mg/m^2).[15] Three toxic deaths occurred, 2 of them caused by hepatic veno-occlusive disease. Thus, it appears that high-dose busulfan may increase the risk of severe hepatic toxicity in patients with MM.

Finally, the association of high-dose melphalan (140 mg/m^2) and TBI appears to be the most effective regimen that can be tolerated in MM patients. However, the CR rate is only 30–50 per cent with this regimen, when applied to patients in first remission after conventional chemotherapy. These results justify the development of new strategies incorporating successive non-cross-resistant high-dose regimens in order to decrease the tumor mass before the TBI/melphalan-prepared autograft. These regimens would be expected to yield an effective antitumoral effect with sparing of the normal hematopoietic stem cells. Administration of high-dose dexamethasone and cyclophosphamide/etoposide may be of significance in this perspective.

The source of autologous hematopoietic stem cells

In considering marrow-ablative therapy requiring autologous hematopoietic stem cell support, the source of stem cells, i.e. bone marrow or peripheral blood, is a major concern, especially for a marrow-derived malignancy such as myeloma. The best source of hematopoietic stem cells should meet the two following criteria: (1) it should contain no or few tumor stem cells, and (2) it should induce a rapid and complete hematopoietic reconstitution.

Bone marrow graft, even when collected in first response to conventional chemotherapy, contains many tumor cells. However, most of these cells are terminally differentiated cells with a low proliferative potential. The plasma cell contamination of blood is generally low. However, there is good evidence, from *in vitro* culture studies, that myeloma progenitor cells do circulate.[16–19] Thus, the relative frequency of myeloma progenitor cells in blood and bone marrow in relationship to frequency of normal hematopoietic stem cells remains unclear.[20] In an analysis of the French registry of autologous transplantation for myeloma, no significant difference was observed between patients receiving a bone marrow graft (*n* = 81) or a peripheral blood graft (*n* = 51) with regard to response rate, duration of response, and survival.[21]

Early mortality after autotransplant is mainly due to myelosuppression. Thus, the speed of hematopoietic reconstitution after stem cell infusion is another important parameter to consider when comparing blood versus bone marrow. After bone marrow graft, we previously reported that the median number of days with severe neutropenia (< 500/mm^3) was 17 and the median duration of severe thrombocytopenia (< 50 000/mm^3) was 29.[22] Fernand et al. reported comparable results after blood stem cell graft collected without hematopoietic growth factor, the median duration of severe neutropenia and thrombocytopenia being respectively 15 and 29 days.[23] Gianni et al. demonstrated that the administration of hematopoietic growth factor (HGF) after high-dose cyclophosphamide improved the mobilization and collection of blood stem cells, resulting in a shorter duration of aplasia after autograft.[24,25] Jagannath et al. adopted a similar approach to collect blood stem cells. They reported a very short duration of neutropenia and thrombocytopenia (15 days), when blood stem cells plus bone marrow was reinfused, GM-CSF being also used after autologous transplantation.[26] However, duration of aplasia after the only use of a bone marrow graft, collected under HGF and followed by the administration of HGF, is unknown.

From the above, it appears that prospective randomized trials are required to compare blood versus bone marrow as source of hematopoietic stem cells after myeloablative therapy.

The role of autologous graft purging

Since there is a potential detriment with reinfusion of myeloma cells and myeloma progenitor cells when autologous stem cell support is used, removal of tumor cells from the graft would appear desirable. However, such strategies are presently of limited clinical interest for two major reasons. First, there is as yet no evidence that purging eradicates the malignant clone from the graft. Second, *in vivo* cytoreductive strategies are not sufficiently effective to avoid expansion of residual tumor burden.

Three groups of investigators have evaluated autologous bone marrow purging. The Vancouver group has used chemotherapeutic purging with 4-HC in fourteen patients, however the residual disease in purged bone marrow was not evaluated.[15] The Bologna group has used immunotoxins containing monoclonal antibodies against B-cell antigens linked to the ribosome-inactivating protein momordin in fourteen patients.[27] The mean plasma cell numbers in bone marrow graft before and after purging were 34 per cent and 0.5 per cent, respectively. The Dana Farber Institute group has used an antibody cocktail of PCA1, CD20, CD10 to purge marrow in eleven patients.[28] This approach induced an effective depletion of PCA1+ CD10+ CD20+ cells. However, in all these studies, it is not clear if the 'clonogenic' cells have been removed since the phenotype of this cell is undefined. Furthermore, none of these studies have investigated the residual disease using molecular techniques to detect immunoglobulin gene rearrangements in the purged bone marrow. Thus, whether these techniques 'clear' the graft of tumor cells belonging to the malignant clone is still unknown.

After allogeneic BMT, only 40 per cent of Stage II–III MM patients achieve CR, demonstrating that currently used conditioning regimens are insufficient to eradicate the malignant clone. After autologous transplantation, 30–40 per cent of patients treated with the best current high-dose regimens can be expected to achieve CR,[20] which does not appear to be significantly different from results obtained with allogeneic BMT. Thus, early after autologous transplantation, a residual tumor burden, probably unrelated to graft contamination but sufficient to explain future disease progression, is detectable in 70 per cent of patients. These considerations may explain why, in two previously reported studies, the extent of tumor cell contamination in unpurged bone marrow graft was not found to be related to duration of response or survival.[21,29] These results clearly indicated that the benefit, if any, of such purging strategies will not be evaluable even in large prospective randomized trials, until a higher CR rate is achieved with new high-dose regimens.

Clinical results of autologous transplantation

UNPURGED AUTOLOGOUS BONE MARROW TRANSPLANTATION

The first studies of unpurged autologous BMT in patients with advanced myeloma were reported from the M. D. Anderson Hospital in the early 1980s.[4] In 1990, Jagannath et al. reported a follow-up analysis of 55 patients (median age, 54 years; range, 20–66 years) receiving TBI at 850 cGy with either melphalan (140 mg/m^2) or thiotepa (750 mg/m^2).[29] Patients with VAD-refractory relapse (fourteen patients) were not found to benefit from this treatment approach. In fact a high early mortality rate of 36 per cent was observed, and none of the patients achieved CR; the median relapse-free and overall survival was only 7 months. By contrast, among the forty-one remaining patients, only one early death was observed, the CR rate was 30 per cent with a median relapse-free survival of 18 months and a projected survival rate at 4 years of 80 per cent. A subsequent analysis of these VAD-responsive patients demonstrated that melphalan/TBI was superior to thiotepa/TBI, with higher CR rates (37 per cent versus 13 per cent), longer median durations of relapse-free survival (20 months versus 8 months), and longer overall survival (80 per cent at 4 years versus median, 23 months).[6] In this study, the degree of marrow plasmocytosis did not affect response rate, relapse-free or overall survival. Thus, melphalan/TBI supported with unpurged autologous BMT appeared to be an effective and safe strategy for patients with non-refractory relapse. However, median remission duration was only 18 months in this 'good-risk' group of patients.

In order to improve these results, other investigators have evaluated the role of autologous BMT as consolidation following initial induction therapy. The Royal Marsden group[12] has reported on fifty previously untreated patients (median age, 51 years; range, 30–69 years), who received the VAMP (for explanation of abbreviations, see Table 11.1 on p. 109) regimen as induction therapy. Patients achieving remission (< 30 per cent plasma cells in the bone marrow) and with adequate renal function were intended to receive autologous BMT prepared with melphalan alone at a dose of 200 mg/m^2. Patients not fulfilling these criteria were to receive melphalan at the dose of 140 mg/m^2 without bone marrow graft. Twenty-eight patients received melphalan 200 mg/m^2 plus autologous BMT, 11 received melphalan 140 mg/m^2, and 11 either refused treatment or developed complications during induction treatment. Twenty-five patients (50 per cent) achieved CR (as defined by the disappearance of monoclonal gammapathy on standard protein electrophoresis). The median survival was 3.5 years. The median duration of response was significantly longer for patients achieving CR than for patients achieving PR (2 years versus 10 months, $p = 0.0005$). However, although an unprecedented rate of CR was achieved, there was no evidence of a plateau in response duration.

The reasons for this insufficient duration of response could be related to several factors. First, progressions may be due to reinfused myeloma cells. This has led investigators to explore the role of bone marrow purging or peripheral blood stem cell transplantation. Second, once CR is achieved, maintenance treatment may be required to observe long-term disease-free survivors. This has led some investigators to explore the role of interferon maintenance therapy. Third, current cytoreductive regimens may be insufficient to eradicate the malignant clone, leading some investigators to develop more intensive programs with successive high-dose therapy regimens.

PURGED BONE MARROW TRANSPLANTATION

There is only a limited experience of autologous BMT with purged marrow. Only three groups of investigators have published data with more than ten patients.

The Dana Farber group used an antibody cocktail directed at antigens expressed by pre-B (CD10), B (CD20), and plasma cells (PCA1), together with rabbit complement to purge marrow.[28] Eleven patients

(mean age, 46 years), with minimal disease (defined as less than 10 per cent bone marrow plasma cells) sensitive to conventional therapy, received high-dose melphalan (70 mg/m^2 on 2 successive days) and TBI (1200 cGy) and were rescued with their purged marrow. Among 9 patients assessable for response (excluding one patient in CR before autologous BMT and one early death), 6 CR and 3 PR were obtained. Of these 6 CR obtained after autologous BMT, 3 had relapsed (3, 5, 16 months after autologous BMT) and 3 continued in CR from 12 to 19 months at the time of reporting. Engraftment was prompt in 10 patients, however one patient failed to engraft and succumbed to a central nervous system hemorrhage. Thus, despite the use of an intensified regimen (TBI, 1200 cGy) and the selection of patients with minimal disease before BMT, the administration of a purged graft did not avoid occurrence of early relapse.

The Bologna group has reported on 14 patients (median age, 47 years) with Stage III disease, previously treated with conventional chemotherapy (refractory, 6 patients; sensitive, 4 patients; relapse, 4 patients).[27] The preparative regimen included TBI (1000 cGy in a single fraction) and cyclophosphamide (with added melphalan or BCNU) or melphalan 100 mg/m^2. The bone marrow harvest was purged using immunotoxins containing monoclonal antibodies against B-cell antigens linked to the ribosome-inactivating protein momordin. Marrow purging resulted in a marked reduction of plasma cells (mean before purging, 34 per cent; after purging, 0.5 per cent). A poor hematopoietic recovery was observed. Indeed, of 13 evaluable patients, 2 did not recover granulopoiesis and 5 did not recover thrombopoiesis. Five early deaths occurred (infections, 2 cases; interstitial pneumonitis, 2 cases; veno-occlusive disease, 1 case). Tumor reduction ranged between 80 per cent and 99 per cent in 10 patients, and 2 patients were refractory to high-dose therapy. The median survival was only 4 months after autologous BMT. Thus, in this study, 6 out of 13 evaluable patients experienced troubles in hematopoietic recovery, suggesting damage to normal hematopoietic stem cells.

The Vancouver group has reported on fourteen patients (median age, 48 years), responsive to VAD (< 10 per cent marrow plasma cells and > 50 per cent reduction in paraprotein).[15] The bone marrow harvest was purged with a fixed dose of 4-hydroperoxycyclophosphamide (100 ug/ml). Patients were prepared with busulfan (16 mg/kg), cyclophosphamide (120 mg/kg), and melphalan (90 mg/m^2). Three early deaths occurred including 2 veno-occlusive disease. Of the remaining 11 patients, 6 achieved CR and 3 of them relapsed, while 5 achieved PR and 4 of them relapsed. The median duration of response was 17 months (range, 5–30 months).

Thus, although data are too limited to evaluate the impact of bone marrow purging on response duration, it appears that relapses occur as early as 6 months after autologous BMT, even in patients achieving CR. Furthermore, although most of these patients had a minimal residual disease before BMT, overall duration of response appears similar to that observed after unpurged autologous BMT. Finally, none of these studies demonstrated that the malignant clone was eradicated from the purged graft.

PERIPHERAL BLOOD STEM CELL (PBSC) TRANSPLANTATION

Since the pioneering studies of Richman[30] and To,[31] demonstrating an increased number of circulating stem cells during the recovery from cytotoxic drug-induced marrow aplasia, PBSC have been usually mobilized and collected after myelotoxic chemotherapy. In multiple myeloma, various protocols have been investigated including high-dose cyclophosphamide alone or combined with Adriamycin in the CHOP regimen,[23] high-dose etoposide,[8] and melphalan.[32] When prior exposure to cytotoxic drugs did not exceed 1 year and when melphalan was not used as the mobilizing agent, PBSC collection was regularly successful.[31] As a result of these findings, several groups have evaluated PBSC transplantation in multiple myeloma.

Fernand et al. attempted to collect blood stem cells after a regimen including intermediate dose cyclophosphamide (4 g/m^2), Adriamycin (100 mg/m^2), vincristine, and methyl prednisolone.[23,33] Patients with successful PBSC collection (> 2 × 10^4 CFU-GM/kg) received additional courses of chemotherapy to achieve remission, followed by autologous transplantation with PBSC, using an intensified chemotherapy/TBI regimen as conditioning (melphalan, 140 mg/m^2; cyclophosphamide, 60 mg/kg; etoposide, 750 mg/m^2; CCNU, 120 mg/m^2; TBI, 1200 cGy). Seventy-three patients (mean age, 46 years) with newly diagnosed myeloma entered the study. Three toxic deaths occurred after CHOP, 12 patients had an unsuccessful PBSC collection, 13 patients were not considered for the planned strategy, and 2 additional patients died of disease progression before transplantation. Results of the only 43 patients (59 per cent), who received the whole strategy, appeared promising. Hematopoietic recovery was satisfactory. All patients achieved good partial or complete response. Four toxic deaths and 10 relapses, 4 of whom died of progressive disease, occurred, resulting in an estimated 4-year survival rate from PBSC collection of 73 per cent. However, it remains unclear whether these promising results may be due to PBSC use or

to some other factors including younger age of patients, selection criteria before transplantation, or reinforced conditioning regimen.

Marit et al. treated twenty-four recently diagnosed MM patients (median age, 49 years) with high-dose cyclophosphamide (7 g/m^2) to recruit PBSC.[33] Patients received melphalan (140 mg/m^2) and TBI (1350 cGy) as conditioning regimen. Seventeen patients received PBSC cells alone but 7 also received additional bone marrow because of inadequate PBSC collection. No toxic deaths occurred. The median time to reach absolute neutrophil count > 500/mm^3 and platelet count > 50 000/mm^3 was 13 and 28 days, respectively. Ten patients achieved a partial remission and 13 a complete response. At the time of reporting (median follow-up, 17 months), 5 relapses had occurred. No difference in relapse rate was observed between patients receiving additional autologous BMT or not.

Studies by Gianni et al. indicate that the combined use of bone marrow and blood cells may accelerate hematopoietic engraftment when GM-CSF is used together with high-dose cyclophosphamide or etoposide for blood cell mobilization and again after transplantation.[24] Jagannath et al. have recently applied this strategy to a cohort of seventy-five previously treated MM patients (median age, 50 years).[26] PBSC collection was attempted after the administration of high-dose cyclophosphamide (6 g/m^2) with or without GM-CSF. Bone marrow collection was also performed. Patients were subsequently intended to receive melphalan (200 mg/m^2) supported by both autologous bone marrow and PBSC, GM-CSF being administered or not after transplant. Out of 75 patients enrolled, 60 received the autologous transplant. Good mobilization of PBSC was achieved when prior chemotherapy did not exceed 1 year and when GM-CSF was used post high-dose cyclophosphamide. After transplantation, the median times to recover neutrophils greater than 500/mm^3 and platelets greater than 50 000/mm^3 were 15 days and 19 days, respectively. The cumulative response rate for all 75 patients was 68 per cent, with 12-month event-free and overall survival projections of about 85 per cent. Thus, although follow-up is too short to evaluate the impact of this strategy on response duration, these results suggest that the combined use of bone marrow and PBSC may speed up hematopoietic recovery after myeloablative therapy. Furthermore, this combined strategy may allow the collection of a sufficient number of stem cells to support a double myeloablative procedure. Finally, if once regarded as alternative sources of hematopoietic stem cells, it appears that the combined use of autologous BMT and PBSC may allow more intensive strategies to be designed without increased mortality.

INTERFERON MAINTENANCE POST-AUTOLOGOUS BMT

Interferon has been reported to prolong the duration of response when used as maintenance treatment after conventional chemotherapy.[35] Furthermore, it has been suggested that IFN maintenance therapy was all the more effective when the residual tumor mass was relatively small. Therefore, the maximal reduction of tumor induced by high-dose therapy could possibly enhance the efficacy of IFN. Thus, the ability of IFN in prolonging duration of response post-autologous transplant was clearly of interest.

We have recently reported the results of a non-randomized study of maintenance IFN after unpurged autologous BMT.[22] Thirty-five consecutive patients (median age, 54 years) with previously untreated Stage III MM were to receive the following three-phase treatment: VAD or VMCP as initial therapy, then autologous BMT prepared with melphalan (140 mg/m^2) and TBI (800 cGy), followed by IFN maintenance therapy. Thirty-one patients (89 per cent), with good performance status and normal renal function after initial chemotherapy, received autologous BMT. Thirty patients received IFN, 3 mU/m^2 3 times a week as soon as the neutrophil and platelet counts had stabilized at over 0.5 and 75 $\times$ 109/l, which occurred at a median of 2.6 months post-BMT. Fifteen patients reached CR (43 per cent), and 14 reached PR (40 per cent) with a 90 per cent decrease in paraprotein among 6 of them. The 33-month, post-autologous BMT probability of progression-free survival was 53 per cent, this probability being significantly higher for patients achieving CR than for patients who did not: 85 per cent versus 24 per cent, respectively. This result suggested that the small residual tumor mass of patients achieving CR after autologous BMT could be highly sensitive to IFN maintenance treatment.

Preliminary results from a randomized study of interferon maintenance have been reported from the Royal Marsden Hospital.[36] Eighty-four patients were randomized to receive IFN, 3 mU/m^2 3 times weekly or no treatment following induction therapy with C-VAMP, consolidated with high-dose melphalan (200 mg/m^2) plus autologous BMT. Median progression-free survival from autologous BMT was 27 months in the control group versus 39 months in the IFN group ($p < 0.02$). Overall survival was also significantly longer for the IFN group ($p < 0.05$). This favourable effect of IFN was only observed for the sixty-two patients who achieved CR after autologous BMT ($p < 0.01$), and 53 per cent of patients in CR who received IFN remained in remission 4 years after treatment. However, for PR and non-responders to

HDM there was no significant prolongation of progression-free or overall survival.

These results suggest that IFN could be useful in prolonging duration of complete response achieved after high-dose therapy. The design of new strategies to increase this complete response rate remains a major objective.

SUCCESSIVE HIGH-DOSE THERAPY REGIMENS

After one course of intensive therapy, only 30 per cent of patients will achieve CR. In order to further increase the CR rate and to prolong duration of response obtained after the first course of high-dose therapy, Harousseau et al. investigated the impact of a double-intensive strategy.[37] Patients (median age, 51 years) with *de novo* ($n = 53$) or advanced ($n = 44$) MM were intended to receive the following two-phase treatment: HDM (140 mg/m^2) without stem cell rescue and without hematopoietic growth factor, followed by a second course of HDM or HDM/TBI supported with unpurged autologous BMT, the bone marrow graft being collected after the first course of HDM. Of the 97 patients who received the first HDM cycle, only 35 (36 per cent) could proceed to the second course of high-dose therapy, the main reasons for exclusion being relapse (10 cases), poor hematologic recovery (7 cases), infections or poor clinical status (11 cases), and early death (8 cases). After the first cycle of HDM, 24 patients (25 per cent) achieved CR. Of the 22 patients in PR who received the second application of high-dose therapy, 11 converted to CR. However, the median survival for the 97 patients was only 24 months and the median duration of response remained 20 months. Thus, although this study demonstrated that a second application of HDM could convert PR to CR, improvement in overall survival or duration of response was not observed. The reason for this failure to improve overall results could be related to several factors. First, because of severe myelosuppression due to the first HDM, only 36 per cent of patients could receive the second course of high-dose therapy. Second, melphalan was the only drug used in this regimen and monotherapy may not be the optimal strategy to target the various subpopulations of cells that belong to the malignant clone.

In order to improve the feasibility of this double-intensive therapy, Jagannath et al. investigated a three-phase strategy involving high-dose cyclophosphamide with GM-CSF for blood stem cell mobilization, followed by melphalan (200 mg/m^2) as the first and melphalan (200 mg/m^2) or melphalan/TBI as the second preparatory regimen, each supported by both

PBSC and autologous BMT.[38] One hundred and thirty-three previously treated patients were enrolled. Seventy-six patients received the double transplant procedure, 38 remained eligible for the second transplant, whereas 19 (14 per cent) were not eligible because of insufficient stem cell collection or other medical reasons. Only one toxic death occurred. A 40 per cent CR rate was observed among patients treated within 12 months from diagnosis for sensitive disease. Median event-free survival and overall survival have not been reached at 30 months. Thus, this strategy appeared feasible with an extremely low rate of toxic death.

In order to further improve these results by targeting the various subpopulations of malignant cells, Barlogie et al. designed the 'total therapy', a strategy incorporating all non-cross-resistant currently available treatment principles.[6] Patients first receive three cycles of VAD. Then, high-dose cyclophosphamide/GM-CSF is introduced to collect PBSC and to induce further cytoreduction. The EDAP regimen, targeting incipient high-grade transformed cells, concludes the induction phase. Double transplantation with melphalan at 200 mg/m^2 supported by both PBSC and autologous BMT follows the induction phase. Maintenance treatment with IFN is given after transplantation. Preliminary results indicate that 'total therapy' is feasible, relatively well tolerated, and associated with a major cytoreduction. However, whether such a strategy prolongs duration of response in unselected patients will require longer follow-up and controlled studies.

Prognostic factors with high-dose therapy

When considering the indication for high-dose therapy in an MM patient, the potential risk and gain associated with autologous transplantation has to be weighed against this patient prognosis with standard therapy. Thus, identification of factors, associated with early death risk as well as factors associated with 'good' response and prolonged survival after autologous transplantation, is a major objective. Two large studies have addressed this issue. Barlogie et al. analyzed 170 patients receiving high-dose therapy with ($n = 80$) or without ($n = 90$) transplant.[32] Harousseau et al. analyzed 167 patients from the French registry receiving autologous BMT ($n = 81$) or PBSC ($n = 51$).[39]

The risk of toxic death was significantly related to performance status (PS) and status of disease at time of transplant. Indeed, Barlogie reported that this risk was 9 per cent for PS $\leqslant$ 1 versus 31 per cent for PS

> 1 and 2 per cent for responsive patients versus 18 per cent for refractory patients. The CR rate was significantly higher when pre-BMT β_2-microglobulin or LDH levels were low in Barlogie's study, and when age was < 50 years or disease was sensitive to conventional therapy at time of BMT in the French study. Survival was significantly longer when β_2-microglobulin or LDH levels were low, PS < 1 and age < 50 years in Barlogie's study, and when β_2-microglobulin was low and disease was sensitive to conventional therapy in Harousseau's study. Neither duration of response nor survival were affected by the source of hematopoietic stem cells in the French study.

Thus, little benefit can be expected from high-dose therapy for patients with disease refractory to conventional therapy. Indeed, in such a setting, a high risk of toxic death is observed, whereas the probability of major cytoreduction resulting in prolonged disease control of more than 1 year is low. When myeloma is still sensitive to standard doses of therapy, patients with low β_2-microglobulin level may preferentially benefit from high-dose therapy because they achieve a high rate of CR and a sustained duration of response, similar to observations with standard dose therapy. Thus, prospective trials testing the effect of high-dose therapy in MM should not be restricted to patients with high β_2-microglobulin level, and should be randomly compared to conventional chemotherapy.

The choice of treatment in the management of multiple myeloma

Currently, the choice of allogeneic rather than autologous transplant remains difficult. After allogeneic BMT, 35 per cent of patients transplanted after first-line therapy will remain alive and in complete remission at 5 years,[40,41] wherease after autologous transplant no series have shown any evidence of a plateau in remission duration. However, transplant-related mortality is much higher after allogeneic BMT (30–40 per cent) than after autologous transplant (5 per cent). Thus, whether the increased morbidity and mortality rate of allogeneic BMT might be offset by the possibility of curing the disease through the putative graft-versus-myeloma effect remains difficult to assess. Furthermore, allogeneic BMT can only be applied to a minority of MM patients, aged under 50 years with HLA-identical siblings. Thus, for the majority of patients aged under 65–70 years, the crucial question remains to know whether high-dose

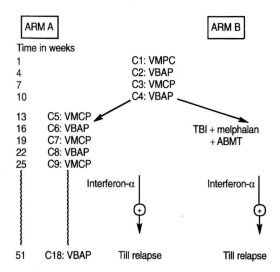

Fig. 15.1 General outline of the IFM 90 trial.

(VMCP/BVAP) or high-dose therapy (melphalan, 140 mg/m^2 and total body irradiation, 800 cGy), supported with unpurged autologous bone marrow collected after two cycles of VMCP/BVAP. Maintenance treatment with IFN-α (3×106 U/m^2; 3 times a week) was used in both arms. In May 1993, 200 patients had been randomized, the enrollment was stopped, and the analysis of the first 100 patients with a follow-up of 1 year or more from diagnosis was performed on an intention-to-treat basis.

Patient characteristics of each group were similar (Table 15.3), and no significant differences were found with regard to age (57.1 years; SD = 5.9), stage DS (II = 29; III = 71), immunoglobulin iso-type (IgG = 58; IgA = 28; Bence Jones = 12; IgD = 2), β$_2$-microglobulin (4.8 mg/l; SD = 4.9), and bone marrow plasmocytosis (38.7 per cent; SD = 26.5).

Thirty-eight of 50 patients (76 per cent) enrolled in the high-dose therapy arm have received autologous BMT and 12 patients did not for the following reasons: 3 for bad performance status, 2 for renal failure, 3 for insufficient bone marrow collection, and 4 for early death. Only one toxic death occurred after autologous BMT (*Streptococcus* infection).

Response rate was significantly higher in the autologous BMT arm than in the conventional arm ($p <$ 0.01). Indeed, 12 of 50 (24.5 per cent) patients achieved CR, 27 of 50 (55.1 per cent) PR, 3 of 50 (6.1 per cent) minimal response (MR) and 7 of 50 (14.3 per cent) remained with a progressive disease (PD) in the autologous BMT arm versus 1 of 50 (2 per cent) CR, 29 of 50 (58 per cent) PR, 8 of 50 (16 per

therapy supported with autologous transplantation is justified by marked prolongation of disease-free and overall survival compared to conventional chemotherapy. As yet, no prospective studies testing the effect of autologous transplantation versus conventional chemotherapy have been reported.

In September 1990, the 'Intergroupe Français du Myelome' (IFM) initiated such a prospective study (Fig. 15.1). Patients aged under 65, with Stage II/III DS previously untreated MM, were randomized, at diagnosis, to receive conventional chemotherapy

Table 15.3 Patient characteristics of the first 100 patients enrolled in the IFM 90 trial

	Arm A chemotherapy	Arm B autologous BMT	Total
No. of patients	50	50	100
Sex (M/F)	26/24	28/22	54/46
Mean age in years (SD)	57.7 (5.9)	56.4 (6)	57.1 (5.9)
Stage DS			
II/III	14/36	15/35	29/71
A/B	45/5	47/3	92/8
M-protein			
IgG	28	30	58
IgA	15	13	28
Bence Jones	6	6	12
IgD	1	1	2
Mean hemoglobin in g/dl (SD)	10.6(2)	11(2.4)	10.8(2.2)
Mean calcemia in μmol/l (SD)	2.56(0.4)	2.46(0.4)	2.5(0.4)
Mean albuminemia in g/l (SF)	38.7(10)	39.1(8.2)	38.7(26.5)
Mean LDH in IU/L (SD)	205(140)	271(161)	239(155)
Mean medullary plasmocytosis in % (SD)	36.3(25)	41.3(28)	38.7(26.5)
Mean β$_2$-microglobulin in mg/l (SD)	5.2(5.1)	4.6(4.9)	4.9(4.9)

None of the characteristics differed significantly between treatment groups.

Table 15.4 Response rate of the first 100 patients enrolled in the IFM 90 trial

	Arm A chemotherapy ($n = 50$)	Arm B autologous BMT ($n = 50$)	p value
Response rate (%)			< 0.01
CR[*]	1(2)	12(24.5)	
Very good partial response ($\geq 90\%$)[*]	7(14)	8(16.3)	
Partial response ($\geq 50\%$)[*]	22(44)	19(38.8)	
Minimal response ($\geq 25\%$)[*]	8(16)	3(6.1)	
Progressive disease[*]	12(24)	7(14.3)	
No. of patients treated with autologous BMT (%)		38*(76)	

[*] Twelve patients were excluded for the following reasons: early death, 4 cases; insufficient bone marrow collection, 4 cases; bad performance status, 3 cases; renal failure, 1 case.

cent) MR, and 12 of 50 (24 per cent) PD in the conventional arm (Table 15.4). Duration of response (> 50 per cent) was significantly longer in the autologous BMT arm than in the conventional arm. Indeed, the 30-month post-diagnosis probability of remaining progression-free was 80 per cent (95 per cent CI = 58–94) in the autologous BMT arm versus 19 per cent (95 per cent CI = 1–37) in the conventional arm ($p < 0.01$). Autologous BMT was also found to improve progression-free survival. The 30-month post-diagnosis probability of progression-free survival was 67 per cent (95 per cent CI = 49–81) in the autologous BMT arm versus 10 per cent (95 per cent CI = 1–42) in the conventional arm ($p < 0.01$) (Fig. 15.2). With a median follow-up of 560 days post-diagnosis, the 30-month post-diagnosis probability of survival was 83 per cent (95 per cent CI = 64–95) in the autologous BMT arm versus 60.5 per cent (95 per cent CI = 35–83) in the conventional arm ($p < 0.09$).

Finally, the first analysis of the IFM trial strongly suggests that response rate, duration of response, and progression-free survival could be improved by auto-

logous BMT. More follow-up is required to appreciate the impact of autologous BMT on survival. If confirmed, these results would justify the use of autologous BMT as front-line therapy in MM patients under the age of 65 years. Furthermore in this trial, complete response after autologous BMT appears to be strongly related with long duration of response and survival. This result clearly indicates that achievement of CR is the first important step toward the improvement of survival in MM. In this regard, the impact of successive applications of non-cross-resistant high-dose therapy regimens supported by autologous BMT and/or PBSC should be investigated in future prospective randomized trials.

Conclusions

During the last 10 years, intensive cytotoxic chemotherapy has proven feasible in elderly patients (up to 65–70 years) with multiple myeloma when supported with autologous transplantation. Indeed for patients with good performance status and normal renal function, the early mortality rate is now under 5 per cent and thus not significantly different to that recorded with conventional chemotherapy. Furthermore, utilization of hematopoietic growth factor both for the procurement of hematopoietic stem cells and after transplantation as well as the combined use of autologous BMT plus PBSC may further decrease the morbidity and mortality rate associated with autologous transplantation.

A marked tumor reduction, including 30 per cent of complete remissions, is usually achieved after autologous transplantation, especially when high-dose therapy is applied in remission and when tumor burden is low. However, even in this favourable situation and when maintenance treatment with IFN is used, there is no evidence that a plateau in

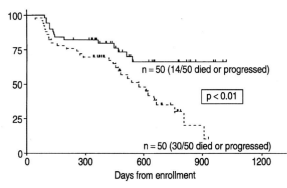

Fig. 15.2 Progression-free survival of the first 100 patients enrolled in the IFM 90 trial. ————, autologous BMT; -----, chemotherapy.

remission duration can be reached. Thus, whether the high response rate induced by such strategies is associated with improved disease-free and overall survival compared to conventional chemotherapy remains to be answered. The first analysis of the IFM 90 trial, designed to address this issue, suggests that response rate, duration of response, and disease-free survival were improved by high-dose therapy. Moreover, this analysis suggested that complete response was an important step towards cure in multiple myeloma. Thus, new strategies incorporating successive applications of non-cross-resistant high-dose regimens could further improve the prognosis of MM patients. If a higher rate of CR is achieved using these combined high-dose strategies, the role of graft purging or positive selection of CD34 normal hematopoietic stem cells could be investigated. Finally, advances in the understanding of cellular and humoral factors involved in the myeloma growth control, may give opportunities for post-transplant maintenance treatment.

References

1. McElwain TJ, Powles RL. High-dose intravenous melphalan for plasma-cell leukaemia and myeloma. *Lancet* 1983; **2**: 822.
2. Selby PH, McElwain TJ, Nandi AC et al. Multiple myeloma treated with high dose intravenous melphalan. *British Journal of Haematology* 1987; **66**: 55.
3. Barlogie B, Hall R, Zander A et al. High dose melphalan with autologous bone marrow transplantation for multiple myeloma. *Blood* 1986; **67**: 1860.
4. Barlogie B, Alexanian R, Dicke K et al. High dose chemoradiotherapy and autologous bone marrow transplantation for resistant multiple myeloma. *Blood* 1987; **70**: 869.
5. Lenhard RE, Oken MM, Barnes JM et al. High-dose cyclophosphamide. An effective treatment for advanced refractory multiple myeloma. *Cancer* 1984; **53**: 1456.
6. Jagannath S, Barlogie B. Autologous bone marrow transplantation for multiple myeloma. *Hematology and Oncology* 1992; **6**: 437.
7. Gockerman J, Bartolucci A, Nelson M. Phase II evaluation of etoposide in refractory multiple myeloma. *Cancer Treatment Reports* 1986; **70**: 801.
8. Gianni AM, Bregni M, Siena S et al. Granulocyte–macrophage colony-stimulating factor or granulocyte colony-stimulating factor infusion makes high-dose etoposide a safe outpatient regimen that is effective in lymphoma and myeloma patients. *Journal of Clinical Oncology* 1992; **10**: 1955.
9. Dimopoulos MA, Delasalle KA, Champlin R. Cyclophosphamide and etoposide therapy with GM-CSF for VAD-resistant multiple myeloma. *British Journal of Haematology* 1993; **83**: 240.
10. Barlogie B, Epstein J, Selvanayagam P et al. Plasma cell myeloma – new biological insights and advances in therapy. *Blood* 1989; **73**: 865.
11. Attal M, Huguet F, Schlaifer D et al. Maintenance treatment with recombinant alpha interferon after autologous bone marrow transplantation for aggressive myeloma in first remission after conventional induction chemotherapy. *Bone Marrow Transplantation* 1991; **8**: 125.
12. Gore ME, Viner C, Meldrum M. Intensive treatment of multiple myeloma and criteria for complete remission. *Lancet* 1989; **14**: 879.
13. Besinger WI, Buckner CD, Clift R et al. Allogeneic marrow transplantation for patients with multiple myeloma. In: Pileri A, Boccadoro M eds *Multiple myeloma from biology to therapy*. III International Workshop, Torino, Italy, 1991: 123.
14. Mansi J, Da Costa F, Vinr C et al. High-dose Busulfan in patients with myeloma. *Journal of Clinical Oncology* 1992; **10**: 1569.
15. Reece DE, Barnett MJ, Connors JM et al. Treatment of multiple myeloma with intensive chemotherapy followed by autologous BMT using marrow purged with 4-hydroperoxycyclophosphamide. *Bone Marrow Transplantation* 1993; **11**: 139.
16. Berenson J, Wong R, Kim K et al. Evidence of peripheral blood B lymphocyte but not T lymphocyte population in human plasma cell myeloma. *Blood* 1987; **70**: 1550.
17. Mellstedt H, Hammorstrom S, Holm G. Monoclonal lymphocyte population in human plasma cell myeloma. *Clinical and Experimental Immunology* 1974; **17**: 371.
18. Bast E, Van Camp B, Reynaert P et al. Idiotypic peripheral blood lymphocytes in monoclonal gammapathy. *Clinical and Experimental Immunology* 1982; **47**: 677.
19. Caligaris-Capio F, Bergui L, Gaidana GL et al. Circulating malignant precursor cells in monoclonal gammapathies. *European Journal of Haematology* 1989; **45** (suppl. 51): 27.
20. Barlogie B, Gahrton G. Bone marrow transplantation in multiple myeloma. *Bone Marrow Transplantation* 1991; **7**: 71.
21. Harousseau JL, Attal M, Divine M et al. Comparison of autologous bone marrow transplantation and peripheral blood stem cell transplantation after first remission induction treatment in multiple myeloma. *Bone Marrow Transplantation* 1995; **15**: 963.
22. Attal M, Huguet F, Schlaifer D et al. Intensive combined therapy for previously untreated aggressive myeloma. *Blood* 1992; **79**: 1130.
23. Fernand JP, Chevret S, Levy Y. The role of autologous blood stem cells in support of high-dose therapy for multiple myeloma. *Hematology and Oncology* 1992; **6**: 451.
24. Gianni AM, Siena S, Bregni M et al. Granulocyte–macrophage colony-stimulating factor to harvest circulating haemopoietic stem cells for autotransplantation. *Lancet* 1989; **2**: 580.
25. Siena S, Bregni M, Brando V et al. Circulation of CD34 (+) hematopoietic stem cells in the peripheral blood of high-dose cyclophosphamide-treated

patients: enhancement by intravenous recombinant human granulocyte–macrophage colony-stimulating factor. *Blood* 1989; **74**: 1905.

26. Jagannath S, Vesole DH, Glenn L et al. Low-risk therapy for multiple myeloma with combined autologous bone marrow and blood stem cell support. *Blood* 1992; **80**: 1666.

27. Gobbi M, Tazzari PL, Cavo M. Autologous bone marrow transplantation with immunotoxin purged marrow for multiple myeloma. Long term results in 14 patients with advanced disease. In: Pileri A, Boccadoro M eds *Multiple myeloma from biology to therapy*. III International Workshop, Torino, Italy, 1991: 137.

28. Anderson KC, Barut BA, Ritz J et al. Monoclonal antibody-purged autologous bone marrow transplantation therapy for multiple myeloma. *Blood* 1991; **77**: 712.

29. Jagannath S, Barloge B, Dicke K. Autologous bone marrow transplantation in multiple myeloma: identification of prognostic factors. *Blood* 1990; **76**: 1860.

30. Richman CM, Weiner RS, Yanhee RA. Increase in circulating stem cell following chemotherapy in man. *Blood* 1976; **47**: 1031.

31. To LB, Haylock DN, Kimber RJ et al. High levels of circulating haemopoietic stem cells in very early remission from acute non-lymphoblastic leukaemia and their collection and cryopreservation. *British Journal of Haematology* 1984; **58**: 399.

32. Barlogie B, Jagannath S. Autotransplant in multiple myeloma. *Bone Marrow Transplantation* 1992; **10**: 37.

33. Fernand JP, Levy Y, Gerota J et al. Treatment of aggressive multiple myeloma by high-dose chemotherapy and total body irradiation followed by blood stem cells autologous graft. *Blood* 1989; **73**: 20.

34. Marit G, Pico JL, Boiron JM. Autologous blood stem cell transplantation (Absct) in high risk myeloma (MM). In: Pileri A, Boccadoro M eds *Multiple myeloma from biology to therapy*. III International Workshop, Torino, Italy, 1991: 146.

35. Mandelli F, Avvisati G, Amadori S et al. Maintenance treatment with alpha-2b recombinant interferon significantly improves response and survival duration in multiple myeloma patients responding to conventional induction chemotherapy. Results of an Italian randomized study. *New England Journal of Medicine* 1990; **322**: 1430.

36. Cunningham D, Powles R, Malpas JS et al. A randomised trial of maintenance therapy with intron-A following high dose melphalan and ABMT in myeloma. *Journal of Clinical Oncology* 1993; **12**: 364 (abstract).

37. Harousseau JL, Milpied N, Laporte JP et al. Double-intensive therapy in high-risk multiple myeloma. *Blood* 1992; **79**: 2827.

38. Jagannath S, Glenn L, Vesole D et al. Double autotransplants (Tx) for multiple myeloma (MM). *Journal of Clinical Oncology* 1993; **12**: 404 (abstract).

39. Harousseau JL, Attal M, Divine M et al. Autologous stem cell transplantation after first remission induction treatment in multiple myeloma: a report of the French Registry on Autologous Transplantation in Multiple Myeloma. *Blood* 1995; **85**: 3077.

40. Gahrton G, Tura S, Ljungman P et al. Allogeneic bone marrow transplantation in multiple myeloma. *New England and Journal of Medicine* 1991; **325**: 1267.

41. Samson D. The current position of allogeneic and autologous BMT in multiple myeloma. *Leukemia and Lymphoma* 1992; **7**: 33.

CHAPTER 16

Bone marrow transplantation with syngeneic or allogeneic marrow in multiple myeloma

GÖSTA GAHRTON, PER LJUNGMAN

Introduction	194	BMT from HLA-matched sibling donors – survival	
Conditioning treatment	195	and relapse-free survival	197
Syngeneic bone marrow transplantation	195	BMT from HLA-matched non-sibling donors	199
Allogeneic bone marrow transplantation	196	Complications and post-BMT treatment	200
Response to BMT	196	Indications for BMT – prospects for the future	201
		References	202

Introduction

Bone marrow transplantation (BMT) has been increasingly used for the treatment of hematologic malignancies[1,2] since the first successful reports in the early 1970s. The rationale behind bone marrow transplantation is that bone marrow ablative therapy may have the potential to eradicate the malignant disease. After high-dose chemotherapy and total body irradiation the patient is saved by bone marrow infusion. Of the three main types of transplantation, i.e. autologous (Chapter 15), syngeneic, and allogeneic BMT, only allogeneic transplantation has the potential for a graft-versus-tumor effect, which may help to eradicate the tumor. The exact nature of this effect is not known. It is fairly well-documented for acute leukemia[3,4] but has been difficult to detect in other disorders, such as multiple myeloma.[5] The graft-versus-tumor effect seems to be associated with graft-versus-host disease (*see below*). Syngeneic transplantation lacks a graft-versus-tumor effect, but has the advantage over allogeneic transplantation that it also lacks graft-versus-host disease.

The first attempts to perform bone marrow transplantation in multiple myeloma were made in the early 1980s. Fefer et al.[6] and Osserman et al.[7] used syngeneic marrow, while the Royal Marsden group used autologous marrow to rescue patients following the marrow-ablative therapy.[8,9] The first promising reports of allogeneic transplantation were published in the mid-1980s.[10–13] Out of three patients who were reported from the Huddinge group,[12] one patient who was resistant to melphalan + prednisolone treatment went into complete remission following BMT. She remained without signs of multiple myeloma for 4 years but then relapsed. Of the patients reported by the Bologna group, one patient was in sustained remission 67 months after transplantation.[14] Following these reports both syngeneic and allogeneic transplants were reported from several groups,[14–21]

Table 16.1 Conditioning regimens in multiple myeloma

TBI	(fractions)	Cyclophosphamide	Busulfan	Other drugs		Reference
10 Gy	(1)	60 mg/kg × 2				Gahrton et al., 1986 [12]
8 Gy	(1)	50 mg/kg × 2				Ozer et al., 1984 [11]
10 Gy	(3)	60 mg/kg × 2				Gallamini et al., 1987[15]
10 Gy	(4)	60 mg/kg × 2				Nikoskelainen et al., 1988 [16]
12 Gy	(6)	60 mg/kg × 2				Buckner et al., 1989 [22]
10 Gy	(1)	60 mg/kg × 2		BCNU 5 mg/kg and/or oral melphalan 1 mg/kg	× 1 × 5	Tura et al., 1986 [13]
13 Gy	(2)	50 mg/kg × 3		HDMP 1g/m^2 and CCNU 200 mg/m^2	× 3 × 1	Samson et al., 1990 [17]
		60 mg/kg × 2	4 mg/kg × 4			Copelan and Tutschka, 1988 [21]
		60 mg/kg × 2	3.5 mg/kg × 4			Buckner et al., 1989 [22] Samson et al., 1990 [17]
		50 mg/kg × 4	4 mg/kg × 4			Tura et al., 1989 [23]
		147 mg/kg (total)	3.5 mg/kg × 4			Bensinger et al., 1992 [19]
		1.8 g/m^2 × 3		CCNU 200 mg/m^2 Ara-C 2 g/m^2 VP-16 1 g/m^2	× 1 × 2 × 1	McCarthy et al., 1990 [24]

most of them from Bologna,[14] Huddinge,[5,18] and Seattle.[19]

Conditioning treatment

The aim of the conditioning treatment before bone marrow transplantation is to eradicate the malignant cell population. The myeloma cell is highly sensitive both to irradiation and to many cytotoxic drugs, particularly the alkylating agents melphalan and cyclophosphamide.

Most allogeneic transplants of patients with multiple myeloma have used conditioning regimens including total body irradiation. This regimen has been combined either with cyclophosphamide, mainly according to the original Seattle protocol, or with other drug combinations (Table 16.1).

In the most recent update of patients reported to the European Group for Blood and Marrow Transplantation (EBMT)[25] total body irradiation (TBI) + cyclophosphamide was the most common combination with 55 out of 162 patients received this conditioning. Other conditioning methods were TBI + melphalan, TBI + cyclophosphamide + other drugs or TBI + melphalan + other drugs. A minority of patients did not receive TBI but only melphalan or a combination of busulfan and cyclophosphamide. This latter regimen was originally used by Tutschka and associates[26] for conditioning of patients with leukemia.

In the EBMT study[25] there was no detectable difference between any of the conditioning methods. However, the great variation concerning doses and schedule, etc., made comparison difficult. The busulfan + cyclophosphamide combination did not appear to be superior to TBI + cyclophosphamide. If anything, it was slightly inferior. This corroborates with other studies that have shown that busulfan + cyclophosphamide might be hampered by more complications, such as hemorrhagic cystits and veno-occlusive disease.[27] Also, for some patients busulfan + cyclophosphamide seems to be inferior to TBI.[27]

The Seattle group has used busulfan + cyclophosphamide in twenty patients with myeloma who received an allogeneic BMT. In their hands, busulfan + cyclophosphamide appears to be a relatively promising method.[19] Attempts to find new effective conditioning schedules for bone marrow transplantation in multiple myeloma seems important since transplant-related mortality which might to some extent depend on the conditioning method is still higher in allogeneic transplants than in autologous transplants.

Syngeneic bone marrow transplantation

Following the first reports of syngeneic BMT in multiple myeloma[6,7] the Seattle group[22] reported a series of six transplants with marrow from syngeneic donors. At the time of transplantation, 7 of these recipients were in a progressive state after chemotherapy. Still, 3 of the patients were alive 726–3560 days after transplantation. Two died of interstitial pneumonits. Four died with multiple myeloma, 2 of whom survived for 1596 and 1750 days, respectively. One patient who was reported to have a small persistent serum monoclonal component at 9.5 years is now living and well 13 years after transplantation and the monoclonal immunoglobulin has disappeared (A Fefer, personal communication).

In the EBMT study, 6 patients have received marrow from syngeneic donors.[28] Three were on second- or third-line treatment at the time of transplantation. Five were in Stage III and 2 patients had not responded to previous treatment. Three were in partial remission and one was in a complete remission. The results were promising. One patient died on day 483 but 5 patients were alive 177, 693, 709, 1045, and 1128 days after BMT. However, all but one of these patients survived with signs of multiple myeloma. Probably, the disease will progress in these patients.

Progression of disease and relapse (*see below*) may well be more frequent in recipients of syngeneic marrow than in those who receive allogeneic marrow. Thus, transplantation using syngeneic marrow should probably be done earlier in the course of the disease but only if the patient has a low tumor burden and has not been heavily pretreated.

Allogeneic bone marrow transplantation

Allogeneic BMT is technically performed just as syngeneic transplantation with one important difference, i.e. that in allogeneic BMT preventive treatment following transplantation is necessary in order to prevent graft-versus-host disease (GVHD). In most BMT so far done worldwide, HLA-compatible sibling donors have been used, but increasing numbers of BMTs are performed with marrow from matched unrelated donors. In multiple myeloma, only a few such transplants have been made.[28]

In comparison to other hematologic malignancies, the number of allogeneic BMTs in multiple myeloma is relatively small. In the registry for the European Group for Bone Marrow Transplantation, there are now 300 reports of such transplants, and 162 of these that were reported until early 1993 have been analyzed.[25] The International Bone Marrow Transplant Registry (IBMTR) has 244 patients in their registry.[29] However, a large proportion of these patients are also reported to the EBMT registry. The Seattle group has

performed about 100 BMTs that are not reported to these registries. An estimated number of allogeneic BMTs worldwide is thus about 400–500 patients.

The relatively small number of BMTs performed in multiple myeloma is probably due to the fact that patients with this disorder are relatively old. Only about 7 per cent are below the age of 50.[30] Also, results of autologous transplantation in multiple myeloma have been promising (Chapter 15). Many centers prefer to perform an autologous transplantation.

Results of BMT presented in this chapter are mainly based on data from the EBMT registry.[25]

Response to BMT

Since BMT is a relatively new treatment method and the guidelines for transplantation have been unclear, the heterogeneity among patients who have undergone BMT is great. Among the patients who have been reported to EBMT prognosis has in most cases been considered poor with standard conventional chemotherapy. Still, as seen from Table 16.2, 72 out of 162 patients entered a complete remission following BMT. However, only 121 could be evaluated for response either because of early death or because they were not yet evaluable. Out of the evaluated patients, 60 per cent entered a complete remission. The complete remission rate was significantly better in patients who were diagnosed in Stage I irrespective of the stage at which they were later transplanted.

Table 16.3 shows that transplantation is significantly more successful if carried out before second-line treatment has been initiated. Of the 64 patients who were on first-line treatment when transplanted 36, i.e. 56 per cent, entered complete remission. The response rate was significantly poorer if the patient already had received second-line or later lines of treatment. Table 16.4 shows that patients who were in a progressive state of the disease did worse than those who were already in complete remission at conditioning for BMT. Patients who were in partial remission or had stable but unresponsive disease entered a complete remission in about 50 per cent

Table 16.2 Bone marrow transplantation in multiple myeloma: complete remission (CR) by stage at diagnosis

Stage at diagnosis	No. of patients	CR following BMT	
		No. of patients	%
I	22	15	68
II	30	11	37
III	109	45	41

Table 16.3 Bone marrow transplantation in multiple myeloma: complete remission (CR) following BMT by number of lines of treatment before BMT

No. of lines treatment before BMT	No. of patients	CR following BMT	
		No. of patients	%
One	64	36	56
Two	50	19	38
Three or more	46	16	35

Table 16.4 Bone marrow transplantation in multiple myeloma: complete remission (CR) by status at conditioning

Status at conditioning	No. of patients	CR following BMT	
		No. of patients	%
CR	18	15	83
Partial remission	66	28	42
Stable disease	22	11	50
Primary refractory	14	5	36
Progressive disease	25	6	24
Relapse	14	6	43

of cases, while of those who were in a progressive state only 25 per cent entered complete remission.

An interesting point in this study was that IgG myelomas seem to respond less well than IgA and light-chain myeloma (Table 16.5). However, this has to be confirmed in other studies.

In the EBMT registry complete remission was defined as the disappearance of the monoclonal immunoglobulin from serum or the light chains from the urine and disappearance of apparent myeloma cells from the marrow, but disregarding the radiographic bone lesion. In fact, the radiographic bone lesions usually did not change significantly following BMT. Only 7 patients showed improvement: major lytic lesions changed to minor lesions in 3 patients, and minor lesions disappeared in 4 patients. It is not known whether these radiographic bone lesions still contain living myeloma cells in complete remission patients who have survived without relapse following BMT.

In a study by the Seattle group[19] 12 complete remissions following BMT were induced among 20 patients who received allografts after preparation with busulfan + cyclophosphamide. Most of these patients were in an advanced stage of the disease. Thus, this study also supports the view that although the best results are obtained in early transplantation before the

Table 16.5 Bone marrow transplantation in multiple myeloma: complete remission (CR) by subtype

Diagnosis	No. of patients	CR following BMT	
		No. of patients	%
IgG	80	27	34
IgA	33	17	52
Light chain	31	18	58

patient has received several lines of treatment, complete remission may be obtained in a relatively high proportion of patients who are in an advanced stage.

BMT from HLA-matched sibling donors – survival and relapse-free survival

The overall actuarial survival following BMT was 32 per cent at 4 years and 28 per cent at 7 years in the EBMT material. Patients who were diagnosed in Stage I appeared to have a better survival than other patients irrespective of the stage at which they were transplanted (Fig. 16.1). Also, patients that were on first-line treatment at the time of transplantation did better than those who were on second- or third-line treatment (Fig. 16.2). Patients who were in complete

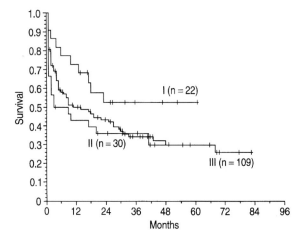

Fig. 16.1 Actuarial survival after bone marrow transplantation according to the stage of the disease at diagnosis. The Kaplan–Meier curves show significantly better survival among patients with Stage I disease at diagnosis than among patients with Stage II and III disease ($p = 0.05$). (Reproduced from ref. 25 with permission.)

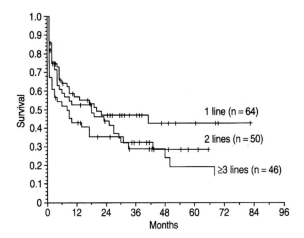

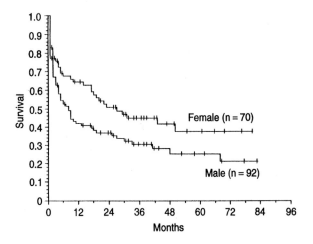

Fig. 16.2 Actuarial survival after bone marrow transplantation according to the number of lines of treatment regimens used before transplantation. The Kaplan–Meier curves show significantly better survival among patients who had received only one line of treatment as compared to those who had received three or more lines of treatment ($p = 0.02$).

Fig. 16.3 Actuarial survival after bone marrow transplantation according to the sex of the patient. The Kaplan–Meier curves show significantly better survival among females than among males ($p = 0.04$). (Reproduced from ref. 25 with permission).

remission already before BMT did better than those who had more advanced disease or were unresponsive to conventional chemotherapy. However, most important was that a fraction of patients that were considered unresponsive to treatment before BMT became long-term survivors.

One interesting observation was that females did significantly better than males (Fig. 16.3). There was a slight adverse effect of female-to-male transplantation, as has been shown for leukemia.[31] However, in the EBMT study males transplanted for myeloma did less well than females, irrespective of whether they received marrow from a male or a female.

The age within the age bracket 25–55 years was not of great importance, although there was a tendency for patients who were older than 40 years of age to do somewhat worse than younger patients. However, it was not statistically significant and would support the view that in adult patients with hematological malignancies there is no clear difference in survival between age groups up to the age of 50–55.[32]

IgA myelomas seem to do better than other types of myeloma (Fig. 16.4). This corroborates with the finding that IgA myelomas seem to respond better to BMT than IgG myelomas. However, the material was relatively small and the heterogeneity precludes firm conclusions concerning the influence of subtype on survival.

A high serum β2-microglobulin value has an adverse prognostic impact, as following chemotherapy (Fig. 16.5).

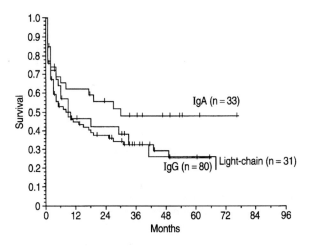

Fig. 16.4 Actuarial survival after bone marrow transplantation according to Ig subtype. The Kaplan–Meier curves shows a tendency for a better survival among patients with IgA myeloma than among patients with IgG ($p = 0.08$), or those with light-chain myeloma ($p = 0.28$). (Reproduced from ref. 25 with permission.)

The relapse-free survival of the seventy-two patients who entered a complete remission is seen in Fig. 16.6. Although it is obvious that relapses do occur continuously in patients who enter remission, it is of interest that there are now nine patients in complete remission more than 4 years following BMT with allogeneic marrow. Whether the myeloma cell clone has been eradicated in these patients is not known. However, in one recent study[33] it has been

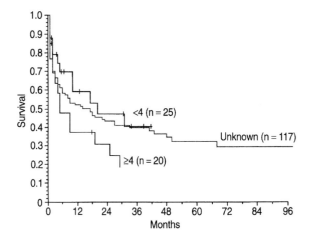

Fig. 16.5 Actuarial survival after bone marrow transplantation according to the value of β_2-microglobulin at diagnosis. There was a slight but not significant trend towards better survival among patients who had β_2-microglobulin values of less than 4 mg/l than in those who had a β_2-microglobulin value of 4 mg/l or higher. (Reproduced from ref. 25 with permission.)

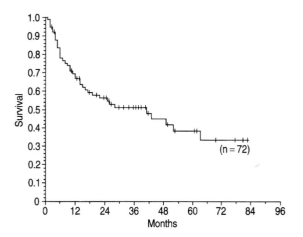

Fig. 16.6 Actuarial relapse-free survival of seventy-two patients who were in complete remission following bone marrow transplantation. (Reproduced from ref. 25 with permission.)

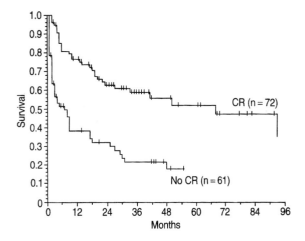

Fig. 16.7 Actuarial survival after bone marrow transplantation according to whether patients were in complete remission after engraftment (CR). The Kaplan–Meier curves show significantly better survival among patients who entered complete remission after engraftment than among those who did not ($p = 0.01$). (Reproduced from ref. 25 with permission.)

shown that using a PCR-based technique, clonal cells were not detectable in the marrow in 3 out of 5 patients who were in sustained hematologic complete remission following BMT. Thus, it may be possible that the myeloma cell clone can be eradicated in some patients.

The rate of disappearance of the myeloma cell clone varies considerably between patients. The median time for disappearance of monoclonal immunoglobulins was 4 months, while some patients lost their monoclonal component as long as 2 years following BMT without any additional treatment.

Figure 16.7 shows that the most important factor for survival was to enter complete remission following transplantation. Patients who engrafted and who did not enter complete remission had a survival that was significantly poorer, although about 20 per cent of these patients survived for more than 3 years. Thus, although complete remission is the most important factor for survival, patients who do not enter complete remission can survive for a significant period of time. Some of these patients had stable disease for some years with only a small monoclonal component in the serum or light chains in the urine. Thus, high-dose therapy followed by allogeneic transplantation may improve survival for some patients who do not enter complete remission.

BMT from HLA-matched non-sibling donors

Only about 25 per cent of patients are expected to have an HLA-matched sibling donor. Thus, about 75 per cent of patients who otherwise would have had an indication for BMT cannot be transplanted unless a suitable donor can be found outside the sibling entourage.[34,35]

Because of the lack of donors and the increasing use of BMT, numbers of countries in the Western world have now established registries of volunteer donors.[36] About 2 million donors are now registered, mostly in Europe and the United States.[37] These donors have been typed for HLA-A and B or for HLA-A, B, and DR. Bone marrow from such unrelated donors is now increasingly used in transplantation, particularly in hematological malignancies.[38–40] Although results may be somewhat less promising than results of sibling donor transplantation, recent studies suggest that if only well-matched donors are selected, results with unrelated marrow transplantation may approach the results of sibling donor transplantation.[41]

In multiple myeloma unrelated donors have only been used in a few patients. In the last review of the EBMT registry, only 6 patients had received marrow from non-sibling donors and only 3 of them were unrelated.[28] The patients were in a poor condition at the time of BMT and the transplant-related death rate was high. Five of the 6 patients died within 75 days and only one patient, who received a graft from an HLA-identical related non-sibling donor, was alive 200 days after BMT. This patient still had signs of disease at the time of the report.

It is too early to predict the outcome of non-sibling and unrelated donor BMT in multiple myeloma. Most probably the poor results are due to the selection of patients with poor prognostic parameters. Thus, based on the promising results in other hematologic malignancies, particular in chronic myelocytic leukemia,[41–43] unrelated donor transplantation should probably be tried in some patients who are not in a poor condition but are judged to respond poorly to other treatment options.

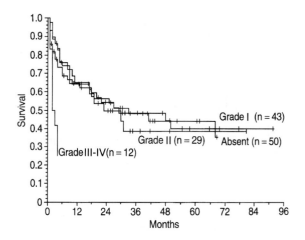

Fig. 16.8 Actuarial survival after bone marrow transplantation according to the grade of graft-versus-host disease following transplantation. The Kaplan–Meier curves show significantly poorer survival for patients with graft-versus-host disease Grades III–IV. (Reproduced from ref. 25 with permission.)

deplete the marrow of T cells.[47,48] T-cell depletion can also be combined with methotrexate, cyclosporine, or methotrexate + cyclosporine treatment. It is the most effective method to prevent GVHD. However, in studies of BMT in chronic myelocytic leukemia it has been shown that effective T-cell depletion may also induce an increased relapse rate.[49] For that reason there has been an overall decline in the use of T-cell depletion in later years. The regimen now generally preferred is the methotrexate + cyclosporine regimen that was originally used by the Seattle group.[45]

Complications and post-BMT treatment

GRAFT-VERSUS-HOST DISEASE

Graft-verus-host disease (GVHD) is an important cause of death in BMT. Figure 16.8 shows the deleterious effect on survival by severe acute GVHD (Grades III–IV). There are several possible ways in which GVHD can be reduced by prophylactic treatment. The treatment modalities generally used have been the original Seattle regimen with methotrexate,[1] cyclosporine[44] or more lately methotrexate + cyclosporine.[45,46] The rationale is that these drugs inhibit the T-cell-dependent GVHD reaction.

Since T cells appear to be most important for GVHD induction, attempts have been made to

OTHER TRANSPLANT-RELATED COMPLICATIONS AND SIDE-EFFECTS

BMT for hematologic malignancies including myeloma is hampered by both transplant-related complications and recurrence of the malignant disorder. Except for GVHD, infectious complications are the most important ones. During the leukopenic phase following BMT, bacterial and fungal infections are common,[50] while later viral infections predominate.[51] Early instigation of broad-spectrum antimicrobial therapy has diminished the mortality from bacterial infections.[52] Prophylaxis with fluconazole can reduce the frequency and mortality in fungal infections.[53] The use of amphotericin attached to liposomes has reduced the toxicity of antifungal therapy which seriously hampers the use of of amphotericin B and thereby probably improves the outcome of disseminated fungal infections in BMT recipients.[54]

The most serious viral infection is cytomegalovirus (CMV) pneumonia. During the last few years major advances have been reached in prophylaxis and therapy of CMV infections. High-dose acyclovir prophylaxis can reduce CMV viremia and infectious mortality.[55] Prophylaxis with ganciclovir reduces the frequency of CMV disease.[56,57] New diagnostic techniques such as antigenemia and PCR allow very early instigation of antiviral therapy (pre-emptive therapy) and preliminary studies have shown that this strategy is effective in reducing severe CMV infections.[58] The outcome of patients with verified CMV pneumonia is still poor despite the use of ganciclovir combined with high-dose intravenous immunoglobulin, with a mortality of approximately 50 per cent.[59–62] Other viral infections that can cause fatal infections in BMT recipients include adenovirus, respiratory syncytial virus, parainfluenza viruses, and human herpesvirus 6. Common late infections include herpes zoster, in particular in patients with chronic GVHD, as well as infections caused by pneumococci and *Haemophilus influenzae*.

RELAPSE FOLLOWING BMT

Relapse is still the most important cause of failure following BMT of patients with multiple myeloma. Patients seemingly in complete remission have relapsed as late as 9 years following transplantation. Interferon is now used in a dose-finding study as a first step to investigate whether post-BMT interferon treatment might diminish the recurrence rate.[63] The rationale behind post-transplant interferon treatment is that the majority of patients in complete remission still have minimal residual disease, which could be responsive to interferon. However, so far, the role of interferon post-BMT treatment is not clear, although it appears that interferon treatment post-autologous bone marrow transplantation might have a value,[64,65] (*see also* Chapter 15).

CAUSE OF DEATH

Although treatment-related mortality following allogeneic BMT in other hematologic malignancies has improved with time,[66] mortality following allogeneic BMT is still significant. Although the recurrence of the disease is the major problem, treatment-related mortality is relatively high, i.e. in the order of 40 per cent.

The causes of death in multiple myeloma are mainly the same as for other hematologic malignancies, i.e. interstitial pneumonitis, GVHD, bacterial

Table 16.6 Allogeneic bone marrow transplantation in multiple myeloma: primary causes of death in 103 patients (EBMT registry)

Cause of death	%
Original disease	26
Bacterial and fungal infections	18
Interstitial pneumonitis	17
Acute graft-versus-host disease	10
Bleedings	7
Organ failure	6
Adult respiratory distress syndrome (ARDS)	4
Chronic graft-versus-host disease	3
Veno-occlusive disease	3
Other causes	7

and fungal infections, hemorrhage, veno-occlusive disease, etc. (Table 16.6).

Indications for BMT – prospects for the future

BMT appears to be an option only for younger patients with multiple myeloma. Patients above 55 years of age should probably not be considered for allogeneic BMT. Less than 10 per cent of the patients are below this age.[30] Among these younger patients it is difficult to tell who is a candidate for BMT. The competing treatment modalities are conventional chemotherapy, more intensive chemotherapy, perhaps combined with cytokines, and autologous stem cell transplantation. For patients with Stage III disease and many with Stage II disease it seems clear that intensification of treatment is superior to conventional low-dose chemotherapy. Conventional chemotherapy does not have the potential to cure multiple myeloma. It results in a median survival of less than 3 years. Nor does intensification of chemotherapy without eradication of the bone marrow cells seem to have the potential to cure the disease. With autologous marrow or peripheral blood stem cells, it appears the eradication of the disease may be more difficult than with allogeneic marrow, since there is a risk of reinfusing malignant cells. However, the recent observations using PCR-based technique to detect minimal residual disease indicate that not only allogeneic but also double autologous transplantation might have the potential to eradicate the disease.[67] Thus, the selection of patients for one or the other method is difficult.

The arguments in favor of allogeneic transplantation are the potential of eradicating the disease without reinfusing malignant cells during transplantation, and a possible help by a graft-versus-myeloma effect. The arguments in favor of autologous transplantation are a lower transplant-related mortality and, perhaps with high intensification or double autologous transplantation, the possibility of eradicating the disease in some patients. However, autologous transplantation still seems to have less chance for disease eradication.

With this background, the following requirements for performing allogeneic BMT should be fulfilled.

1. The patients should be treatment-requiring. Patients with Stage I and Stage II disease frequently have very slowly progressing myeloma. They are not candidates for BMT.
2. The patients should be under about 55 years of age. Older patients are poor candidates for BMT and have high transplant-related mortality.
3. The patients should be aware of all the possible complications and still willing to undergo BMT.
4. Patients responding to first-line treatment are probably good candidates. However, here autologous transplantation is also a clear option if the patient enters complete remission
5. Patients who are on first-line treatment but respond poorly seem to be good candidates.
6. Patients that have responded to first-line treatment but have later failed to do so and are up for second-line treatment are candidates. These patients are not in complete remission and will probably be poor candidates for autologous transplantation.
7. Patients who respond to later lines of treatment are clear candidates.
8. Some patients who do not respond to later lines of treatment and are in a relatively good condition may be candidates, although the response rate might be low. Occasional patients enter complete remission and some patients will be long-term survivors. With other treatment methods these patients usually fail.

These indications for performing an allogeneic BMT in multiple myeloma are not clear-cut. Improvement of the BMT technology is continuously worked on. Post-transplant infectious prophylaxis, treatment of veno-occlusive disease, etc., are improving and this might change indications in favour of allogeneic transplantation. Combinations with interferon or other new methods such as anti-IL-6 antibodies post-transplant may also change indications. It is also possible that new technology, such as gene transfer, could be a means to improve the immunogenicity of the multiple myeloma cell and thus 'vaccination' post-transplant might be an option. Recently, it has been shown that gene transfer can successfully be done to multiple myeloma cells.[68] Hopefully, these new methods will lead to the cure

of an increasing number of patients with multiple myeloma.

References

1. Thomas ED, Storb R, Clift RA et al. Bone-marrow transplantation. *New England Journal of Medicine* 1975; **292**: 832–43.
2. Thomas ED, Storb R, Clift RA et al. Bone-marrow transplantation. *New England Journal of Medicine* 1975; **292**: 895–902.
3. Weiden PL, Sullivan KM, Flournoy N et al. Anti-leukemic effect of chronic graft-versus-host disease: contribution to improved survival after allogeneic marrow transplantation. *New England Journal of Medicine* 1981; **304**: 1529–33.
4. Horowitz MM, Gale RP, Sondel PM et al. Graft-versus-leukemia reactions after bone marrow transplantation. *Blood* 1990; **75**: 555–72.
5. Gahrton G, Tura S, Ljungman P et al. for the European Group for Bone Marrow Transplantation. Allogeneic bone marrow transplantation in multiple myeloma. *New England Journal of Medicine* 1991; **325**: 1267–73.
6. Fefer A, Greenberg PD, Cheever MA et al. Treatment of multiple myeloma with chemoradiotherapy and identical twin bone marrow transplantation (abstract). *ASCO Proceedings* 1982; **1**: C731.
7. Osserman ED, Dire LB, Sherman WH et al. Identical twin marrow transplantation in multiple myeloma. *Acta Haematologica* 1982; **68**: 215–23.
8. McElwain TJ, Powles RL. High-dose intravenous melphalan for plasma cell leukaemia and myeloma. *Lancet* 1983; **1**: 822–4.
9. Selby P, McElwain TJ, Nanci AC et al. Multiple myeloma treated with high dose intravenous melphalan. *British Journal of Haematology* 1987; **66**: 55–62.
10. Highby DJ, Brass C, Fitzpatrick J, Henderson ES. Bone marrow transplantation in multiple myeloma: a case report with protein studies (abstract). *ASCO Proceedings* 1982; C747.
11. Ozer H, Han T, Nussbaum-Blumenson A et al. Allogeneic BMT and idiotypic monitoring in multiple myeloma. *AACR Abstract* 1984; **84**: 161.
12. Gahrton G, Ringdén O, Lönnqvist B et al. Bone marrow transplantation in three patients with multiple myeloma. *Acta Medica Scandinavica* 1986; **219**: 523–7.
13. Tura S, Cavo M, Baccarani M et al. Bone marrow transplantation in multiple myeloma. *Scandinavian Journal of Haematology* 1986; **36**: 176–9.
14. Tura S, Cavo M, Rosti G et al. Allogeneic bone marrow transplantation for multiple myeloma. *Bone Marrow Transplantation* 1989; **4**: 106–8.
15. Gallamini A, Buffa F, Baciagalupo A et al. Allogeneic bone marrow transplantation in multiple myeloma. *Acta Haematologica* 1987; **77**: 111–14.
16. Nikoskelainen J, Kokela K, Katka E et al.

Allogeneic bone marrow transplantation in multiple myeloma: a report of four cases. *Bone Marrow Transplantation* 1988; **3**: 495–500.

17. Samson D, Kanfer E, Taylor J et al. Allogeneic BMT for multiple myeloma: the Riverside experience (abstract). *British Journal of Haematology* 1990; **74**(suppl. 1): 53.

18. Gahrton G, Tura S, Flesch M et al. Bone marrow transplantation for multiple myeloma: Report from the European Cooperative Group for Bone Marrow Transplantation. *Blood* 1987; **69**: 1262–4.

19. Bensinger WI, Buckner CD, Clift RA et al. A phase I study of busulfan and cyclophosphamide in preparation for allogeneic marrow transplant for patients with multiple myeloma. *Journal of Clinical Oncology* 1992; **10**: 1492–7.

20. Ben-Yehuda A, Or R, Naparstek E et al. T-cell depleted bone marrow transplantation for plasma cell myeloma. *Blut* 1988; **56**: 229–31.

21. Copelan EA, Tutschka PJ. Marrow transplantation following busulphan and cyclophosphamide for multiple myeloma. *Bone Marrow Transplantation* 1988; **3**: 363–5.

22. Buckner CD, Fefer A, Bensinger WI et al. Marrow transplantation for malignant plasma cell disorders: summary of the Seattle experience. *European Journal of Haematology* 1989; **43**(suppl. 51): 186–90.

23. Tura S, Cavo M, Gobbi M et al. High-dose chemo-radiotherapy and allogeneic bone marrow transplantation in multiple myeloma. *European Journal of Haematology* 1989; **43**(suppl. 51): 191–5.

24. McCarthy D, Kanfer E, Samson D et al. BMT after conditioning with lomustine, cytosine arabinoside, etoposide and cyclophosphamide. *Bone Marrow Transplantation* 1990; **5**(suppl. 2): 124.

25. Gahrton G, Tura S, Ljungman P et al. Prognostic factors in allogeneic bone marrow transplantation for multiple myeloma. *Journal of Clinical Oncology* 1995; **13**: 1312–22.

26. Tutschka PJ, Copelan EA, Klein JP. Bone marrow transplantation for leukemia following a new busulfan and cyclophosphamide regimen. *Blood* 1987; **70**: 1382–8.

27. Ringdén O, Ruutu T, Remberger M et al. A randomized trial comparing busulfan with total body irradiation as conditioning in allogeneic marrow transplant recipients with leukemia: a report from the Nordic Bone Marrow Transplant Group. *Blood* 1994; **83**: 2723–30.

28. Gahrton G, Ljungman P, Tura S et al. Allogeneic bone marrow transplantation in multiple myeloma. In: Champlin RE, Gale RP eds *New strategies in bone marrow transplantation*. UCLA Symposia on molecular and cellular biology, new series. New York: Wiley-Liss 1991: 395–404.

29. Durie BGM, Gale RP, Horowitz MM. Allogeneic and twin transplants for multiple myeloma: an IBMTR analysis. *Multiple myeloma. From biology to therapy. Current concepts.* INSERM, Mulhouse, 24–26 October, 1994 (abstract).

30. National Board of Health and Welfare. The Cancer Registry. Cancer Incidence in Sweden 1988. Stockholm, Sweden, 1991.

31. Zwaan FE, Hermans J, Barrett AJ, Speck B. Bone marrow transplantation for acute nonlymphoblastic leukaemia: a survey of the European Group for Bone Marrow Transplantation (E.G.B.M.T.). *British Journal of Haematology* 1984; **56**: 645–53.

32. Ringdén O, Horowitz MM, Gale RP et al. Outcome after allogeneic bone marrow transplant for leukemia in older adults. *Journal of the American Medical Association* 1993; **270**: 57–60.

33. Bird JM, Russell NH, Samson D. Minimal residual disease after bone marrow transplantation for multiple myeloma: evidence for cure in long-term survivors. *Bone Marrow Transplantation* 1993; **12**: 651–4.

34. Hows JM, Bradley BA. The use of unrelated marrow donors for transplantation. *British Journal of Haematology* 1990; **76**: 1–6.

35. Gahrton G. Bone marrow transplantation with unrelated volunteer donors. Comments and Critique. *European Journal of Cancer* 1991; **27**: 1537–9.

36. James DCO. Organisation of a hospital bone marrow panel. In: Smit Sibinga CT, Cas PC, Opelz G eds *Transplantation and blood transfusion*. Boston: Martinus Nijhoff, 1984: 131–9.

37. van Rood JJ, for the Editorial Board of the BMWD. *Bone marrow donors world wide*, 6th ed. June, 1991.

38. Beatty PG. The world experience with unrelated donor transplants. *Bone Marrow Transplantation* 1991; **7**(suppl. 1): 54–8.

39. Howard MR, Hows JM, Gore M et al. Unrelated donor marrow transplantation between 1977 and 1978 at four centres in the United Kingdom. *Transplantation* 1990; **49**: 547–53.

40. Kernan NA, Bartsch G, Ash RC et al. Analysis of 462 transplantations from unrelated donors facilitated by the National Marrow Donor Program. *New England Journal of Medicine* 1993; **328**: 593–602.

41. Ringdén O, Remberger M, Persson U et al. Similar incidence of graft-versus-host disease using HLA-A, -B and -DR identical unrelated bone marrow donors as with HLA-identical siblings. *Bone Marrow Transplantation* 1995; **15**: 619–25.

42. McGlave PB, Beatty P, Ash R, Hows JM. Therapy for chronic myelogenous leukemia with unrelated bone marrow transplantation: Results in 102 cases. *Blood* 1990; **75**: 1728–32.

43. Beatty PG, Hansen JA, Longton GM et al. Marrow transplantation from HLA-matched unrelated donors for treatment of hematologic malignancies. *Transplantation* 1991; **51**: 443–7.

44. Powles RL, Clink HM, Spence D et al. Cyclosporin A to prevent graft-versus-host disease after allogeneic bone marrow transplantation. *Lancet* 1980; **1**: 327–9.

45. Storb R, Deeg HJ, Pepe M et al. Methotrexate and cyclosporin versus cyclosporin alone for prophylaxis of graft-versus-host disease in patients given HLA identical marrow grafts for leukemia: long

term follow-up of a controlled trial. *Blood* 1989; **73**: 1729–34.

46. Aschan J, Ringdén O, Sundberg B et al. Methotrexate combined with cyclosporin A decreases graft-versus-host disease, but increases leukemic relapse compared to monotherapy. *Bone Marrow Transplantation* 1991; **7**: 113–19.

47. Prentice HG, Blacklock HA, Janossy G et al. Use of anti-T cell monoclonal antibody OKT3 to prevent graft-versus-host disease in allogeneic bone marrow transplantation for acute leukaemia. *Lancet* 1982; **1**: 700–3.

48. Filipovitch AH, McGlave PB, Ramsay NK et al. Pretreatment of donor bone marrow with monoclonal antibody OKT3 for prevention of graft-versus-host disease in allogeneic histocompatible bone marrow transplantation. *Lancet* 1982; **1**: 1266–9.

49. Apperley JF, Jones L, Hale G et al. Bone marrow transplantation for patients with chronic myeloid leukaemia: T-cell depletion with Campath-1 reduced the incidence of graft-versus-host disease but may increase the risk of leukaemia relapse. *Bone Marrow Transplantation* 1986; **1**: 53–66.

50. Meyers JD. Infections in marrow recipients. In: Mandell GI, Douglas RG, Bennett JE eds *Principles and practice of infectious diseases* 2nd edn. New York: John Wiley & Sons, 1985: 1674–6.

51. Zaia JA. Viral infections associated with bone marrow transplantation. *Hematology Oncology Clinics of North America* 1990; **4**: 603–23.

52. Wingard JR. Prevention and treatment of bacterial and fungal infections. In: Forman SJ, Blume KG, Thomas ED eds *Bone marrow transplantation*. Oxford: Blackwell Scientific, 1994: 363–75.

53. Goodman I, Winston D, Greenfield R et al. A Controlled trial of fluconazole to prevent fungal infections in patients undergoing bone marrow transplantation. *New England Journal of Medicine* 1992; **326**: 905–7.

54. Ringdén O, Meunier F, Tollemar J et al. Efficacy of amphotericin B encapsulated in liposomes (AmBisome) in the treatment of invasive fungal infections in immunocompromised patients. *Journal of Antimicrobial Chemotherapy* 1991; **28**(suppl. B): 73–82.

55. Meyers JD, Reed ECC, Shepp DH et al. Acyclovir for prevention of cytomegalovirus infection and disease after allogeneic marrow transplantation. *New England Journal of Medicine* 1988; **318**: 70–5.

56. Goodrich J, Bowden R, Fisher L et al. Ganciclovir prophylaxis to prevent cytomegalovirus disease after allogeneic marrow transplant. *Annals of Internal Medicine* 1993; **118**: 173–8.

57. Winston D, Ho W, Bartoni K et al. Ganciclovir prophylaxis to prevent cytomegalovirus infection and disease in allogeneic bone marrow transplant recipients. *Annals of Internal Medicine* 1993; **118**: 179–84.

58. Ljungman P, Aschan H, Boström L et al. Early ganciclovir therapy of cytomegalovirus viremia diagnosed by polymerase chain reaction in allogeneic bone marrow transplant patients. In: Einhorn J, Nord C, Norrby S eds *18th International Congress of Chemotherapy, Stockholm, Sweden.* American Society for Microbiology, 1993: 637–8.

59. Emanuel D, Cunningham I, Jules EK et al. Cytomegalovirus pneumonia after bone marrow transplantation successfully treated with the combination of ganciclovir and high-dose intravenous immune globulin. *Annals of Internal Medicine* 1988; **109**(10): 777–82.

60. Ljungman P, Engelhard D, Link H et al. Treatment of interstitial pneumonitis due to cytomegalovirus with ganciclovir and intravenous immune globulin: experience of European Bone Marrow Transplant Group. *Clinical Infectious Diseases* 1992; **14**(4): 831·5.

61. Reed EC, Bowden R, Dandiker PS et al. Treatment of cytomegalovirus pneumonia with ganciclovir and intravenous cytomegalovirus immunoglobulin in patients with bone marrow transplants. *Annals of Internal Medicine* 1988; **109**(10): 782–8.

62. Schmidt GM, Kovacs A, Zaia JA et al. Ganciclovir/immunoglobulin combination therapy for the treatment of human cytomegalvirus-associated interstitial pneumonia in bone marrow allograft recipients. *Transplantation* 1988: **46**(6): 905–7.

63. Samson D. Maintenance with recombinant alphal2b interferon after allogeneic BMT for multiple myeloma – a pilot study. EBMT Chronic Leukaemia Working Party Multiple Myeloma Study Group.

64. Cunningham D, Milan S, Millar B et al. Strategies for the management of myeloma with conventional chemotherapy and high dose melphalan (HDM) and ABMT – possible roles for verapamil and maintenance interferon. *Proceedings of the III International Workshop on Multiple Myeloma*, 1991: 133.

65. Attal M, Huget F, Schlaifer D et al. Maintenance treatment with recombinant alpha interferon after autologous bone marrow transplantation for aggressive myeloma in first remission after conventional induction chemotherapy. *Bone Marrow Transplantation* 1991; **8**: 125–8.

66. Gratwohl A, Hermans J, van Biezen A et al. Allogeneic BMT for CML. The EBMT experience. Paper presented at the 20th Annual Meeting of the European Group for Bone Marrow Transplantation, Harrogate, UK, 13–17 March 1994. Abstract, p. 95.

67. Björkstrand B, Ljungman P, Bird JM et al. Double high-dose chemo-radiotherapy with autologous stem cell transplantation can induce molecular remissions in multiple myeloma. *Bone Marrow Transplantation* 1995; **15**: 367–71.

68. Björkstrand B, Dilber, MS, Li KJ et al. Retroviral-mediated gene transfer into human myeloma cells. *British Journal of Haematology* 1994; **88**: 325–31.

CHAPTER 17

Complications and supportive treatment

DIETRICH PEEST, HELMUTH DEICHER

Introduction	205	Infections	208
Hypercalcemia and bone destruction	205	Neurologic complications	209
Renal dysfunction	206	Hyperviscosity	209
Light-chain amyloidosis, light-chain deposition disease,		Cryoglobulinemia	210
and light- and heavy-chain deposition disease	207	Concludings remarks	210
Bone marrow dysfunction	208	References	210

Introduction

Most patients with multiple myeloma (MM) have a high tumor cell burden at diagnosis which exceeds 1.2×10^{12} cells/m^2 body surface in Stage III patients.[1] However, symptoms and signs presented by the patients are usually not the consequence of the tumor mass extent, but originate from individual complications induced by each myeloma plasma cell clone. The plasma cells produce and secrete monoclonal immunoglobulins, cytokines, and other biologically and physically active factors which interfere with bone metabolism, renal functions, hematopoiesis, immune mechanisms, and other organ systems. Different patterns of such complications contribute to the heterogeneity of MM patients in terms of symptoms, treatment strategy, and prognosis.

Hypercalcemia and bone destruction

Active bone destruction in MM caused by excessive osteoclast stimulation (*see* Chapter 5), may result in hypercalcemia due to increased release of calcium from the skeleton.[2] Hypercalcemia leads to impaired tubular concentrating ability due to decreased sodium reabsorption in Henle's loop, and to interference with ADH action. Consequences are increased natriuresis and loss of water. Calcium hyperfiltration causes further inhibition of sodium reabsorption leading to a progressive decrease of the extracellular fluid volume (ECV). This in turn stimulates the renin–aldosterone circuit which leads to hypokalemia because of the increased sodium level in the distal nephron. Glomerular filtration is further impaired by hypercalcemia-induced constriction of afferent glomerular arterioles. Furthermore, hypercalcemia causes transiently increased hydrogen ion secretion, e.g. in the stomach, thus producing alkalemia. Histologically, tubular injury of hypercalcemia is first seen in the distal loop of Henle and in collecting ducts;

Table 17.1 Hypercalcemia syndrome

Symptoms	Treatment
Mild form	
	Adequate chemotherapy
Serum calcium > 2.6 – < 3.5 mmol/l	Intravenous bisphosphonates (e.g. clodronate 300–600 mg/d1, 2, 3 or pamidronate 15–90 mg/dl)
Polyuria, polydipsia	Oral hydration with up to 3 l low-calcium mineral water
Constipation, nausea	Oral potassium substitution
Malaise, neuromuscular weakness	Prednisolone
Toxic form	
	Adequate chemotherapy
Serum calcium > 3.5 mmol/l	Forced diuresis using 3 – > 5l/day 0.15 M NaCl solution
Polyuria followed by dehydration, oliguria anuria, acute renal failure	Furosemide 40 – >200 mg/day, potassium substitution
Cardiac ventricular arrhythmia, shortening of QT interval	Intravenous bisphosphonates (*see above*)
Somnolence followed by coma	Prednisolone

progressive tubular necrosis and scarring finally leads to renal tubular acidosis and renal failure.

In the presence of renal failure and dehydration, impaired calcium excretion and enhanced calcium reabsorption in the proximal tubuli constitute additional mechanisms leading to hypercalcemia.[3] Clinically, oligo- or asymptomatic mild hypercalcemia (> 2.6 – < 3.5 mmol/l) can be distinguished from the severe, or toxic form requiring more intensive treatment (Table 17.1). Adequate early chemotherapy remains the most important treatment measure (*see* Chapter 11). Reconstitution of decreased ECV is mandatory; in elderly patients, monitoring of central venous pressure may be necessary to prevent heart failure. Since the advent of bisphosphonates which reliably inhibit bone resorption by osteoclasts through an as yet unknown mechanism involving osteoblasts and macrophages[4,5] as well as via their binding to bone mineral, other drugs such as corticosteroids (which inhibit intestinal calcium absorption and stimulate renal sodium reabsorption), calcitonin (which inhibits the tubular reabsorption of calcium but to a small extent), or mithramycin (formerly held to inhibit osteoclast proliferation), are no longer needed in the management of tumor-induced hypercalcemia in multiple myeloma.

Besides hypercalcemia, bone destruction also induces pathological fractures of vertebrae and other bones, as well as severe incapacitating bone pain (*see* Chapter 10). Merlini et al.[6] were first to show that prolonged intermittent intravenous/intramuscular administration of bisphosphonates in addition to the usual care of MM patients, significantly reduced bone pain, the appearance of new osteolytic lesions, and the incidence of pathological fractures. A recent prospective study[7] using oral clodronate reported similarly favorable results concerning decreased progression of osteolytic bone lesions and disappearance of bone pain. However, in view of the inconstant and poor resorption of orally administered bisphosphonates,[4] and because solid data concerning optimal dosage for long-term application of bisphosphonates are still lacking, further studies are needed to firmly establish this new and promising treatment principle.

Progress of osteolytic lesions, particularly of vertebral fractures, as well as of osteoporotic bone lesions threatening skeleton stability, can be treated efficiently by local radiation providing sufficient doses are applied. Based on long-term observation of patients with solitary myeloma, doses less than 25 Gy have been shown to be insufficient to prevent local progress, whereas excellent sustained local responses have been obtained with doses between 40 and 50 Gy,[8,9] thus demonstrating dose requirements for efficient local irradiation of MM. Pain relief and duration of pain control is significantly better in MM patients receiving local radiation during concomitant standard chemotherapy than in those irradiated in a period without systemic treatment.[10]

Renal dysfunction

Approximately 20–35 per cent of all MM patients present with impairment of kidney function at diagnosis.[11,12] These patients have a significantly worse prognosis than those without. Many factors may be involved in damaging the kidneys. Hypercalcemia (*see above*), dehydration, excess of uric acid, infections, nephrotoxic drugs (e.g. diclofenac, antibiotics) are unspecifically harmful. Amyloid light-chain (AL), light-chain deposition disease (LCDD), and light- and heavy-chain deposition disease (LHCDD) are characteristic complications of monoclonal gammopathies such as MM. Radiographic contrast agents may induce acute renal failure, particularly in dehydrated MM patients.

The morphology of the myeloma kidney is characterized by eosinophilic or polychromatophilic, sometimes laminated casts in the distal and collecting tubules.[13] The distal tubules are often dilated and present an atrophic, flattened epithelium. Multinucleated syncytial epithelial cells and fibrotic changes of the interstitium may be found. The casts contain immunoglobulin light chains, complete immunoglobulin molecules, or albumin.[14,15] Glomeruli are usually affected only if amyloidosis is present. In nephrocalcinosis calcium is deposited in the epithelial cells of the tubules.[13]

Renal failure in MM is usually associated with the presence of Bence Jones proteinuria indicating a central role of immunoglobulin light chains in kidney injury. Healthy persons produce an excess of free polyclonal immunoglobulin light chains which has been calculated to be filtered through the glomeruli at a rate of about 5 mg/kg body weight/day.[16] While only a small fraction (0.04 mg/kg body weight/day) will be excreted in the urine, most of the immunoglobulin light chains will be reabsorbed in the proximal tubules and catabolized by the epithelial cells. In monoclonal gammopathies with pronounced production of free monoclonal immunoglobulin light chains, the normal capacity of reabsorption and catabolism is often not sufficient to eliminate the protein. Consequently, Bence Jones proteinuria will develop and casts of monoclonal protein and toxic catabolism products may accumulate in the epithelial cells with the outcome of renal damage.[17] Although the mechanisms are not fully understood, individual physicochemical properties of the monoclonal immunoglobulin light chains are discussed to explain the various degrees of renal involvement. Using an animal model it is possible to test the individual potential of a given human Bence Jones protein to induce deposits and renal damage.[15]

Early clinical signs of myeloma kidney are impairment of tubular renal functions such as capability to concentrate and to acidify the urine. An adult Fanconi syndrome with specific reabsorption defects of glucose, amino acids, phosphate, and electrolytes can be observed in some patients.[18] Later stages are characterized by the loss of nephrons with the consequence of progressive reduction of glomerular filtration rates. A non-selective

proteinuria sometimes associated with a nephrotic syndrome is suggestive of amyloidosis.

Hydration and administration of sufficient chemotherapy is essential for MM patients with renal failure. In a prospective trial it has been shown that fluid intake of at least 3 l/day can improve or even reverse renal failure in many patients.[19] Furthermore, there are patients with oliguric renal failure whose renal function improves after an intermittent period of dialysis and appropriate chemotherapy, diminishing the tumor clone thus decreasing Bence Jones protein production. In studies with small numbers of patients, it has been shown that plasmapheresis combined with dialysis and chemotherapy successfully reverses acute progressing renal failure in more MM patients than dialysis and chemotherapy alone.[20,21] It is well worth while offering chronic hemodialysis to MM patients with terminal renal failure provided that chemotherapy successfully controlling the tumor is available.[22,23]

Light-chain amyloidosis, light-chain deposition disease, and light- and heavy-chain deposition disease

Light-chain amyloidosis (AL), light-chain deposition disease (LCDD), and light- and heavy-chain deposition disease (LHCDD) are pathologically similar complications associated with monoclonal gammopathies such as MM, and may be summarized under the term monoclonal immunoglobulin deposition diseases (MIDD).[24] Differences among MIDDs can only be defined by histochemical or immunohistochemical methods. In AL, monoclonal immunoglobulin light chains or light-chain fragments are linked together forming amyloid fibrils with ß-pleated sheet formation, which can be stained with Congo red. Additionally an amyloid P-component is present. These characteristics are missing from the amorphous deposits of LCDD and LHCDD.

The amyloidogenic capacity of an individual monoclonal immunoglobulin light chain is determined by physicochemical properties of the molecule. Sequence analyses have failed to identify a distinct chemical structure responsible for amyloid formation. Almost all of the known light-chain V-region subgroups have been found in AL. Proteolysis of light chains may be involved,[25] because proteolytic cleavage of certain Bence Jones proteins resulted in the formation of amyloid fibrils *in vitro*. Tissue-specific proteolytic enzymes may then determine the pattern of organ involvement in individual patients.

The deposits which develop in less than 10 per cent of MM patients can affect nearly all organs (e.g. kidney, heart, liver, nervous system, gut, skin, vessels, etc.).[26] Clinical signs include weakness, weight loss, edema, dyspnea, parestesias, carpal tunnel syndrome, macroglossia, liver enlargement, or slight splenomegaly. Skin bleeding may occur due to fragile vessels and a deficiency of clotting factor X as a result of factor X binding to amyloid.

Examination of biopsies of clinically affected organs will allow the diagnosis in most cases.[24] In addition to standard staining procedures (Congo red staining, polarization microscopy) immunohistochemical analyses must be applied. Rectal biopsies which have to include submucosal tissue have a sensitivity of about 80 per cent while bone marrow or gingival biopsies are less valuable (sensitivity < 50 per cent). The sensitivity of subcutaneous fat aspiration was found to be 50–70 per cent, provided suitable techniques were used.[27–29]

Chemotherapy as necessary for MM is also used for treatment of MIDD, although the results are unsatisfactory and the prognosis remains poor. Chemotherapy may stop or slow progression of deposits; however, an apparent impact on survival has not been established so far.[30] Other drugs such as colchicine have been applied without convincing success.[31,32]

Bone marrow dysfunction

Hematopoiesis is frequently impaired in MM. Twenty-five per cent of Stage II and 62 per cent of Stage III patients (classification according to tumor cell mass[1]) present with an anemia of < 10 g hemoglobin/dl at diagnosis (unpublished results of GMTG trial MM02[33]), indicating a correlation between degree of anemia and tumor cell load. The mechanisms inducing anemia in MM are unknown. Renal failure and damage of hematopoietic precursors by radiation or numerous chemotherapy cycles in advanced-stage patients may aggravate anemia. In newly diagnosed MM patients anemia will rapidly improve after several cycles of chemotherapy successfully reducing the tumor load. However, red blood cell transfusions are frequently needed in advanced MM patients to achieve a better quality of life. A hemoglobin level > 10 g/dl should be maintained particularly in elderly patients and those with cardiovascular diseases. In some MM patients, including those without renal failure, anemia can be sufficiently corrected by the application of recombinant erythropoietin at a dose of 150–200 IU/kg body weight 3 times per week.[34]

A mild reduction of granulocyte and platelet counts is often found in MM patients at diagnosis; severe reductions are rare. Patients with platelet counts more than 150000/µl have a significantly better prognosis than those with less thrombocytes.[11,35] For treatment of chemotherapy-induced granulocytopenia *see below*.

An increased incidence of acute leukemias of non-lymphocytic type has been reported for MM patients. Such leukemias often develop after a period of myelodysplasia with pancytopenia in the peripheral blood. In a Canadian MM trial the risk of converting into leukemia has been determined with 17 per cent at 50 months treatment with alkylating agents, more than 200 times higher than in the normal population.[36] Although at a much lower incidence (3 per cent after 5 years), another group demonstrated a significant association to the duration of melphalan treatment not found in the control group treated with cyclophosphamide.[37] Since the occurrence of leukemias has also been reported in untreated patients with monoclonal gammopathies, it has been postulated that the development of leukemia may be a variant of the natural course of MM, and might be explained by a defect of early hematopoietic stem cells in MM.[38] Chemotherapy and radiation probably escalate this intrinsic risk. The prognosis of MM patients developing secondary leukemias is poor. Induction of remissions with standard regimens for leukemia should be attempted, though elevated toxicity and complication rates have to be expected due to pretreatment induced reduction of stem cell reserve, age of the patients, and myeloma related impairment of different organ functions.

Infections

Many MM patients suffer from frequent bacterial infections[39] and death rates due to infections in the range of 15 per cent to 50 per cent of all MM cases have been observed.[40,41] Tumor-induced immunosuppression aggravated by chemotherapy and/or irradiation forms the basis for a complex immunodeficiency syndrome in these patients. Besides impaired cellular defence due to neutropenia, reduced T-lymphocyte functions,[42] and natural killer cell activity,[43] the most prominent defect concerns the production of polyclonal immunoglobulins.[39] Nearly all MM patients show reduced concentrations of polyclonal serum immunoglobulins, and impaired antibody responses to vaccination with different antigens have been demonstrated.[44–46] Pre-B and B-lymphocyte compartments are reduced in MM.[47] Furthermore, monocytes from MM patients inhibit *in vitro* mitogen-induced polyclonal immunoglobulin production when tested in co-cultures with normal or MM B lymphocytes.[48,49] Soluble suppressor activity could be measured in supernatants of cultured mononuclear cells isolated from the bone marrow of MM patients.[50] Such supernatants inhibit the proliferation of normal polyclonal B lymphocytes while the differentiation of the lymphoblastoid B-cell line CESS into immunoglobulin secreting cells remains undisturbed. These observations in men are compatible with studies in mice showing that plasmacytoma cells produce a soluble factor (PC factor) which stimulates macrophages to produce a second factor (PIMS), capable of inhibiting primary antibody response of spleen B lymphocytes.[51]

The most common infections are those of the respiratory tract, i.e. bronchitis and pneumonia; infections of the urinary tract, septicemia, and skin infections are less common.[39,46,52] In non-neutropenic MM patients with stable plateau phase, predominantly infections due to *Streptococcus pneumoniae* or *Haemophilus influenzae* were reported, also seen in other groups of patients with humoral immunodeficiency. However, during periods of neutropenia following intensive chemotherapy and in advanced disease infections due to *Staphylococcus aureus* and Gram-negative bacteria were predominantly observed in MM patients.

MM patients presenting with bacterial infection should be handled as other immunodeficient patients. Early complete diagnosis (e.g. bacterial cultures from blood and other specimens, chest X-rays, bronchoscopy with BAL) followed by intensive bactericidal antibiotic therapy considering the predominant spectrum of bacteria in these patients (*see above*) is mandatory.

Hematopoietic growth factors are available to reduce the period of chemotherapy-induced neutropenia in order to prevent or support treatment of infections. Although, granulocyte–macrophage colony-stimulating factor (GM-CSF) as well as granulocyte colony-stimulating factor (G-CSF) can stimulate the growths of myeloma cells *in vitro*,[53] this effect is probably not relevant *in vivo*.[54] Such cytokines may therefore be applied to MM patients and seem to be of benefit (personal observations). However, reliable data which validate the safety and efficiency of GM-CSF or G-CSF in MM are still lacking.

Immunoglobulin preparations for i.v. application have been tested in controlled trials to prevent infections in multiple myeloma,[46,52] where doses between 10 g every 3–4 weeks and 0.4 g/kg bodyweight monthly have been administered. A significant reduction of both frequency and severity of infections as compared to control groups could be observed. Patients not responding to vaccination with pneumococcal vaccine, i.e those with a pronounced immunodeficiency, have profited most from i.v. immunoglobulin treatment.[46] This corresponds to

the observation that patients with low or negative lipid A antibody titers have significantly more infections than those with titers within the normal range.[55] Therefore, MM patients with a relevant defect in the antibody response to bacterial antigens, i.e. patients clinically identified by presenting with recurrent infections, should be regularly substituted with i.v. immunoglobulins.

Neurologic complications

Radicular pain, typically aggravated by coughing, is an early sign of spinal cord or nerve root compression by fractured vertebrae or extramedular parts of the MM tumors. Sensoric or motoric defects, paraplegia, and loss of sphincter control are late symptoms. Early diagnosis is needed to prevent permanent damage. Magnetic resonance imaging is presently the most suitable method because it can detect paravertebral tumors and osteolytic destruction before conventional X-rays are positive.[56] Usually magnetic resonance imaging can replace a myelogram, which may only be needed in a few exceptional cases for diagnosing imminent spinal cord compression.

Since MM tumors are very sensitive to X-rays, radiation with a local dose of 40 Gy is the therapy of choice. This should be applied even before symptoms occur if diagnostic methods indicate a risk of neurologic complications. High doses of steroids, e.g. 16 mg dexamethasone/day p.o., should be given as a supplement in patients with a spinal cord compression syndrome. If the diagnosis of local MM growth is certain, decompression laminectomy is rarely necessary.

Polyneuropathy is a rare complication in MM.[57–59] It may be induced by amyloidosis or hyperviscosity. Polyneuropathy is also part of an uncompletely understood MM associated syndrome characterized by polyneuropathy, organomegaly, endocrinopathy, monoclonal gammopathy, and skin changes (POEMS syndrome).[60] Furthermore, individual monoclonal proteins may have an affinity to neural constituents, although most of them belong to the IgM class.[61–63] Specific therapy is lacking. Improvement of polyneuropathic symptoms have been observed after tumor reducing chemotherapy in many patients. However, therapy remains disappointing in others.

Hyperviscosity

A hyperviscosity syndrome is a rare complication of MM, occurring more often in immunocytoma

Table 17.2 Hyperviscosity syndrome

Symptoms	Treatment

Slowly developing
  vertigo
  lethargy
  fatigue
  blurred vision
  angina pectoris/intestinal angina
  bleeding tendency (platelet 'coatover' effect)
  somnolence, coma
Monoclonal gammopathy (IgM, IgA, IgG3)
Increased plasma viscosity
Fundus paraproteinemicus
	Adequate chemotherapy
	Plasmapheresis

(Waldenström's macroglobulinemia) due to the tendency of monoclonal IgM to form polymers.[64–66] IgA and IgG3 myeloma proteins may also polymerize and thus cause clinical symptoms associated with pathologically increased blood viscosity. However, no straightforward relationship between plasma, or whole-blood viscosity, serum protein concentration, and clinical symptoms has been established for monoclonal gammopathies,[67] so that individual factors such as age, presence of atherosclerotic blood vessel and/or organ lesions, cardiac performance, and anemia may all contribute. Blood being a non-Newtonian fluid, its dynamic viscosity depends to a large extend on the shearing velocity which increases with decreasing blood vessel diameter and reaches its highest value within capillaries. Here, single red blood cells suspended in blood plasma travel through the vessel lumen, so that increased plasma viscosity may impair local oxygen supply. Different methods are being used to measure plasma viscosity, precluding the comparison of results from one laboratory to the next. Hence the term hyperviscosity syndrome remains a comparatively ill-defined clinical entity (Table 17.2), and other causes of the observed symptoms must be excluded in every case. Specific treatment measures are not at hand; adequate chemotherapy will result in decreased M-protein levels and thus alleviate symptoms. In severe cases, plasmapheresis is indicated,[68] but will only bring about short-lived relief unless accompanied by effective chemotherapy.

Cryoglobulinemia

Cryoglobulinemia is a rare complication in MM and commonly induces acrocyanosis, Raynaud's phenomenon or purpura.[69,70] The M-components of these patients bind other immunoglobulin molecules upon cooling resulting in reversible protein precipitation, gelification, increase of serum viscosity, or complement activation by immune complexes. Cryoglobulinemia Type I (containing only the M-component in the precipitate) and Type II (containing the M-component and polyclonal immunoglobulins) are associated with monoclonal gammopathies, while Type III (containing polyclonal components) is not.

Concluding remarks

The quality of life and prognosis of MM patients are significantly influenced by tumor-induced complications. Therefore, a complete diagnosis and an individual strategy for cytoreduction and for treatment and/or prevention of complications have to be established in each single MM patient. Such a strategy should include chemotherapy, radiation, surgery, drug therapy for pain control, blood support, antibiotics, i.v. immunoglobulins, fluid intake, bisphosphonates, hemodialysis, plasmapheresis, hematopoietic growth factors, and other supportive measures.

References

1. Salmon SE, Wampler SB. Multiple myeloma: quantitative staging and assessment of response with a programmable pocket calculator. *Blood* 1977; **49** 379–89.
2. Mundy GR. Hypercalcemia of malignancy revisited. *Journal of Clinical Investigation* 1988; **82**: 1–6.
3. Hosking DJ. Assessment of renal and skeletal components of hypercalcemia. *Calcification Tissue International* 1990; **46**: (suppl.) S11–S19.
4. Fleisch H. Bisphosphonates. Pharmacology and use in the treatment of tumor-induced hypercalcaemic

and metastatic bone disease. *Drugs* 1991; **42**: 919–44.

5. Sahni M, Guenther HL, Fleisch H et al. Bisphosphonates act on rat bone resorption through the mediation of osteoblasts. *Journal of Clinical Investigation* 1993; **91**: 2004–11.

6. Merlini G, Parrinello GA, Piccinini L et al. Long-term effects of parenteral dichloromethylene bisphosphonate (CL2MBP) on bone disease of myeloma patients treated with chemotherapy. *Hematology and Oncology* 1990; **8**: 23–30.

7. Lahtinen R, Laakso M, Palva I et al. Randomised, placebo-controlled multicentre trial of clodronate in multiple myeloma. Finnish Leukaemia Group. *Lancet* 1992; **340**: 1049–52.

8. Holland J, Trenkner DA, Wasserman TH, Fineberg B. Plasmacytoma. Treatment results and conversion to myeloma. *Cancer* 1992; **69**: 1513–17.

9. Bataille R. Localized plasmacytomas. In: Salmon SE ed. *Clinics in haematology* vol.11, no.1. London: W.B. Saunders, 1982: 113–22.

10. Adamietz IA, Schöber C, Schulte RW et al. Palliative radiotherapy in plasma cell myeloma. *Radiotherapy and Oncology* 1991; **20**: 111–16.

11. Peest D, Coldewey R, Deicher H et al. Prognostic value of clinical, laboratory, and histological characteristics in multiple myeloma: improved definition of risk groups. *European Journal of Cancer* 1993; **29A**: 978–83.

12. Durie BGM. Staging and kinetics of multiple myeloma. In: Salmon, SE ed. *Clinics in haematology*, vol.11, no.1. London: W.B. Saunders, 1982: 3–18.

13. Zlotnick A, Rosenmann E. Renal pathologic findings associated with monoclonal gammopathies. *Archives of Internal Medicine* 1975; **135**: 40–5.

14. Levi DF, Williams RC Jr., Lindstrom FD. Immunofluorescent studies of the myeloma kidney with special reference to light chain disease. *American Journal of Medicine* 1968; **44**: 922–33.

15. Solomon A, Weiss DT, Kattine AA. Nephrotoxic potential of Bence Jones proteins. *New England Journal of Medicine* 1991; **324**: 1845–51.

16. Waldmann TA, Strober W, Mogielnicki RP. The renal handling of low molecular weight proteins. II. Disorder of serum protein catabolism with tubular proteinuria, the nephrotic syndrom, or uremia. *Journal of Clinical Investigation* 1972; **51**: 2162–73.

17. Clyne DH, Pollak VE. Renal handling and pathophysiology of Bence Jones proteins. *Contributions to Nephrology* 1981; **24**: 78–87.

18. Maldonado JE, Velosa JA, Kyle RA et al. Fanconi syndrome in adults: A manifestation of a latent form of myeloma. *American Journal of Medicine* 1975; **58**: 354–64.

19. MRC Working Party on Leukaemia in Adults. Analysis and management of renal failure in fourth MRC myelomatosis trial. *British Medical Journal* 1984; **288**: 1411–16.

20. Feest TG, Burge PS, Cohen SL. Successful treatment of myeloma kidney by diuresis and plasmaphoresis. *British Medical Journal* 1976; **1**: 503–4.

21. Pasquali S, Cagnoli L, Rovinetti C et al. Plasma exchange therapy in rapidly progressive renal failure due to multiple myeloma. *International Journal of Artificial Organs* 1985; **8** (suppl 2): 27–30.

22. Coward RA, Mallick NP, Delamore IW. Should patients with renal failure associated with myeloma be dialysed? *British Medical Journal* 1983; **287**: 1575–8.

23. Johnson WJ, Kyle RA, Pineda AA et al. Treatment of renal failure associated with multiple myeloma. Plasmapheresis, hemodialysis, and chemotherapy. *Archives of Internal Medicine* 1990; **150**: 863–9.

24. Buxbaum J. Mechanisms of disease: monoclonal immunoglobulin deposition. Amyloidosis, light chain deposition disease, and light and heavy chain deposition disease. *Hematology and Oncology Clinics of North America* 1992; **6**: 323–46.

25. Linke RP, Zucker-Franklin D, Franklin EC. Morphologic, chemical, and immunologic studies of amyloid-like fibrils formed from Bence Jones proteins by proteolysis. *Journal of Immunology* 1973; **111**: 10–23.

26. Kyle RA. Amyloidosis. In: Salmon SE ed. *Clinics in hematology*, vol.11, no.1, *Myeloma and related disorders*. London: W.B. Saunders, 1982: 151–80.

27. Gertz MA, Li CY, Shirahama T, Kyle RA. Utility of subcutaneous fat aspiration for the diagnosis of systemic amyloidosis (immunoglobulin light chain). *Archives of Internal Medicine* 1988; **148**: 929–33.

28. Breedveld FC, Markusse HM, MacFarlane JD. Subcutaneous fat biopsy in the diagnosis of amyloidosis secondary to chronic arthritis. *Clinical and Experimental Rheumatology* 1989; **7**: 407–10.

29. Duston MA, Skinner M, Meenan RF, Cohen AS. Sensitivity, specificity, and predictive value of abdominal fat aspiration for the diagnosis of amyloidosis. *Arthritis Rheumatism* 1989; **32**: 82–5.

30. Kyle RA, Greipp PR. Primary systemic amyloidosis: comparison of melphalan and prednisone versus placebo. *Blood* 1978; **52**: 818–27.

31. Kyle RA, Greipp PR, Garton JP, Gertz MA. Primary systemic amyloidosis. Comparison of melphalan/prednisone versus colchicin. *American Journal of Medicine* 1985; **79**: 708–16.

32. Schattner A, Varon D, Green L et al. Primary amyloidosis with unusual bone involvement: Reversiblity with melphalan, prednisone, and colchicine. *American Journal of Medicine* 1989; **86**: 347–8.

33. Peest D, Deicher H, Coldewey R et al. A comparison of polychemotherapy and melphalan/prednisone for primary remission induction, and interferon-alpha for maintenance treatment, in multiple myeloma. A prospective trial of the German Myeloma Treatment Group. *European Journal of Cancer* 1994; **31A**: 146–51.

34. Ludwig H, Fritz E, Kotzmann H et al. Erythropoietin treatment of anemia associated with multiple myeloma. *New England Journal of Medicine* 1990; **322**: 1693–9.

35. Cavo M, Galieni P, Zuffa E et al. Prognostic variables and clinical staging in multiple myeloma. *Blood* 1989; **74**: 1774–80.

36. Bergsagel DE, Bailey AJ, Langley GR et al. The chemotherapy of plasma-cell myeloma and the incidence of acute leukemia. *New England Journal of Medicine* 1979; **301**: 743–8.

37. Cuzick J, Erskine S, Edelman D, Galton DA. A comparison of the incidence of the myelodysplastic syndrome and acute myeloid leukaemia following melphalan and cyclophosphamide treatment for myelomatosis. A report to the Medical Research Council's working party on leukaemia in adults. *British Journal of Cancer* 1987; **55**: 523–9.

38. Bergsagel DE. Chemotherapy of myeloma: drug combinations versus single agents, an overview, and comments on acute leukemia in myeloma. *Hematology and Oncology* 1988; **6**: 159–66.

39. Jacobson DR, Zolla-Pazner S. Immunosuppression and infection in Multiple Myeloma. *Seminars in Oncology* 1986; **13**: 282–90.

40. Peest D, Coldewey R, Deicher H. Overall vs. tumor-related survival in multiple myeloma. *European Journal of Cancer* 1991; **27**: 672–2.

41. Twomey JJ, Houston MB. Infections complicating multiple myeloma and chronic lymphocytic leukemia. *Archives of Internal Medicine* 1973; **132**: 562–5.

42. Massaia M, Dianzani U, Bianchi A et al. Defective generation of alloreactive cytotoxic T lymphocytes (CTL) in human monoclonal gammopathies. *Clinical and Experimental Immunology* 1988; **73**: 214–18.

43. Österborg A, Nilsson B, Björkholm M et al. Natural killer cell activity in monoclonal gammopathies: relation to disease activity. *European Journal of Haemotology* 1990; **45**: 153–7.

44. Fahey JL, Scoggins R, Utz JP, Szwed CF. Infection, antibody response and gamma globulin components in multiple myeloma and macroglobulinemia. *American Journal of Medicine* 1963; **35**: 698–707.

45. Krull P, Deicher H. Primäre Antikörperbildung bei Patienten mit malignen lymphoretikulären Systemerkrankungen und metastasierenden Tumoren. *Zeitschift für Immunitätsforschung* 1973; **145**: 70–7.

46. Chapel HM, Lee M, Hargreaves R et al. Randomised trial of intravenous immunoglobulin as prophylaxis against infection in plateau-phase multiple myeloma. The UK Group for Immunoglobulin Replacement Therapy in Multiple Myeloma. *Lancet* 1994; **343**: 1059–63.

47. Duperray C, Bataille R, Boiron JM et al. No expansion of the pre-B and B-cell compartments in the bone marrow of patients with multiple myeloma. *Cancer Research.* 1991; **51**: 3224–8.

48. Broder S, Humphrey R, Durm M et al. Impaired synthesis of polyclonal (non-paraprotein) immunoglobulins by circulating lymphocytes from patients with multiple myeloma. *New England Journal of Medicine* 1975; **293**: 887–92.

49. Peest D, Brunkhorst U, Schedel I, Deicher H. *In vitro* immunoglobulin production by peripheral blood mononuclear cells from multiple myeloma patients and patients with benign monoclonal gammopathy. Regulation by cell subsets. *Scandinavian Journal of Immunology* 1984; **19**: 149–57.

50. Peest D, Hölscher R, Weber R et al. Suppression of polyclonal B cell proliferation mediated by supernatants from human myeloma bone marrow cell cultures. *Clinical Experimental Immunology* 1989; **75**: 252–7.

51. Berman JE, Zolla-Pazner S. Control of B cell proliferation: Inhibition of responses to B cell mitogens induced by plasma cell tumors. *Journal of Immunology* 1985; **134**: 2872–8.

52. Schedel I. Application of immunoglobulin preparations in multiple myeloma. In: Nydegger UE, Morell A eds. *Clinical use of intravenous immunoglobulins.* London: Academic Press, 1986: 123–32.

53. Klein B, Bataille R. Cytokine network in human multiple myeloma. *Hematology Oncology Clinics of North America* 1992; **6**: 273–84.

54. Barlogie B, Jagannath S, Dixon DO et al. High-dose melphalan and granulocyte-macrophage colony-stimulating factor for refractory multiple myeloma. *Blood* 1990; **76**: 677–80.

55. Stoll C, Schedel I, Peest D. Serum antibodies against common antigens of bacterial lipopolysaccharides in healthy adults and in patients with multiple myeloma. *Infection* 1985; **13**: 115–19.

56. Ludwig H, Frühwald F, Tscholakoff D et al. Magnetic resonance imaging of the spine in multiple myeloma. *Lancet* 1987; **II**: 364–6.

57. Kelly JJ, Kyle RA, Miles JM, O'Brien PC, Dyck PJ. The spectrum of peripheral neuropathy in myeloma. *Neurology* 1981; **31**: 24–31.

58. Read DJ, Vanhegan RI, Matthews WB. Peripheral neuropathy and benign IgG paraproteinaemia. *Journal of Neurology, Neurosurgery and Psychiatry* 1978; **41**: 215–9.

59. Delauche MC, Clauvel JP, Seligmann M. Peripheral neuropathy and plasma cell neoplasias: a report of 10 cases. *British Journal of Haematology* 1981; **48**: 383–92.

60. Resnick D, Greenway GD, Bardwick PA et al. Plasma-cell dyscrasia with polyneuropathy, organomegaly, endocrinopathy, M-protein, and skin changes: the POEMS syndrome. *Radiology* 1981; **140**: 17–22.

61. Dellagi K, Dupouney P, Brouet JC et al. Waldenström's macroglobulinemia and peripheral neuropathy: a clinical and immunological study of 25 patients. *Blood* 1983; **62**: 280–5.

62. Wehmeier U, Rilke H, Patzold U et al. Crossreacting idiotypes of kappa-monoclonal immunoglobulins M in sera of patients with Waldenström's macroglobulinemia. *Immunobiology* 1987; **176**: 144–53.

63. Hoppe U, Dräger HS, Patzold U et al. Polyneuropathy in Waldenström's macroglobulinaemia. Passive transfer from man to mouse. *Acta Neurologica Scandinavica* 1987; **75**: 112–6.

64. Fahey JL, Barth WF, Solomon A. Serum hyperviscosity syndrome. *Journal of the American Medical Association* 1965; **192**: 464–7.

65. McGrath, MA, Penny R. Paraproteinemia: blood

hyperviscosity and clinical manifestations. *Journal of Clinical Investigation* 1976; **58**: 1155–62.

66. Preston FE, Cooke KB, Foster ME et al. Myelomatosis and the hyperviscosity syndrome. *British Journal of Haematology* 1978; **38**: 517–30.

67. Crawford J, Cox EB, Cohen HJ. Evaluation of hyperviscosity in monoclonal gammopathies. *American Journal of Medicine* 1985; **79**: 13–22.

68. Isbister JP, Biggs JC, Penny R. Experience with large volume plasmapheresis in malignant paraproteinemia and immune disorders. *Australian and New Zealand Journal of Medicine* 1978; **8**: 154–64.

69. Osterland CK, Espinoza LR. Biological properties of myeloma proteins. *Archives of Internal Medicine* 1975; **135**: 32–6.

70. Brouet JC, Clauvel P-C, Danon F et al. Biological and clinical significance of cryoglobulins. *American Journal of Medicine* 1974; **57**: 775–88.

Index

adhesion *see* cell adhesion molecules (CAMs)
Ag-dependent/independent phases of B cells, 23
agarose gel electrophoresis, detection and quantification
    of M-component, 45
alkylating agents
    with/without prednis(ol)one, 112–13
    see also chemotherapy
amyloidosis, 85, 101, 207–8
anemia, 98
    erythropoietin therapy (*table*), 173
aneuploidies, 57, 92
antibody diversity, normal generation, 39
antigens, chronic antigen stimulation (CAS), 17–18
apoptosis, loss, *bcl–2* expression, 91
asbestos, 17
atomic bomb survivors, excess absolute risk (EAR), all
    leukemias and MM, 15
autoimmune diseases, 18
autologous transplantation *see* bone marrow
    transplantation

ß$_2$-microglobulin, serum marker, 102, 105
B cells
    Ag-dependent/independent phases, 23
    B1 and B2, CD5 molecule, 24
    clonal B cell excess (CBE), 41
    idiotype-bearing, 40–2
        Ig surface expression, 42–3
    late peripheral, 25–6
    malignant clone, hypothesis for growth and
        progression of MM, 32
    and plasma cells, 23
    and surface markers, 23
bacterial infections, 208–9
bcl–1 oncogene, t(11;14)(q13:q32) translocation, 59
bcl–2 oncogene, 65, 67
bcl–6 locus, 65, 67
Bence Jones, Henry, 4–6
Bence Jones proteinuria
    historical aspects, 8–9, 45
    renal dysfunction, 85, 100

see also monoclonal gammopathies of undetermined
    significance (MGUS)
benzene, 17
biclonal gammopathies, M-component isotypes, 40
bone changes in MM, 51–4
    bone remodeling, 52
    cytokines, 52–3
    destruction, hypercalcemia, 204–5
    osteoclastic resorption, 51–2
bone gla-protein (BGP), uncoupled bone remodeling, 52
bone marrow
    cell differentiation, 29–31
    dysfunction, 208
    stromal cells, 29–31
        adhesion molecules (*table*), 30
bone marrow transplantation (allogeneic BMT), 189–91,
    196
    complications, 200–1
    conditioning treatment, 195
    graft-versus-host disease, 200
    HLA-matched non-sibling donors, 199–200
    HLA-matched sibling donors, 197–9
    indications, 201–2
    response, 196–7
bone marrow transplantation (autologous, ABMT),
    182–93
    clinical results, unpurged/purged, 185–7
    graft purging, 185
    interferons following, 154, 188
    peripheral blood stem cells (PBSC), 187
bone marrow transplantation (syngeneic BMT),
    195–6
busulfan, 184

C$_1$ esterase-inhibitor deficiency, 85
C-*myc* oncogene, 65–6
C-reactive protein, serum marker, 92, 103
capillary leak syndrome, 85
carboquone, 122
case-control studies, causes of plasma cell neoplasms, 16
CD5, B1 and B2 cells, 24

CD40, expression, 93–4
CD56, expression, 87
cell adhesion molecules (CAMs), 89
  bone marrow stromal cells (*table*), 30
  CD56 expression, 87
  and ECM proteins, 29
chemotherapy, 108–22
  clinical studies, 114, 116
  combination, 113–19, 182
    ABCM, 115
    tables, 109, 114
    VBMCP (M–2) protocol, 113
    VMCP and VMCP/VBAP, 113–14
  compared with interferons, 152
  continuous infusion, 118
  drug resistance, 120–1, 139–40
  high-dose therapy, 119–20, 182–4, 188–9
    prognostic factors, 189
  indications and aims for treatment, 108–12
  induction therapy, success rate, 130–1
  intermediate dose melphalan, 120
  maintenance therapy, 119
  new drugs, 121–2
  objective response (*table*), 110–12
  and PBSC, 187
  randomized studies (*table*), 114
  refractory disease, 133–6
    drug resistance, 139–40
    myeloablative therapy, 136–7
    systemic radiotherapy, 137–9
  regimens (*table*), 109
  single alkylating agents, with/without
    prednis(ol)one, 112–13
  steroids, 118
  VAD, 117–18
chemotherapy regimens, 109
chlorodeoxyadenosine (CDA), 121
chromosome abnormalities in MM, 55–63
  aneuploidies, 57
  banding studies, 56–7
    published (*table*), 56
  chromosome–1 deletions, 58, 59
  clinical association to karyotypes, 60
  clonal evolution, 59–60
  donor chromosomes for 14q+ markers, 59
  history, 56
  karyotypes, published (*table*), 57
  single abnormalities (*table*), 57, 59
  t(11;14)(q13:q32) translocation, 58–9
chromosomes
  and oncogenes, 60–1
    bcl–1 oncogene, 59
chronic antigen stimulation (CAS), 17–18
clinical features of MM, 98–107
  diagnostic criteria, 101–2
  skeletal symptoms, 99–100
clinical staging of MM, 103–5
  Durie and Salmon staging system (*table*), 103–4
  ECOG, 104
clinical studies
  case-control studies, 16
  chemotherapy (*tables*), 114, 116

cohort studies, 15–16
clonal B cell excess (CBE), immunofluorescence, 41
cohort studies, causes of plasma cell neoplasms, 15–16
cold agglutinin disease, 85
complementarity-determining region (CDR), 26
complications of MM, 205–13
cryoglobulinemia, 85, 210
cyclophosphamide
  continuous-infusion, 118
  high-dose, 183
  oral, 112
cytokines, 73–82, 159–81
  associated with lytic bone lesions (*table*), 52–3
  bone changes in MM, 52–3
  and cytokine receptors, 88–9
  erythropoietin, 172–4
  in model of growth and progression of MM, 30
  myeloid growth factors, 171–2
  plasma cells (MM), role in disease, 29
  see also growth factors; interferons; interleukins
cytotoxic therapy *see* chemotherapy

deoxycoformycin, 121
dexamethasone, combined therapy with interferons,
  166–7
diagnostic criteria
  MGUS, 84, 102
  MM, 84, 101–2
  other monoclonal gammopathies, 85
DiSC assay, 121
drugs
  investigational drugs, 121–2
  multidrug resistance, 120–1, 139–40
  new, 121–2
  see also chemotherapy

ECOG (Eastern Cooperative Oncology Group), 115
epidemiology, 12–21
erythrocyte sedimentation rate (ESR), raised, 44–5
erythropoietin, 172–4
etoposide, high-dose, 183
excess absolute risk (EAR), radiation exposure, 15
extracellular plasmacytoma, 85

familial plasmacytoma, 18
Fanconi syndrome, 85
Finnish Leukemia Group, 133
France, IFM study (*tables*), 190, 191

G–CSF gene, 104
  homology with IL–6 gene, 77
  see also cytokines
Gaucher's disease, 18
genetic factors, 18–19
German study, interferons, 153
glucocerebroside, Gaucher's disease, 18
p-glycoprotein, marker of multidrug resistance (MDR),
  139
gp130 IL–6 transducer, activation by cytokines, 79–80
growth factors, 73–82, 209
  requirements of plasma cells (MM), 28–9
  see also cytokines

heavy-chain disease, 85
heavy-chain gene, rearrangement, isotype switching, 38
hematopoiesis, dysfunction, 208
hematopoietic growth factors, 28–9, 73–82, 209
historical aspects of MM, 1–11, 45, 56, 148
HLA associations, 18, 19, 43
Hodgkin's lymphoma, and smoking, 16
humoral anti-idiotypic immunity, 43
hypercalcemia syndrome, 101, 205
    destruction of bone, 204–5
hyperviscosity syndrome, 100, 209–10

idarubicin, chemotherapy, 121
idiotype-bearing B cells, 40–4
idiotype-specific immunity induction of T cells, 46
IFM study, France, 190, 191
immune regulation, action of interferons, 149–50
immune response
    and plasma cell generation, 23
    plasma cell properties, 24
immunocytoma *see* Waldenstrom's macroglobulinemia
immunofluorescence, B cells, clonal B cell excess
    (CBE), 41
immunoglobulins, 36–50
    generation of antibody diversity, 39
    heavy-chain gene rearrangement, and isotype
        switching, 38
    idiotype-bearing B cells, 40–4
    idiotypic immune network regulation, 42–4
    IgM, secretion by plasma cells, 24
    isotype switching, 38–9
    M-component
        detection and quantification, 44–6
        and familial plasmacytoma, 19
        isotypes, and diagnosis (*table*), 39
        structure, 36
    normal gene loci, 37
    normal gene rearrangement, 37–9
    therapy with, 209
    variable, diversity, and joining (VDJ) segments, 23,
        37–8
    see also M-component
incidence and prevalence of MGUS, 12–13
infections, 208–9
interferons, 148–58, 159–67
    action
        antigrowth effect, 149
        immune regulation, 149–50
        on myeloma cells, 150
    biology of, 149, 159–61
    English study, 155
    German study, 153, 155
    history, 148
    IFN-α, 159–60
        inducing production of IL–6, 77
    IFN-ß, 155
    IFN-γ, 155
        blocking expression of IL–6, 77
    Italian study, 153, 155
    pharmacology and pharmacokinetics, 149
    recombinant, 149
    Swedish study, 153–4, 155

SWOG study, 154, 155
    therapy
        combined with chemotherapy, 151–2, 162–3, 166–7
        compared with chemotherapy, 152
        indications, 155–6
        maintenance treatment, 153–4, 163–7; post
            autologous BMT, 154, 188
        patients in relapse, 152–3
        previously untreated patients, 150–2, 161–2
interleukins
    IL–1 and IL–1ß, 76, 89
    IL–2, 167–9
    IL–3, 75
    IL–4, 169
    IL–5, 75
    IL–6, 74–5
        action of GM–CSF, 75
        autocrine production, 77
        cytokines using IL–6 transducer chain, 75–6
        gp130 IL–6 transducer, growth factors, 79–80
        IL–6 gene, homology with G–CSF gene, 77
        inhibition, 76–7, 169–71
        major myeloma cell growth factor, *in vivo* and *in
            vitro*, 79
        site of production, 75
        soluble IL–6 receptors, 88, 92–3
        treatment with anti-IL–6 antibodies, 78–9
    IL–10, 77–8, 79–80
irradiation *see* radiotherapy
isotype switching, heavy-chain gene rearrangement, 38
Italian study, interferons, 153

Kahler, Otto, 7–8
Karnofsky index, 101
karyotypes, published (*table*), 57

labeling index *see* myeloma cells; plasma cells, plasma
    cell labeling index (PCLI)
LAK cells, 91
leukemias, EAR, atomic bomb survivors, 15
leukocyte alkaline phosphatase (LAP), serum marker, 92
light-chain amyloidosis, 85, 207–8
light-chain deposition disease, 207–8
lymphokine-activated killer (LAK) cells, 91
lymphoproliferative disease, monoclonal IgM-
    associated, 85

M-component
    and age, 13
    antigen reactivity, 40
    detection and quantification, 44–6
        agarose gel electrophoresis, 45
    and familial plasmacytoma, 19
    Ig, quantification, 45–6
    Ig isotypes, monoclonal gammopathies, 39–40
    Ig isotypes and diagnosis (*table*), 39
    Ig molecules, schema, 36–7
    immunization studies, 46
    isotypes, and diagnosis (*table*), 39
    and MGUS, 12
    prevalence, 12
    and race, 13

M-component (continued)
  structure, 36
major histocompatibility complex (MHC)
  HLA associations, 18, 19
  T cells, recognition, 43
MDR (multidrug resistance), 120–1, 139–40
Medical Research Council (MRC) Vth Myeloma Study, 110, 115, 116
melphalan
  action, 139
  high-dose, 136–7, 182–4
    with TBI, 184
  intermediate-dose, 120
  see also chemotherapy
memory B cells, plasma cell precursors (MM), 26
MGUS *see* monoclonal gammopathies of undetermined significance
ß₂-microglobulin, serum marker, 102, 105
Molony leukemia virus integration-IV (MLVI–4) locus, 66
monoclonal gammopathies
  list, 85
  M-component isotypes, 39–40
  with peripheral neuropathy, 85
monoclonal gammopathies of undetermined significance (MGUS), 83–97
  agarose gel electrophoresis, 45
  age of patient, 13
  diagnostic criteria, 84, 102
  differentiation from MM, 83–95
    apoptosis, 91
    CD40, 93
    criteria, 84, 102–3
    differences, 84–9
    oncogenes, 89–90
    serum markers, 91–2
    soluble IL–6R, 92–3
    summary, 94
    T cells and NK cells, 91
  incidence and prevalence, 12–13
  and M-component, 36
  and race, 18
  racial differences, 13–14
monoclonal IgM-associated lymphoproliferative disease, 85
monoclonal immunoglobulin depositional disease (MIDD), 207
mortality, trends, 14
mouse models of plasmacytoma, 18
MRC (Medical Research Council) Vth Myeloma Study, 110, 115, 116
MTF (Myeloma Task Force), 110, 111, 131
multidrug resistance (MDR), 120–1, 139–40
multiple myeloma
  circulating plasma cell precursors, 26–7
  complications, 205–13
  diagnostic criteria, 84, 102
  differentiation from MGUS, 83–95
  epidemiology, 12–21
  familial, 19
  growth and progression
    hypothesis, 32

  model, 31
plasma cell cytokines, 29
plasma cell growth factor requirements, 28–9
plasma cell phenotype, 27–8
plasma cell precursors, 25–6
  circulating, 26–7
refractory disease, 130–47
  classification, 132–3
smoldering (SMM), 84, 85, 87
myeloablative therapy, refractory MM, 136–7
myeloid growth factors, 171–2
  G–CSF gene, 77, 104
Myeloma Task Force (MTF), 110, 111, 131
myelomatous plasma cells
  cytokines, role in disease, 29
  effects of interferons, 156
  growth factor requirements, 28–9
  interactions with bone marrow stromal cells, 76
  phenotypes, 27–8
  plasma cell labeling index (PCLI), 84, 86–7, 103
    (*tables*), 88, 105
  plasma cell precursors, 25–6

neopterin, serum marker, 92
neurologic complications of MM, 209
non-Hodgkin's lymphoma, *bcl–6* locus, 67
Nordic Myeloma Study Group (NMSG), 135

occupational risk, 17
odds ratio, 16
oncogenes, 64–72
  bcl–1 oncogene, 59, 65
  oncogenesis, 64–5
  Rb1 deletions, 60
osteocalcin, serum marker, 91
osteoclast activating factors (OAF), 22, 32
osteosclerotic myeloma, 85

p53 tumor suppressor gene, 65, 68–9
  in MM, MGUS, and PCL (*table*), 68, 89–90
p-glycoprotein (PGP), marker of multidrug resistance (MDR), 139–40
PCR, detection of monoclonal B cells, 41–2
peripheral blood stem cell (PBSC) transplantation, 187
plasma cell neoplasms
  case-control studies, 16
  cohort studies, 15–16
  epidemiology, 12–21
  induction, 15
  see also myelomatous plasma cells; plasmacytoma
plasma cells, 22–35
  antigrowth effects of interferon, 149
  bone marrow environment, 29–31
  description, 22
  differences in MM and MGUS, 86–7, 89–90
  generation, 23
  heterogeneity, 24
  idiotypic immune network regulation, 42–4
  in life history of B-lineage cells, 23
  malignancy major features, 22
  plasma cell labeling index (PCLI), 84, 86–7, 88, 103, 105

plasma cells (continued)
  precursors, bone marrow-seeking, 25
  proliferation, benign *see* MGUS
  properties, and immune response, 24
  types
    lymphoplasmacytoid, 24
    Marschalko, 24
plasma cells (myelomatous) *see* myelomatous plasma
  cells
plasmacytoma
  familial, 18
  localized, 100
  mouse models, 18
  solitary, extracellular, 85
POEMS syndrome, 101, 209
  polyneuropathy, 101
  tamoxifen, 122
polyneuropathy, POEMS syndrome, 101, 209
prednis(ol)one chemotherapy, 112–13
prognostic factors, 102–3
  parameters (*table*), 70
purine analogs in chemotherapy, 121

racial differences
  M-component, 13–14
  MM and MGUS, 18
radiation exposure, excess absolute risk (EAR), 15
radiotherapy, 108–10, 122–3, 137–9
  double hemi-body irradiation (DHBI), 122–3, 138,
    183–4
  local, 122
  for skeletal symptoms, 99
  systemic radiotherapy in refractory disease, 137–9
  total body irradiation (TBI), 122–3, 183–4
    with chemotherapy, 136–7
ras oncogenes, 65, 67–8, 89–90
  in MM, MGUS, and PCL (*table*), 68
  and p53, prognostic parameters (*table*), 70
Rb1 tumor suppressor gene, 60, 65, 69, 90
renal dysfunction, 100, 206–7
rheumatoid arthritis, 17–18
Rustizky's disease, 7

scleromyxedema, 85
serum markers
  ß$_2$-microglobulin, 102, 105
  C-reactive protein, 92, 103
  history, 45
  leukocyte alkaline phosphatase (LAP), 92
  in MGUS, 91–2
  neopterin, 92
  osteocalcin, 91
  p-glycoprotein, 139
  usefulness, 91

see also M-component
smoking, 16
  Hodgkin's lymphoma, 16
smoldering multiple myeloma (SMM), 84, 85, 87
  diagnostic criteria, 101–2
socioeconomic status, 16
solitary plasmacytoma, 85
South West Oncology Group (SWOG), 110, 111, 131
  interferon study, 154
Spain, PETHEMA group, 116, 134
spinal cord compression, 99
SSCP analysis, 67–8
staging *see* clinical staging of MM
stem cells
  for autologous bone marrow transplantation
    peripheral blood (PBSC), 187
    purging, 185
    source, 184–5
  early myeloma cell clone, 42
  plasma cell precursors, 25–6
  tumorigenesis, 74
  see also myeloma cells; plasma cells
steroids, combined with interferons, 166–7
Swedish study, interferons, 153–4
SWOG *see* South West Oncology Group
systemic capillary leak syndrome, 85

T cells
  idiotype-specific, 44
    immunity induction, 46, 91
    recognition of major histocompatibility complex, 43
    regulatory, and with NK phenotype, 91
    response to tumor clone, 43
tamoxifen, POEMS syndrome, 122
terminology, historical aspects, 7
transplantation *see* bone marrow transplantation;
    peripheral blood (PBSC) transplantation
tumor necrosis factor, inducing production of IL-6, 77
tumor suppressor genes
  p53, 65, 68–9
  Rb1 deletions, 60
tyrosine kinase, serum marker, 92

variable, diversity, and joining (VDJ) segments
  Ig, 23
  isotype switching, 38
  number of combinations, 39
  number of genes, 38

Waldenstrom's macroglobulinemia
  genetics, 56
  hyperviscosity syndrome, 100
  M-component, 19, 36, 45